AF614748

Methods in Molecular Biology™

Series Editor
John M. Walker
School of Life Sciences
University of Hertfordshire
Hatfield, Hertfordshire, AL10 9AB, UK

For further volumes:
http://www.springer.com/series/7651

Gene Regulation

Methods and Protocols

Edited by

Minou Bina

Department of Chemistry, Purdue University, West Lafayette, Indiana, USA

Editor
Minou Bina
Department of Chemistry
Purdue University
West Lafayette, Indiana, USA

ISSN 1064-3745 ISSN 1940-6029 (electronic)
ISBN 978-1-62703-283-4 ISBN 978-1-62703-284-1 (eBook)
DOI 10.1007/978-1-62703-284-1
Springer New York Heidelberg Dordrecht London

Library of Congress Control Number: 2013931173

Printed on acid-free paper

Humana Press is a brand of Springer
Springer is part of Springer Science+Business Media (www.springer.com)

Preface

In recognition of monumental impact of gene expression in producing normal and abnormal cellular states, we are pleased to offer the scientific community a volume on Gene Regulation, Methods and Protocols. The goal is to provide scientists in academia, food, and pharmaceutical industry, as well as public institutions, complementary technologies for investigating various facets of regulatory systems that contribute to the control of protein-coding genes in mammalian cells.

The technologies are broad in their scope. They include biochemical assays, methods in molecular biology, spectroscopic techniques, and high-throughput approaches for delineating key processes that contribute to the regulation of gene expression. The chapters are organized to offer a comprehensive, integrated, and coherent view of control systems and their associated components. The introductory chapter draws attention to the interconnectivity of regulatory circuits and provides examples of regulatory hubs. This chapter underscores the importance of protein networks in dynamics of gene activation, gene repression, and histone modifications.

Several techniques encompass chromatin structural features that influence gene activation. Two chapters detail methods for DNase hypersensitivity analysis, for high-throughput DNA sequencing to study chromatin accessibility and to identify regulatory DNA sequences on a global scale. Another chapter focuses on DNA derived from nucleosome-free regions for functional assays, to localize and study the activity of *cis*-regulatory modules dispersed within promoters and enhancers of genes.

A chapter highlights the utility of Heavy methyl-SILAC (Stable Isotope Labeling of Amino Acids in Cell Culture) to distinguish preexisting and newly generated methyl marks on histones. Coupling this technology with quantitative liquid chromatography and mass spectrometry (LC–MS) makes possible monitoring changes in site-specific histone methylation patterns.

Several chapters offer protocols for identifying regulatory sequences for functional assays and studies of protein–DNA interactions. Two chapters feature luciferase-based assays, including a dual luciferase reporter system to localize transcription factor binding sites within promoter segments of genes. Another chapter covers a functional assay that can be implemented for de novo identification of endogenous transcriptional regulatory modules in cultured mammalian cells.

Three chapters are devoted to technologies designed to examine genome-wide association of transcription factors with regulatory regions of genes. A protocol focuses on isolation, purification, and immunoprecipitation of DNA fragments associated with a transcription factor of interest, with the intention of massive parallel sequencing. A method (HaloCHIP) utilizes a HaloTag protein fusion and corresponding interaction resin (HaloLink) to capture crosslinked protein–DNA complexes directly from cell lysates. This approach effectively yields the DNA fragments bound to a protein of interest, circumvents the need for using antibodies, and facilitates downstream analyses including DNA amplification, use of microarrays, and massive DNA sequencing.

A powerful genetic method, a modified yeast one-hybrid system, is presented for analysis of DNA–protein interactions on a genome-wide scale. When compared to other methods, the modified system offers several advantages including low-cost, large-scale output, and ease of reagent handling.

Several chapters cover in vitro techniques for studies of protein–DNA interactions. A new application of mRNA display provides a method for in vitro selection of DNA-binding proteins. In a single experiment, using DNA as bait, the selection system can identify various DNA–protein complexes including those that contain hetero-oligomers of transcription factors bound to DNA.

A technology (SILAC-based quantitative proteomics) is included for identifying specific interactions between proteins and functional DNA elements in an unbiased manner. Another technique, electrophoretic mobility shift assay (EMSA) is useful for characterization of nucleoprotein complexes formed with short DNA fragments. This technique is particularly convenient for determining whether a DNA fragment of interest includes binding site(s) for transcription factors present in nuclear extracts prepared from a given cell type. When used in conjunction with specific antibodies, EMSA offers a strategy (supershift assay) for identification of transcription factors that associate with a DNA fragment.

Several chapters cover spectroscopic techniques for studying protein–DNA and protein–protein interactions. Fluorescence Resonance Energy Transfer (FRET) is useful for measuring changes in DNA conformation due to protein binding because small changes in the distance between two fluorophores (2–10 nm) translate into large changes in energy transfer. With Fluorescence Cross-Correlation Spectroscopy (FCCS), it is possible to assess interaction of target molecules in aqueous condition. When applied to cultured cells, with FCCS one can directly observe dimerization between transcription factors in living cells. Another spectroscopic technique, Fluorescence Anisotropy/polarization Microplate assay (FAMA), provides a powerful tool to investigate the interaction of transcriptional coactivators with transcription factors that bind DNA.

Two schemes are covered for isolation and characterization of relatively large multiprotein complexes. One scheme was designed to purify multisubunit complexes gently and quickly from crude extracts prepared from mammalian cells. An example describes isolation of the mammalian Mediator complex from cultured cells. The other scheme uses a single-step FLAG affinity purification to isolate chromatin modifying complexes for subsequent characterization by sucrose gradient equilibrium centrifugation and mass spectrometry.

Several complementary approaches are offered for studies of histone acetylases (i.e. p300) and deacetylases. A chromatin immunoprecipitation (ChIP) protocol is described for examining p300-dependent regulatory elements in genomic DNA. A mammalian two-hybrid system is presented to detect interactions between two proteins in vivo. This approach overcomes limitations inherent to the yeast two-hybrid system. As an example, the mammalian system was applied to study interactions of p300 with a transcription factor associated with hypoxia (HIF-1α) and a transcription factor relevant to inflammation (NF-κB-p65). Another protocol uses a HDAC inhibitor and takes advantage of different types of p300 to study the interplay of bromodomain and histone acetylation in p300-dependent gene expression. Use of HDAC inhibitor Valproic acid is also described for inducing differentiation of pluripotent stem cells.

To examine gene repression, a chapter details strategies for studies of protein interactions with different DNA methyltransferases. Furthermore a combination of sedimentation and immunoprecipitation assays is presented for analysis and characterization of complexes that repress transcription.

A chapter deals with peptide microarray technology used to identify substrates for recombinant kinases, to discover and evaluate kinase-inhibitors, and to examine changes in activity of kinases in cell lysates and lysates from fresh frozen (tumor) tissue. The method described was developed for examining dynamics of peptide microarrays with real-time read out, and for determining the influence of assay parameters on optimization experiments.

To investigate multicomponent transcriptional complexes, an experimental approach assesses the function of each component on chromatin reconstructed in vitro. Another technique, promoter-independent abortive initiation assay, exploits the intrinsic ability of RNA polymerases to initiate transcription from nicked DNA templates. These assays can be used to measure the effect of transcription factors such as TFIIB and RNA polymerase mutations on abortive transcription. A protocol (nuclear recruitment assay) details a strategy to validate transcription factor interactions in mammalian cells. In addition, a method is described for isolating cell lines with multicopy arrays of reporter transgenes, for real-time high-resolution imaging of transcriptional activation dynamics in single cells.

I hope that the breadth and scope of methodologies offer a beneficial resource for studies of gene regulation. In principle, the protocols and experimental strategies could be applied by researchers in diverse fields including molecular biology, genomics, biochemistry, biomedicine, nutrition, and agricultural sciences.

I thank the authors for their contributions. I thank Professor John Walker, the series editor, for his encouragement.

West Lafayette, IN, USA *Minou Bina*

Contents

Preface . . . . *v*
Contributors . . . . *xiii*

1 Gene Regulation 1
Minou Bina

2 Isolation of Nuclei for Use in Genome-Wide DNase Hypersensitivity Assays to Probe Chromatin Structure 13
Guoyu Ling and David J. Waxman

3 DNase I Digestion of Isolated Nulcei for Genome-Wide Mapping of DNase Hypersensitivity Sites in Chromatin 21
Guoyu Ling and David J. Waxman

4 Isolation and Analysis of DNA Derived from Nucleosome-Free Regions 35
Matthew Murtha, Yatong Wang, Claudio Basilico, and Lisa Dailey

5 Acquisition of High Quality DNA for Massive Parallel Sequencing by In Vivo Chromatin Immunoprecipitation 53
M. van den Boogaard, L.Y.E. Wong, V.M. Christoffels, and P. Barnett

6 Luciferase Assay to Study the Activity of a Cloned Promoter DNA Fragment 65
Nina Solberg and Stefan Krauss

7 Promoter Deletion Analysis Using a Dual-Luciferase Reporter System 79
Yong Zhong Xu, Cynthia Kanagaratham, Sylwia Jancik, and Danuta Radzioch

8 Application of mRNA Display for In Vitro Selection of DNA-Binding Transcription Factor Complexes 95
Seiji Tateyama and Hiroshi Yanagawa

9 Isolation of Intracellular Protein – DNA Complexes Using HaloCHIP, an Antibody-Free Alternative to Chromatin Immunoprecipitation 111
Danette L. Daniels and Marjeta Urh

10 A Modified Yeast One-Hybrid System for Genome-Wide Identification of Transcription Factor Binding Sites 125
Kazuyuki Yanai

11 Identifying Specific Protein–DNA Interactions Using SILAC-Based Quantitative Proteomics 137
Cornelia G. Spruijt, H. Irem Baymaz, and Michiel Vermeulen

12 Electrophoretic Mobility-Shift and Super-Shift Assays for Studies and Characterization of Protein–DNA Complexes 159
Elsie I. Parés-Matos

13 Combination of Native and Denaturing PAGE for the Detection of Protein Binding Regions in Long Fragments of Genomic DNA. 169
Kristel Kaer and Mart Speek

14 Quantitative NanoProteomics Approach for Protein Complex (QNanoPX) Using Gold Nanoparticle-Based DNA Probe. 183
Shu-Hui Chen and Mei-Yin Lin

15 Chromatin Assembly and In Vitro Transcription Analyses for Evaluation of Individual Protein Activities in Multicomponent Transcriptional Complexes 193
Takayuki Furumatsu and Hiroshi Asahara

16 Using FRET to Monitor Protein-Induced DNA Bending: The TBP-TATA Complex as a Model System 203
Rebecca H. Blair, James A. Goodrich, and Jennifer F. Kugel

17 Promoter Independent Abortive Transcription Assays Unravel Functional Interactions Between TFIIB and RNA Polymerase. 217
Simone C. Wiesler, Finn Werner, and Robert O.J. Weinzierl

18 Fluorescence Cross-correlation Spectroscopy (FCCS) to Observe Dimerization of Transcription Factors in Living Cells 229
Hisayo Sadamoto and Hideki Muto

19 Nuclear Recruitment Assay as a Tool to Validate Transcription Factor Interactions in Mammalian Cells 243
C.J.J. Boogerd, V.M. Christoffels, and P. Barnett

20 Preparation of Cell Lines for Single-Cell Analysis of Transcriptional Activation Dynamics 249
Ilona U. Rafalska-Metcalf and Susan M. Janicki

21 Peptide Microarrays for Profiling of Serine/Threonine Kinase Activity of Recombinant Kinases and Lysates of Cells and Tissue Samples 259
Riet Hilhorst, Liesbeth Houkes, Monique Mommersteeg, Joyce Musch, Adriënne van den Berg, and Rob Ruijtenbeek

22 Immunoaffinity Purification of Protein Complexes from Mammalian Cells . . 273
Chieri Tomomori-Sato, Shigeo Sato, Ronald C. Conaway, and Joan W. Conaway

23 Simple and Efficient Identification of Chromatin Modifying Complexes and Characterization of Complex Composition 289
Jeong-Heon Lee and David Skalnik

24 Heavy Methyl-SILAC Labeling Coupled with Liquid Chromatography and High-Resolution Mass Spectrometry to Study the Dynamics of Site-Specific Histone Methylation 299
Xing-Jun Cao, Barry M. Zee, and Benjamin A. Garcia

25 Analysis of p300 Occupancy at the Early Stage of Stem Cell Differentiation by Chromatin Immunoprecipitation 315
Melanie Le May and Qiao Li

26 Mammalian Two-Hybrid Assays for Studies of Interaction of p300 with Transcription Factors. 323
Daniela B. Mendonça, Gustavo Mendonça, and Lyndon F. Cooper

27 Fluorescence Anisotropy Microplate Assay to Investigate the Interaction of Full-Length Steroid Receptor Coactivator-1a with Steroid Receptors 339
Chen Zhang, Steven K. Nordeen, and David J. Shapiro

28 Use of Histone Deacetylase Inhibitors to Examine the Roles of Bromodomain and Histone Acetylation in p300-Dependent Gene Expression 353
Jihong Chen and Qiao Li

29 Histone Deacetylase Inhibitor Valproic Acid as a Small Molecule Inducer to Direct the Differentiation of Pluripotent Stem Cells........... 359
Jihong Chen, Natascha Lacroix, and Qiao Li

30 Sedimentation and Immunoprecipitation Assays for Analyzing Complexes that Repress Transcription.. 365
Ping Lu, Bruce S. Hostager, Paul B. Rothman, and John D. Colgan

31 Methods for Studies of Protein Interactions with Different DNA Methyltransferases.. 385
Jianchang Yang

Index *397*

Contributors

HIROSHI ASAHARA • *Department of Molecular and Experimental Medicine, The Scripps Research Institute, La Jolla, CA, USA*

P. BARNETT • *Heart Failure Research Centre, Academic Medical Centre, section K2, Amsterdam, The Netherlands*

CLAUDIO BASILICO • *Department of Microbiology, NYU School of Medicine, New York, NY, USA*

H. IREM BAYMAZ • *Department of Molecular Cancer Research, University Medical Center Utrecht, Utrecht, The Netherlands*

ADRIËNNE VAN DEN BERG • *PamGene International BV, 's-Hertogenbosch, The Netherlands*

MINOU BINA • *Department of Chemistry, Purdue University, West Lafayette, IN, USA*

REBECCA H. BLAIR • *Department of Chemistry and Biochemistry, University of Colorado, Boulder, CO, USA*

M. VAN DEN BOOGAARD • *Heart Failure Research Centre, Academic Medical Centre, section K2, Amsterdam, The Netherlands*

C.J.J. BOOGERD • *Heart Failure Research Center, Academic Medical Center, section K2, Amsterdam, The Netherlands*

XING-JUN CAO • *Department of Molecular Biology, Princeton University, Princeton, NJ, USA*

SHU-HUI CHEN • *Department of Chemistry, National Cheng Kung University, Tainan, Taiwan*

JIHONG CHEN • *Departments of Pathology and Laboratory Medicine, Faculty of Medicine, University of Ottawa, Ottawa, ON, Canada*

V.M. CHRISTOFFELS • *Heart Failure Research Centre, Academic Medical Centre, section K2, Amsterdam, The Netherlands*

JOHN D. COLGAN • *Department of Internal Medicine, Roy J. and Lucille A. Carver College of Medicine, University of Iowa, Iowa City, IA, USA*

JOAN W. CONAWAY • *Stowers Institute for Medical Research, Kansas City, MO, USA*

RONALD C. CONAWAY • *Stowers Institute for Medical Research, Kansas City, MO, USA*

LYNDON F. COOPER • *Bone Biology and Implant Therapy Laboratory, Department of Prosthodontics, University of North Carolina at Chapel Hill, Chapel Hill, NC, USA*

LISA DAILEY • *Department of Microbiology, NYU School of Medicine, New York, NY, USA*

DANETTE L. DANIELS • *Promega Corporation, Madison, WI, USA*

TAKAYUKI FURUMATSU • *Department of Orthopaedic Surgery, Okayama University Graduate School, Shikata-cho, Okayama, Japan*

BENJAMIN A. GARCIA • *Department of Molecular Biology, Princeton University, Princeton, NJ, USA*

JAMES A. GOODRICH • *Department of Chemistry and Biochemistry, University of Colorado, Boulder, CO, USA*

RIET HILHORST • *PamGene International BV, 's-Hertogenbosch, The Netherlands*

BRUCE S. HOSTAGER • *Department of Internal Medicine, Roy J. and Lucille A. Carver College of Medicine, University of Iowa, Iowa City, IA, USA*
LIESBETH HOUKES • *PamGene International BV, 's-Hertogenbosch, The Netherlands*
SYLWIA JANCIK • *McGill University Health Centre, Montreal General Hospital Research Institute, Montreal, QC, Canada*
SUSAN M. JANICKI • *Molecular and Cellular Oncogenesis Program, The Wistar Institute, Philadelphia, PA, USA*
KRISTEL KAER • *Department of Gene Technology, Tallinn University of Technology, Tallinn, Estonia*
CYNTHIA KANAGARATHAM • *McGill University Health Centre, Montreal General Hospital Research Institute, Montreal, QC, Canada*
STEFAN KRAUSS • *Unit for Cell Signaling, Oslo University Hospital, Oslo, Norway*
JENNIFER F. KUGEL • *Department of Chemistry and Biochemistry, University of Colorado, Boulder, CO, USA*
NATASCHA LACROIX • *Department of Pathology and Laboratory Medicine, University of Ottawa, Ottawa, ON, Canada*
JEONG-HEON LEE • *Department of Pediatrics, Wells Center for Pediatric Research, Indiana University School of Medicine, Indianapolis, IN, USA*
QIAO LI • *Department of Pathology and Laboratory Medicine, Faculty of Medicine, University of Ottawa, Ottawa, ON, Canada; Department of Cellular and Molecular Medicine, Faculty of Medicine, University of Ottawa, Ottawa, ON, Canada*
MEI-YIN LIN • *Department of Chemistry, National Cheng Kung University, Tainan, Taiwan*
GUOYU LING • *Department of Biology, Boston University, Boston, MA, USA*
PING LU • *Department of Internal Medicine, Roy J. and Lucille A. Carver College of Medicine, University of Iowa, Iowa City, IA, USA*
MELANIE LE MAY • *Department of Pathology and Laboratory Medicine, University of Ottawa, Ottawa, ON, Canada; Department of Cellular and Molecular Medicine, Faculty of Medicine, University of Ottawa, Ottawa, ON, Canada*
DANIELA B. MENDONÇA • *Bone Biology and Implant Therapy Laboratory, Department of Prosthodontics, University of North Carolina at Chapel Hill, Chapel Hill, NC, USA*
GUSTAVO MENDONÇA • *Bone Biology and Implant Therapy Laboratory, Department of Prosthodontics, University of North Carolina at Chapel Hill, Chapel Hill, NC, USA*
MONIQUE MOMMERSTEEG • *PamGene International BV, 's-Hertogenbosch, The Netherlands*
MATTHEW MURTHA • *Department of Microbiology, NYU School of Medicine, New York, NY, USA*
JOYCE MUSCH • *PamGene International BV, 's-Hertogenbosch, The Netherlands*
HIDEKI MUTO • *Laboratory of Molecular Cell Dynamics, Hokkaido University, Sapporo, Hokkaido, Japan*
STEVEN K. NORDEEN • *University of Colorado at Denver and Health Sciences Center, Aurora, CO, USA*
ELSIE I. PARÉS-MATOS • *Department of Chemistry, University of Puerto Rico at Mayagüez, Mayagüez, PR, USA*
DANUTA RADZIOCH • *McGill University Health Centre, Montreal General Hospital Research Institute, Montreal, QC, Canada*
ILONA U. RAFALSKA-METCALF • *Molecular and Cellular Oncogenesis Program, The Wistar Institute, Philadelphia, PA, USA*

PAUL B. ROTHMAN • *Department of Internal Medicine, Roy J. and Lucille A. Carver College of Medicine, University of Iowa, Iowa City, IA, USA*
ROB RUIJTENBEEK • *PamGene International BV, 's-Hertogenbosch, The Netherlands*
HISAYO SADAMOTO • *Laboratory of Functional Biology, Tokushima Bunri University, Sanuki, Kagawa, Japan*
SHIGEO SATO • *Stowers Institute for Medical Research, Kansas City, MO, USA*
DAVID J. SHAPIRO • *Department of Biochemistry, University of Illinois at Urbana-Champaign, Urbana, IL, USA*
DAVID SKALNIK • *Biology Department, School of Science, Indiana University—Purdue University Indianapolis, Indianapolis, IN, USA*
NINA SOLBERG • *Unit for Cell Signaling, Oslo University Hospital, Oslo, Norway*
MART SPEEK • *Department of Gene Technology, Tallinn University of Technology, Tallinn, Estonia*
CORNELIA G. SPRUIJT • *Department of Molecular Cancer Research, University Medical Center Utrecht, Utrecht, The Netherlands*
SEIJI TATEYAMA • *Department of Biosciences and Informatics, Keio University, Yokohama, Japan*
CHIERI TOMOMORI-SATO • *Stowers Institute for Medical Research, Kansas City, MO, USA*
MARJETA URH • *Promega Corporation, Madison, WI, USA*
MICHIEL VERMEULEN • *Department of Molecular Cancer Research, University Medical Center Utrecht, Utrecht, The Netherlands*
YATONG WANG • *Department of Microbiology, NYU School of Medicine, New York, NY, USA*
DAVID J. WAXMAN • *Department of Biology, Boston University, Boston, MA, USA*
ROBERT O.J. WEINZIERL • *Department of Life Sciences, Imperial College London, London, UK*
FINN WERNER • *RNAP laboratory, Institute of Structural and Molecular Biology, Division of Biosciences, University College London, London, UK*
SIMONE C. WIESLER • *Department of Life Sciences, Imperial College London, London, UK*
L.Y.E. WONG • *Heart Failure Research Centre, Academic Medical Centre, University of Amsterdam, Amsterdam, The Netherlands*
YONG ZHONG XU • *McGill University Health Centre, Montreal General Hospital Research Institute, Montreal, QC, Canada*
HIROSHI YANAGAWA • *Department of Biosciences and Informatics, Keio University, Yokohama, Japan*
KAZUYUKI YANAI • *Department of Biomolecular Science, Faculty of Science, Toho University, Chiba, Japan*
JIANCHANG YANG • *Human Health and Environment Program, Desert Research Institute (DRI), Las Vegas, NV, USA*
BARRY M. ZEE • *Department of Molecular Biology, Princeton University, Princeton, NJ, USA*
CHEN ZHANG • *Department of Biochemistry, University of Illinois at Urbana-Champaign, Urbana, IL, USA*

Chapter 1

Gene Regulation

Minou Bina

Abstract

This review concisely highlights the complexity of regulatory events. It provides examples of how interconnectivity of regulatory hubs could maintain transcriptional synergy and orchestrate the proper spatial and temporal patterns of gene expression.

Key words: Chromatin structure, Histone modification, Transcription factors, Mediator complex, MLL network, PCR2 network, Gene activation, CBP, p300, Nuclear receptors, Coactivators, Gene repression, Regulatory hubs, Cellular memory

1. Overview

Mammalian genomes are relatively long and associated with numerous proteins to form a highly dynamic and organized structure known as chromatin. The human genome contains about 22,500 protein coding genes (1). A significant fraction of the genome consists of repetitive DNA; only a small proportion (roughly 1.5 %) of genomic sequences corresponds to protein-coding genes (1). A significant proportion of noncoding DNA encompasses sequence elements that control the elaborate patterns of gene expression in diverse cell types (2). In mammalian chromosomes, protein-coding segments are often separated by relatively long intervening sequences.

Typically, protein-coding genes contain three types of regulatory DNA segments: a core promoter that directs the initiation of transcription at correct sites, a nearby (proximal) promoter, and distal DNA sequences that can enhance or silence transcription (3, 4). The core promoter serves as a docking site for the assembly of the basal transcriptional machinery (3). The proximal promoter is defined as the region immediately upstream (up to a few hundred base pairs) from the core promoter. Proximal promoters contain short-sequences (*cis*-elements) that interact with transcription factors

Minou Bina (ed.), *Gene Regulation: Methods and Protocols*, Methods in Molecular Biology, vol. 977,
DOI 10.1007/978-1-62703-284-1_1, © Springer Science+Business Media, LLC 2013

(TFs) that bind DNA. Enhancer segments also include transcription factor binding sites (TFBSs), but unlike proximal promoters, enhancers may occur anywhere in genomic DNA including far upstream, within, and far down-stream of the genes whose expression they control.

2. Proteins and DNA Sequences that Influence the Initiation of Transcription

Protein-coding genes are transcribed by RNA polymerase II (Pol II) producing the templates (mRNAs) for protein synthesis. Initiation of mRNA synthesis involves several multiprotein complexes known as general transcription factors (GTFs). These factors include TFIIA, TFIIB, TFIID, TFIIE, TFIIF, and TFIIH (5). A relatively large complex (the mediator) supports communications of Pol II with the transcription initiation apparatus and with TFs that bind DNA (6–8), summarized in (Fig. 1).

Pol II core promoters are structurally and functionally diverse (9). A relatively small fraction (~25 %) of mammalian promoters contains the sequence motif (TATA box) that directs Pol II to initiate transcription at a correct site. In mammalian promoters, transcription start sites (TSSs) often occur near or within sequences known as CpG islands (CGIs). These islands correspond to genomic DNA segments that contain guanine-tracks and many CpG dinucleotides (10–14). Generally, TATA boxes direct focused transcription initiation: consisting of either a single major transcription start site or several TSSs localized within several nucleotides (9). In contrast, CGIs associated promoters often display dispersed initiation patterns (9, 11). CpG-poor promoters correspond to highly tissue-specific genes (15). CpG-rich promoters are associated with several types of genes including housekeeping (ubiquitously expressed), tissue-specific, and developmental genes (11, 15). CGIs also are associated with genes regulated by transcription factors that respond to upstream signals (16–18). There are also examples of CGIs encompassing genes controlled by a large family of TFs known as nuclear hormone receptors.

3. Chromatin Structure and Organization

Organization of genomic DNA into compact structures is intimately associated with processes that control gene expression (19). The first level of DNA compaction involves nucleosome formation to produce the repeating DNA packaging units in chromatin. In each nucleosome, a DNA segment (147 base-pairs) is wrapped around a protein core consisting of histone H2A, H2B, H3, and H4 (20).

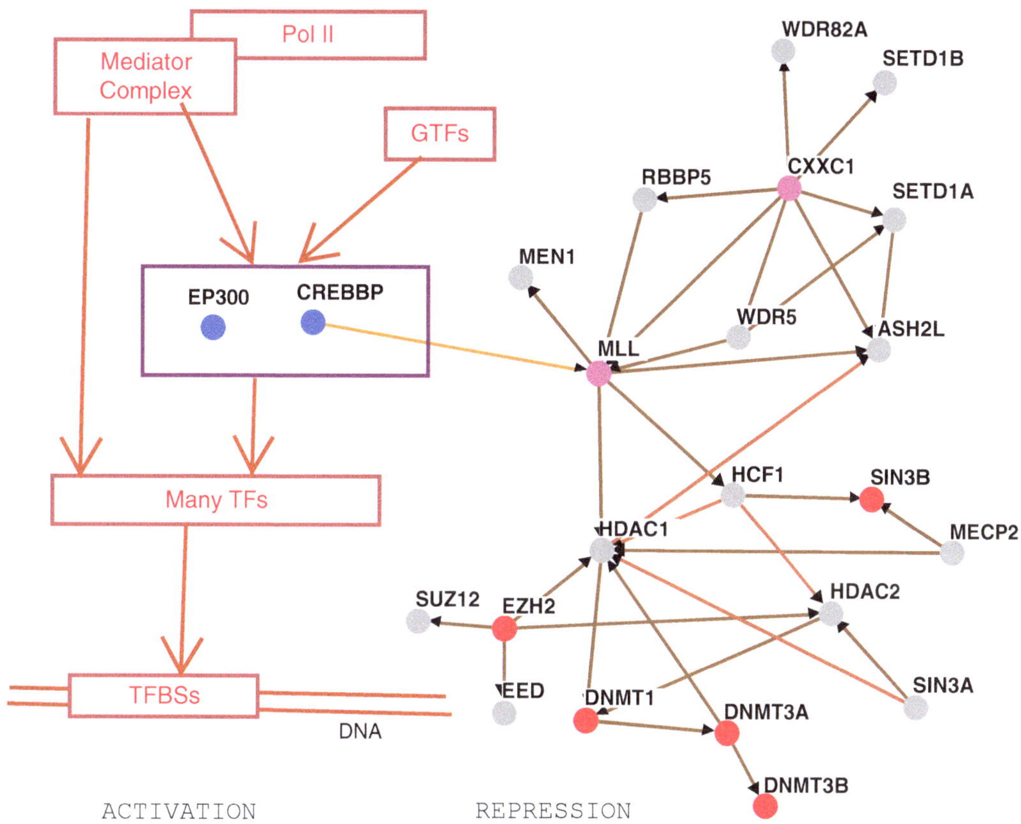

Fig. 1. Examples of hubs and protein networks that regulate transcription. Proteins are depicted as nodes and interactions as edges between nodes (49). The networks were constructed using the Osprey visualization system (49). The interactions were obtained from BioGRID (http://www.thebiogrid.org), a public database that archives and disseminates genetic and protein interaction data from model organisms and humans (50). The interactions were derived from results of low and high throughput experiments including co-crystal structures, yeast and mammalian two-hybrid assays, complex reconstitution, affinity capture and Western analysis, in vitro pull-down, and in vivo coimmunoprecipitation assays (50). The figure does not display all reported interactions but rather focuses on illustrating the underlying connectivity among proteins that regulate gene expression. TFs correspond to transcription factors; TFBSs correspond to TF binding sites in genomic DNA. Naturally occurring human TFBSs often include distinct sequence motifs that appear frequently in promoter regions of Pol II genes (16, 51–54).

Since histone-DNA interactions can influence the accessibility of chromatin to proteins that bind *cis*-elements, the control of gene expression is often viewed in the context of processes that influence the formation of chromatin structures that can support transcription and structures that repress transcription (19).

Various chromatin configurations are driven by proteins that interact with DNA, by enzymes that modify the core histones in nucleosomes, and by proteins that bind modified residues in histones (19). Interactions of genomic DNA with regulators of transcription contribute to formation of transcriptionally poised or active chromatin structures. In such structures, regulatory sequences are highly accessible to cleavage by DNase I (21). Consequently, DNase I hypersensitive regions are hallmarks of transcriptionally poised or active chromatin states (22).

Post-translational modification of core histones contributes to partitioning the genome into distinct domains including euchromatin, where DNA is kept accessible for transcription, and heterochromatin, where transcription is repressed (23). Histone modifications occur on the side chains of specific residues in the histone tails and functionally impact transcription (23). Modifi cations include methylation (me), acetylation (ac), phosphorylation, and ubiquitination (23). Histone methyltransferases often include a domain (SET) that catalyzes the methylation reaction (24). Histone acetyltransferases are known as HATs (25). Methylation may be in one of three different forms: mono-, di-, or trimethyl for lysines as well as mono- or di-methyl for arginines (23).

Enzymes that add or remove modifications in histones operate within a complex protein-network that is closely linked to processes that control initiation, up-regulation, or repression of transcription (Figs. 1 and 2). Functions of acetylation include disruption of contacts between DNA and histones to "unravel" chromatin to produce an open chromatin configuration (23). Functions of acetylated and methylated residues include recruitment of regulatory proteins to chromatin (23).

4. Chromatin States and Regulation of Transcription

In mammalian cells, gene regulation operates in the context of dynamic changes in chromatin structure. Various chromatin configurations and states are driven by the underlying genomic DNA sequence, by DNA methylation, and by histone modification patterns. Chromatin states include active, repressed, and bivalent (19). These states arise from the type of modifications on the core histones in nucleosomes. By convention, methylations are described in the context of the histone that is modified, the residue and position of modification, and the number of methyl groups.

Active promoters are commonly marked by H3K4me3, H3K4me2, and histone acetylation (19). Active enhancers are marked with H3K4me1, H3K4me2, and H3K27ac (19, 22). Repressed promoters show unique patterns of chromatin modifications including DNA methylation, H3K27me3, and H3K9me3; such modifications also are found in constitutively repressed heterochromatin structure (19). Bivalent promoters are marked with both H3K4me3 and H3K27me3 (19). In chromatin-state maps, ~22 % of high-CpG-promoters are bivalent.

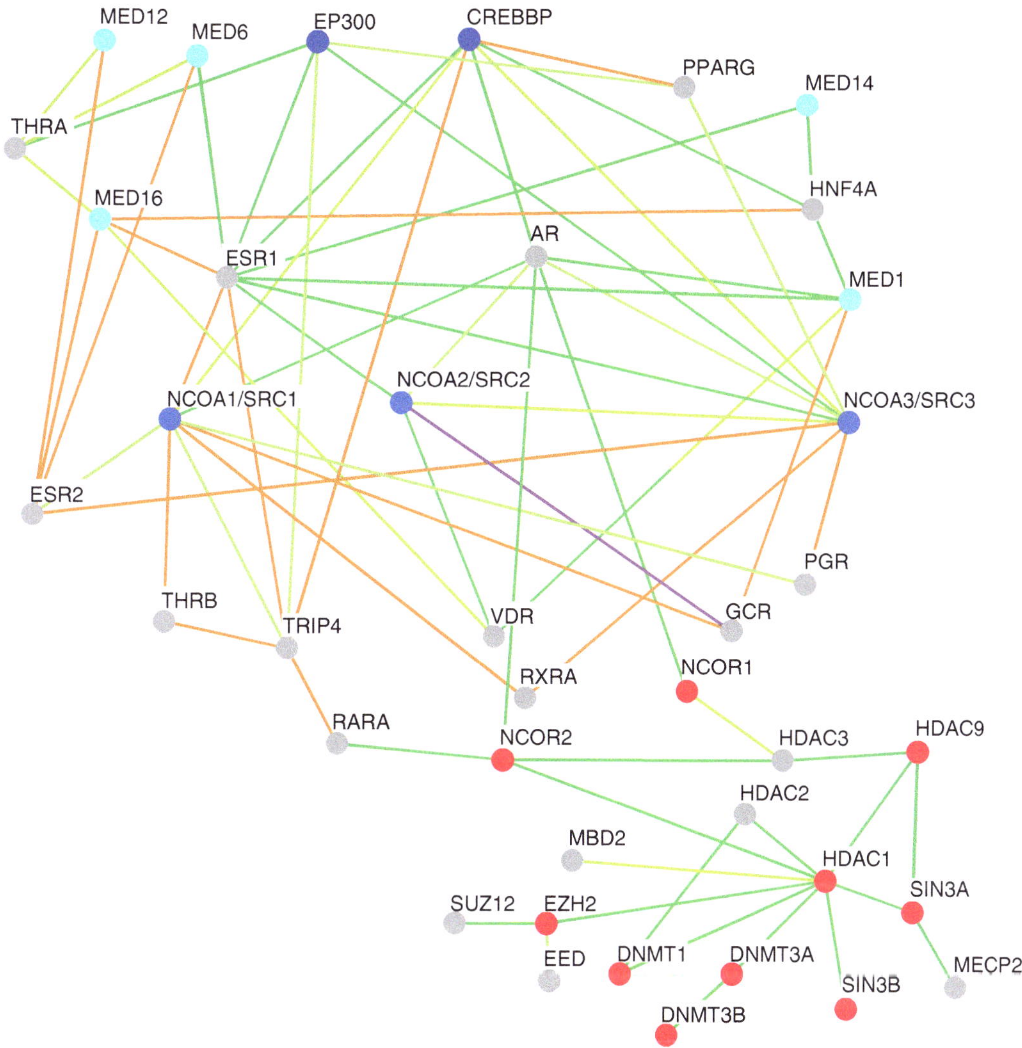

Fig. 2. Examples of hubs and protein networks that regulate transcription by nuclear receptors. See Fig. 1 for the source of interaction data and the system for visualization of protein networks.

Generally, CpG density in genomic DNA correlates positively with the degree of H3K4 trimethylation, indicating that these two properties are mechanistically linked (10, 15). Regions associated with H3K4me3 entail characteristic features of accessible chromatin structure, including acetylated histones and hypersensitivity to DNase I (19). Sharp peaks of H3K4me3 are associated with the TSSs of many transcribed genes (15). The CpGs in these promoters are hypo-methylated (19). CpG-rich promoters may be enriched in Pol II, poised for transcription. In contrast, by default, AT-rich promoters are transcriptionally inactive (19).

5. Protein Networks that Regulate Gene Expression

To uncover the mechanisms that control spatial and temporal patterns of gene expression, it is important to grasp the underlying complexity of protein networks that drive the regulatory machineries. Ongoing in vitro and biochemical assays have identified and dissected key components in these machineries, leading to mechanistic clues. High throughput analyses are discovering the missing components. A global view indicates that regulatory systems are elaborate and dynamic. Their components communicate through intricate and interwoven networks of interactions among proteins that bind DNA, enzymes that modify DNA and histones, transcription factors that direct Pol II to initiate transcription, and proteins that mediate and modulate the level of transcription (Figs. 1 and 2).

The central molecular circuitries that drive coordinated gene expression are largely based on proteins that bind DNA, on enzymes that add or remove modifications from histones, and on protein complexes that transmit regulatory information to other components of the systems. The protein networks can be viewed as composite hubs that receive and transmit information to activate, upregulate, downregulate, or repress the expression of a given gene (Figs. 1 and 2). Modules in hubs often consist of relatively large, multisubunit protein complexes (26). Such complexes may include a core composed of several polypeptide chains. Dynamics of the system arise in part from transient association of the core with other regulators of transcription.

A key hub in protein networks includes the relatively large, multisubunit complex known as the mediator (6, 7, 27, 28). To Pol II, the mediator transmits inputs from GTFs and inputs from transcription factors that bind DNA (6, 7, 28, 29). The mediator subunits are organized into three modules: head, middle, and tail (28, 30). These modules create a core that processes inputs for activation, down-regulation, or repression of transcription (27). The mediator core plays important roles at all stages of transcription, from the recruitment of Pol II to genes in response to diverse signals to controlling transcription initiation and elongation (30). Several of the mediator subunits function in integrating inputs conveyed from the molecular circuitry controlled by nuclear receptors (7, 29) (Fig. 2).

In protein networks, two evolutionarily related histone acetyltransferases (CREBBP and EP300) provide a hub that coordinates multiple events within the transcription apparatus (31–33). CREBBP and EP300 can be viewed as scaffolds upon which multicomponent transcriptional regulatory complexes are built to facilitate assembly of regulatory machineries in a local transcription environment (31, 32).

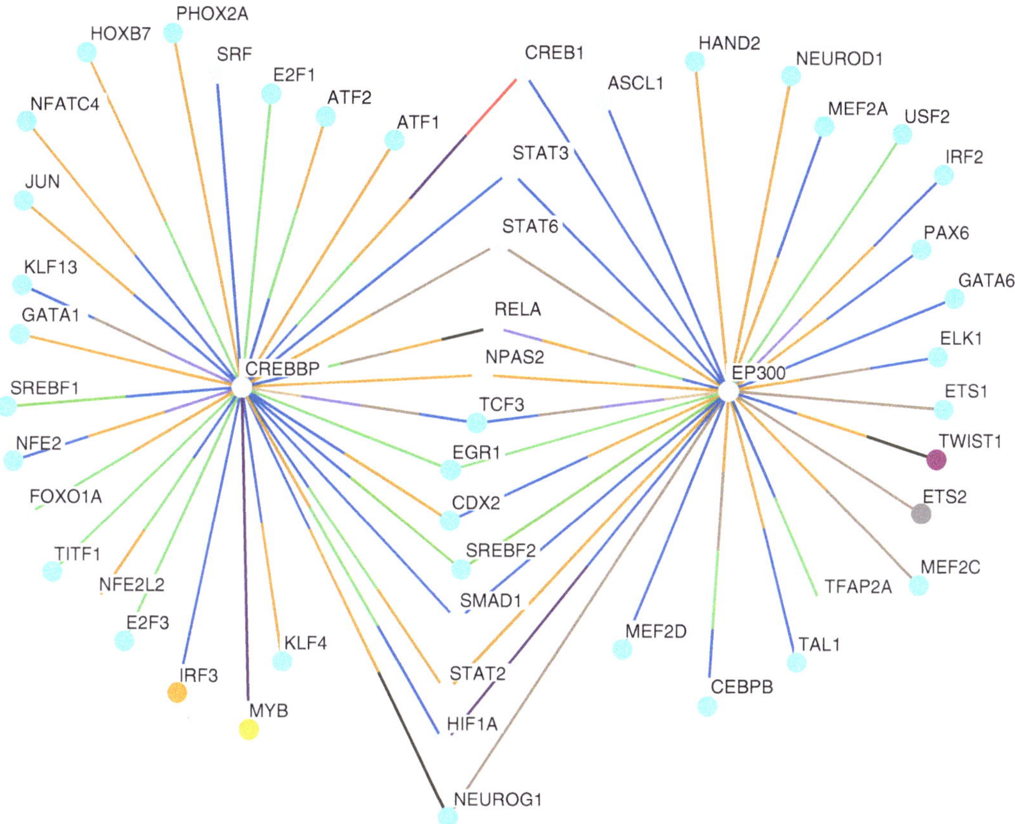

Fig. 3. Examples of transcription factors that bind CREBBP and EP300. See Fig. 2 for examples of nuclear receptors that bind CREBBP and EP300. See Fig. 1 for the source of interaction data and the system for visualization of protein networks.

CREBBP and EP300 play overlapping roles in activation and regulation of transcription (31, 32). Both CREBBP and EP300 associate with a myriad of DNA binding proteins (Fig. 3). Examples include transcription factors that regulate gene expression during development (i.e., HOXB7, KLF13, HAND2, PAX6), transcription factors that control cell-cycle progression (i.e., E2F1, E2F3), transcription factors that response to signaling cues (i.e., ATF1, CREB1, STAT3, EGR1, HIF1), and transcription factors that are ligand-dependent (Fig. 2).

The structure of CREBBP and EP300 share several conserved, modularly organized domains (31, 32). The HAT domain resides in the central region of the proteins. Both the N-and the C-terminal regions of CREBBP and EP300 contribute to activation of transcription. Three domains in EP300 and CREBBP interact with transcription factors that bind DNA. The structure of EP300 and CREBBP also includes a domain (bromo) that binds acetylated lysine to recruit acetylated proteins to a local transcriptional machinery (31).

Both CREBBP and EP300 facilitate communication of the transcription apparatus with nuclear receptors and their coactivators

(Fig. 2). Nuclear hormone receptors are a relatively large family of transcription factors involved in the regulation of genes that play critical roles in numerous biological processes, including development, reproduction, and maintenance of stable metabolic and physiological conditions (34). Nuclear receptors often function as molecular switches, alternating between two states (repression and activation), depending on the absence or presence of cognate ligand (35). Figure 2 provides examples of nuclear receptors including receptor for estrogen (ESR1 and ESR2), progesterone (PGR), androgen (AR), glucocorticoid (GCR), vitamin D (VDR), thyroid hormone (THRA and THRB), and retinoic acid (RARA).

Nuclear receptor coactivators enhance transcriptional activation by liganded, DNA-bound receptors (35). Prominent among these coactivators are those in the SRC family consisting of NCOA1/SCR1, NCOA2/SCR2, and NCOA3/SCR3 (35). The coactivators contribute to enhancement of transcription by modifying histones and through interaction with other components in regulatory networks (Fig. 2).

In protein networks, another hub includes enzymes that methylate H3K4 to create chromatin states that can support transcription (Fig. 1). Several H3K4 methyltransferases are components of relatively large multi-protein complexes that regulate the formation of transcriptionally competent chromatin states (36). These enzymes include SETD1A, SETD1B, and the members of MLL family of regulators of transcription.

In the MLL hub, several proteins provide subunits for formation of core complexes (Fig. 1). These proteins include WDR5, ASH2L, RBP5, HCF1, and MEN1 (37). The core components contribute to communication with other hubs. For example, interaction of MLL with CREBBP links histone H3K4 methylation to histone acetylation and contributes to formation of protein complexes that recruit to chromatin transcription factors that bind DNA (Fig. 1). The MLL hub also includes a protein (CXXC1/Cfp1) that binds unmethylated CpGs to recruit SETD1A and SETD1B to methylate H3K4 (38).

Emerging evidence also supports central roles for MLL in maintaining cellular memory during cell division (39). During cell division, cellular memory is maintained by the Trithorax group (TrxG) and Polycomb group (PcG) of proteins (36). Polycomb proteins preserve the silent state of their associated genes over several cell generations (40). MLL perpetuates active transcription in dividing cells through association with chromatin in a manner favoring genes that were highly transcribed during the interphase stage of cell-cycle (39). Such an association may results through interactions of MLL with specific sequence motifs localized in CGIs (41).

In part, silent transcriptional states are mediated by PRC1 and PRC2: Polycomb repressive complexes 1 and 2 (42). PRC1 catalyzes histone H2A ubiquitination and promotes chromatin compaction.

PRC2 catalyzes tri-methylation of histone H3K27 (42). PCR2 is a multisubunit enzyme (43). The subunits include EZH2, SUZ12, and EED. EZH2 catalyzes the addition of methyl groups to histone H3K27. SUZ12 and EED provide critical stimulatory inputs for the methylation reaction (43).

In mammals, PRC1 and PRC2 target a large proportion of CpG-rich promoters (44). In ES cells, approximately 20 % of these promoters are bound by PRC2 and marked by H3K27me3 modification (19). Evidence supports a causal role for GC-rich sequences in recruitment of PRC2 and implicates a specific subset of CGIs for the initial localization of PRC2 to specific regions in mammalian chromatin (44).

EZH2, the catalytic subunit of PRC2, is a key component in a hub that contributes to repression of transcription (Fig. 1). This hub includes enzymes that deactylate histones and enzymes that methylate DNA. EZH2 plays central roles in recruitment of DNA methyltransferases (DNMT1, DNMT2, and DNMT3) to chromatin to repress transcription (45). Methylated CGIs are strongly and heritably repressed (46).

Gene repression also is tightly linked to histone deacetylation (Figs. 1 and 2). HATs and histone deacetylases (HDACs) exert opposing effects on chromatin structure: HATs to relaxation of compact structures, to permit transcription factors access the DNA, HDACs to chromatin condensation to repress transcription (45). HDACs belong to a relatively large protein family (45). Gene repression is largely achieved by combinatorial action of various enzymatic complexes known as co-repressor complexes (47). HDAC1 and HDAC2 are often associated with distinct repressive complexes known as SIN3, CoREST, and NCoR/SMRT (48). SIN3 complexes include SIN3A or SIN3B, two evolutionarily related proteins with overlapping expression patterns (48). In addition to HDAC1 and HDAC2, the CoREST complex includes a lysine demethylase (47). NCoR/SMRT co-repressors are evolutionarily related (47). NCoR (known as NCOR1) and SMRT (known as NCOR2) contribute to repression of genes regulated by nuclear receptors through association with HDACs (Fig. 2).

Nuclear-receptor networks are rather complex and elaborate (Fig. 2). They communicate with Pol II via distinct subunits in the mediator complex (29). Furthermore, receptor-networks include links to CREBBP and EP300 and links to co-repressor complexes (Fig. 2).

Thus, through networks of protein-protein interactions and combinatorial association of various enzymatic complexes, cells may create a potentially very large number of distinct but related regulatory complexes to maintain cooperatively (Figs. 1–3). The dynamics and the interconnectivity of the networks change over time to produce four-dimensional networks to maintain transcriptional synergy and temporal patterns of gene expression.

References

1. International Human Genome Sequencing Consortium (2001) Initial sequencing and analysis of the human genome. Nature 409:860–921
2. Heintzman ND, Ren B (2009) Finding distal regulatory elements in the human genome. Curr Opin Genet Dev 19:541–549
3. Maston GA, Evans SK, Green MR (2006) Transcriptional regulatory elements in the human genome. Annu Rev Genomics Hum Genet 7:29–59
4. Lemon B, Tjian R (2000) Orchestrated response: a symphony of transcription factors for gene control. Genes Dev 14:2551–2569
5. Sikorski TW, Buratowski S (2009) The basal initiation machinery: beyond the general transcription factors. Curr Opin Cell Biol 21: 344–351
6. Conaway JW, Florens L, Sato S, Tomomori-Sato C, Parmely TJ, Yao T, Swanson SK, Banks CA, Washburn MP, Conaway RC (2005) The mammalian Mediator complex. FEBS Lett 579:904–908
7. Malik S, Roeder RG (2005) Dynamic regulation of pol II transcription by the mammalian Mediator complex. Trends Biochem Sci 30: 256–263
8. Levine M, Tjian R (2003) Transcription regulation and animal diversity. Nature 424: 147–151
9. Juven-Gershon T, Kadonaga JT (2010) Regulation of gene expression via the core promoter and the basal transcriptional machinery. Dev Biol 339:225–229
10. Illingworth RS, Gruenewald-Schneider U, Webb S, Kerr AR, James KD, Turner DJ, Smith C, Harrison DJ, Andrews R, Bird AP (2010) Orphan CpG islands identify numerous conserved promoters in the mammalian genome. PLoS Genet 6
11. Deaton AM, Bird A (2011) CpG islands and the regulation of transcription. Genes Dev 25:1010–1022
12. Saxonov S, Berg P, Brutlag DL (2006) A genome-wide analysis of CpG dinucleotides in the human genome distinguishes two distinct classes of promoters. Proc Natl Acad Sci USA 103:1412–1417
13. Rozenberg JM, Shlyakhtenko A, Glass K, Rishi V, Myakishev MV, FitzGerald PC, Vinson C (2008) All and only CpG containing sequences are enriched in promoters abundantly bound by RNA polymerase II in multiple tissues. BMC Genomics 9:67
14. Gardiner-Garden M, Frommer M (1987) CpG islands in vertebrate genomes. J Mol Biol 196:261–282
15. Mikkelsen TS, Ku M, Jaffe DB, Issac B, Lieberman E, Giannoukos G, Alvarez P, Brockman W, Kim TK, Koche RP et al (2007) Genome-wide maps of chromatin state in pluripotent and lineage-committed cells. Nature 448:553–560
16. Bina M, Wyss P, Ren W, Szpankowski W, Thomas E, Randhawa R, Reddy S, John PM, Pares-Matos EI, Stein A et al (2004) Exploring the characteristics of sequence elements in proximal promoters of human genes. Genomics 84:929–940
17. Pares-Matos EI, Milligan JS, Bina M (2006) Exploring transcription factor binding properties of several non-coding DNA sequence elements in the human NF-IL6 gene. J Mol Biol 357:732–747
18. Bina M, Crowely E (2001) Sequence patterns defining the 5′ boundary of human genes. Biopolymers 59:347–355
19. Zhou VW, Goren A, Bernstein BE (2011) Charting histone modifications and the functional organization of mammalian genomes. Nat Rev Genet 12:7–18
20. Richmond TJ, Davey CA (2003) The structure of DNA in the nucleosome core. Nature 423:145–150
21. Gross DS, Garrard WT (1988) Nuclease hypersensitive sites in chromatin. Annu Rev Biochem 57:159–197
22. ENCODE Project Consortium (2011) A user's guide to the encyclopedia of DNA elements (ENCODE). PLoS Biol 9:e1001046
23. Kouzarides T (2007) Chromatin modifications and their function. Cell 128:693–705
24. Ruthenburg AJ, Allis CD, Wysocka J (2007) Methylation of lysine 4 on histone H3: intricacy of writing and reading a single epigenetic mark. Mol Cell 25:15–30
25. Anamika K, Krebs AR, Thompson J, Poch O, Devys D, Tora L (2010) Lessons from genome-wide studies: an integrated definition of the coactivator function of histone acetyl transferases. Epigenetics Chromatin 3:18
26. Malik S, Roeder RG (2010) The metazoan Mediator co-activator complex as an integrative hub for transcriptional regulation. Nat Rev Genet 11:761–772
27. Blazek E, Mittler G, Meisterernst M (2005) The mediator of RNA polymerase II. Chromosoma 113:399–408

28. Kornberg RD (2005) Mediator and the mechanism of transcriptional activation. Trends Biochem Sci 30:235–239
29. Chen W, Roeder RG (2011) Mediator-dependent nuclear receptor function. Semin Cell Dev Biol 22:749–758
30. Conaway RC, Conaway JW (2011) Function and regulation of the Mediator complex. Curr Opin Genet Dev 21:225–230
31. Khan O, La Thangue NB (2001) p300/CBP proteins: HATs for transcriptional bridges and scaffolds. J Cell Sci 114:2363–2373
32. Chen J, Li Q (2011) Life and death of transcriptional co-activator p300. Epigenetics 6: 957–961
33. Snowden AW, Perkins ND (1998) Cell cycle regulation of the transcriptional coactivators p300 and CREB binding protein. Biochem Pharmacol 55:1947–1954
34. Mangelsdorf DJ, Thummel C, Beato M, Herrlich P, Schutz G, Umesono K, Blumberg B, Kastner P, Mark M, Chambon P et al (1995) The nuclear receptor superfamily: the second decade. Cell 83:835–839
35. Leo C, Chen JD (2000) The SRC family of nuclear receptor coactivators. Gene 245:1–11
36. Schuettengruber B, Martinez AM, Iovino N, Cavalli G (2011) Trithorax group proteins: switching genes on and keeping them active. Nat Rev Mol Cell Biol 12:799–814
37. Ansari KI, Mishra BP, Mandal SS (2009) MLL histone methylases in gene expression, hormone signaling and cell cycle. Front Biosci 14:3483–3495
38. Thomson JP, Skene PJ, Selfridge J, Clouaire T, Guy J, Webb S, Kerr AR, Deaton A, Andrews R, James KD et al (2010) CpG islands influence chromatin structure via the CpG-binding protein Cfp1. Nature 464:1082–1086
39. Blobel GA, Kadauke S, Wang E, Lau AW, Zuber J, Chou MM, Vakoc CR (2009) A reconfigured pattern of MLL occupancy within mitotic chromatin promotes rapid transcriptional reactivation following mitotic exit. Mol Cell 36:970–983
40. Ringrose L, Paro R (2007) Polycomb/Trithorax response elements and epigenetic memory of cell identity. Development 134:223–232
41. Bina M, Wyss P, Novorolsky E, Zulkelfi N, Xue J, Price R, Fay M, Gutmann Z, Folger B (2013) Discovery of MLL binding units and their localization to CpG islands. Submitted for publication
42. Schwartz YB, Pirrotta V (2007) Polycomb silencing mechanisms and the management of genomic programmes. Nat Rev Genet 8:9–22
43. O'Meara MM, Simon JA (2012) Inner workings and regulatory inputs that control Polycomb repressive complex 2. Chromosoma 121:221–234
44. Mendenhall EM, Koche RP, Truong T, Zhou VW, Issac B, Chi AS, Ku M, Bernstein BE (2010) GC-rich sequence elements recruit PRC2 in mammalian ES cells. PLoS Genet 6:e1001244
45. Vire E, Brenner C, Deplus R, Blanchon L, Fraga M, Didelot C, Morey L, Van Eynde A, Bernard D, Vanderwinden JM et al (2006) The Polycomb group protein EZH2 directly controls DNA methylation. Nature 439:871–874
46. Goll MG, Bestor TH (2005) Eukaryotic cytosine methyltransferases. Annu Rev Biochem 74:481–514
47. Perissi V, Jepsen K, Glass CK, Rosenfeld MG (2010) Deconstructing repression: evolving models of co-repressor action. Nat Rev Genet 11:109–123
48. Hayakawa T, Nakayama J (2011) Physiological roles of class I HDAC complex and histone demethylase. J Biomed Biotechnol 2011: 129383
49. Breitkreutz BJ, Stark C, Tyers M (2003) Osprey: a network visualization system. Genome Biol 4:R22
50. Stark C, Breitkreutz BJ, Chatr-Aryamontri A, Boucher L, Oughtred R, Livstone MS, Nixon J, Van Auken K, Wang X, Shi X et al (2011) The BioGRID Interaction Database: 2011 update. Nucleic Acids Res 39:D698–D704
51. FitzGerald PC, Shlyakhtenko A, Mir AA, Vinson C (2004) Clustering of DNA sequences in human promoters. Genome Res 14: 1562–1574
52. Marino-Ramirez L, Spouge JL, Kanga GC, Landsman D (2004) Statistical analysis of over-represented words in human promoter sequences. Nucleic Acids Res 32:949–958
53. Bina M, Wyss P, Lazarus SA, Shah SR, Ren W, Szpankowski W, Crawford GE, Park SP, Song XC (2009) Discovering sequences with potential regulatory characteristics. Genomics 93: 314–322
54. Vinson C, Chatterjee R, Fitzgerald P (2011) Transcription factor binding sites and other features in human and Drosophila proximal promoters. Subcell Biochem 52:205–222

Chapter 2

Isolation of Nuclei for Use in Genome-Wide DNase Hypersensitivity Assays to Probe Chromatin Structure

Guoyu Ling and David J. Waxman

Abstract

DNase hypersensitivity (DHS) analysis coupled with high-throughput DNA sequencing (DNase-seq) has emerged as a powerful tool to analyze chromatin accessibility and identify regulatory sequences in genomic DNA on a global scale. In this method, intact nuclei are isolated from fresh tissue or cultured cells and then subjected to limited digestion using DNase I. The resulting short DNA fragments released by DNase digestion, which correspond to regions of open chromatin structure, are subsequently purified and identified by high throughput next generation DNA sequencing. This chapter describes methods used to isolate intact nuclei from mouse liver suitable for DNase-seq studies. The following chapter presents a detailed protocol for DNase I digestion of liver nuclei followed by the isolation of DNase-released fragments for sequencing and genome-wide mapping of DHS sites.

Key words: DNase I hypersensitivity assay, Next generation DNA sequencing, DNase-seq, Chromatin structure

1. Introduction

In eukaryotic cells, genomic DNA is wrapped around a core of histone proteins, forming nucleosomes, which are further packaged into chromatin. Chromatin controls the accessibility of regulatory proteins to DNA and plays an important role in many critical nuclear processes, including gene regulation, DNA repair and replication (1). DHS assays using the enzyme DNase I have long been used to identify chromatin regions that are open (i.e., accessible to cleavage by DNase I) (Fig. 1). These open chromatin regions are frequently associated with DNA regulatory elements, including gene promoters, distal enhancer sequences, silencers, insulators, and locus control regions (2), and thus encompass genomic sequences of substantial biological importance. Recent advances

Minou Bina (ed.), *Gene Regulation: Methods and Protocols*, Methods in Molecular Biology, vol. 977,
DOI 10.1007/978-1-62703-284-1_2, © Springer Science+Business Media, LLC 2013

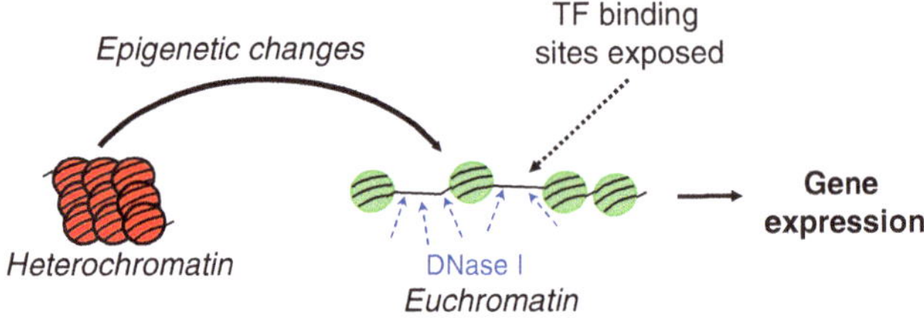

Fig. 1. DNase hypersensitivity identifies open genomic regions in chromatin that facilitate transcription factor (TF) binding to chromatin and induce gene expression. Nucleosomes are compacted in closed, inactive heterochromatin but are more open, exposing sites of DNase hypersensitivity in the euchromatin state.

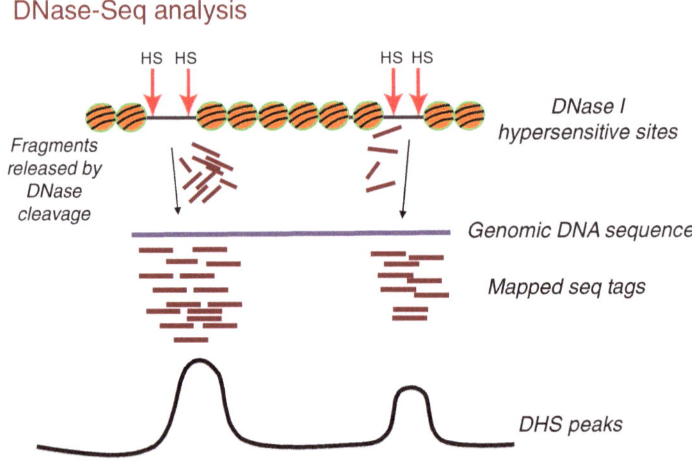

Fig. 2. DNase-seq analysis. Sites of hypersensitivity (HS) are susceptible to cutting by DNase I, which releases many fragments of variable length from each hypersensitive region. The released fragments are then purified, sequenced on one end, and the resultant sequence tags then mapped back to the genome. A peak detection algorithm is used to identify DHS peaks, two of which are shown.

have enabled traditional DHS assays to be coupled with high throughput mapping methods, such as tiling microarrays (DNase-chip) and, more recently, DNA sequencing (DNase-seq), making it possible to obtain high resolution genome-wide maps of DHS sites upon mapping released DNA fragments back to the genome (Fig. 2) (3–6).

DNase-seq can be carried out with nuclei isolated from intact mammalian tissues, as exemplified by studies from this laboratory on sex differences in mouse liver chromatin structure (7). DNase-seq studies using intact nuclei isolated from fresh tissue, such as mouse liver, have the important advantage of providing detailed information about the regulation of chromatin structure under physiological conditions in vivo. However, care must be taken to ensure that the procedure yields high quality nuclei with minimal disturbance of chromatin structure. In the protocol described here, isolation of fresh, high quality nuclei from tissues is facilitated by sucrose ultracentrifugation. In brief, fresh liver tissue is collected, homogenized, and the nuclei are pelleted by ultracentrifugation through a 2 M sucrose cushion, which helps maintain the integrity of the nuclei and chromatin structure. Only intact nuclei are of sufficiently high density to pass through the sucrose cushion (8). Nuclei isolated and purified by this method give reproducible and reliable DNase-seq results (7). The same protocol can also be used to isolate nuclei from tissue culture cells, although in that case a simpler, detergent-based method may work as well (5). We anticipate that the protocol described here can be applied to other tissues that yield high quality nuclei without major modifications.

2. Materials

Prepare all solutions using ultrapure water and analytical grade reagents. Store all buffers at 4 °C unless otherwise noted. Protease inhibitors, spermine, spermidine, and DTT should be added fresh to each solution just prior to their use.

2.1. Materials for Dissection of Liver Tissue

1. Scissors, blades, and forceps for tissue dissection.
2. CO_2 chamber for rodents.
3. Paper towels, Kimwipes, and petri dishes.
4. Ice-cold 1.15% KCl and 1× Phosphate-buffered saline (PBS).
5. 250 ml beaker, 50 ml conical tubes.
6. 1 ml syringes and 27G needles (optional).

2.2. Tissue Homogenization and Isolation of Nuclei

1. Preparative ultracentrifuge and SW28 rotor, or equivalent. Ultra-clear centrifugation tubes (Beckman Coulter, Brea, CA; cat. # 344058).
2. Potter-Elvehjem Tissue Grinder with Teflon Pestle, 10, 15, 30 or 55 ml size (Wheaton Science Products, Millville, NJ; see Note 1), and a power drill.
3. Nuclear homogenization buffer (NEHB): 10 mM HEPES-KOH, pH 7.9, 25 mM KCl, 1 mM EDTA, 2 M sucrose, 10% glycerol, 0.15 mM spermine, 0.5 mM spermidine, 10 mM NaF, 1 mM orthovanadate, 1 mM PMSF, 0.5 mM DTT, and 1× protease inhibitor cocktail (Sigma, St Louis, MO; cat. # P8340) (see Note 2).
4. Nuclear storage buffer: 20 mM Tris–HCl, pH 8.0, 75 mM NaCl, 0.5 mM EDTA, 50% (v/v) glycerol, 1 mM DTT, and 0.1 mM PMSF.

5. 1 M HEPES-KOH, pH 7.9. Dissolve 238.3 g HEPES in 700 ml of ultrapure water. Use potassium hydroxide pellets and 1 M KOH solution to adjust pH to 7.9. Bring up to 1,000 ml with ultrapure water.
6. 1 M Tris–HCl, pH 8.0. Dissolve 121.1 g Tris base in 800 ml of ultrapure water. Add concentrated HCl to bring to pH 8.0. Bring up to 1,000 ml with ultrapure water.
7. 0.5 M EDTA, pH 8.0 (Sigma, St. Louis, MO).
8. Dounce homogenizer with B pestle, chilled on ice (Wheaton Science Products, Millville, NJ).
9. Inverted tissue culture microscope, hemocytometer, and trypan blue solution (Sigma, cat. #T8154 or equivalent) for counting nuclei.
10. 1.5-ml microcentrifuge tubes, chilled on ice.
11. Liquid nitrogen for freezing nuclei.

3. Methods

All steps described here, except animal-related procedures, should be carried out in a 4 °C cold room. Samples should be kept at 4 °C to minimize nonspecific DNA degradation.

3.1. Dissection of Liver Tissue

1. Prepare complete NEHB on ice (see Note 2). Precool the ultracentrifuge, rotor, tubes, and the tissue grinder before starting the experiments.
2. Euthanize the mice using approval protocols, e.g., CO_2 asphyxiation followed by cervical dislocation. Subsequent steps should be carried out promptly to ensure that the minced liver tissue is submerged in ice-cold NEHB (**step 6**) as quickly as possible.
3. If required, remove blood sample from the heart using a 1-ml syringe with 50 μl of heparin and a 27G needle (optional).
4. Remove each liver and rinse in ice-cold PBS. Snap-freeze a small piece of tissue and store at −80 °C, e.g., for RNA isolation at a later time.
5. Place the liver in a 100 × 15 mm petri dish on ice containing ice-cold 1.15% KCl. Cut the liver into small pieces and wash twice with 1.15% KCl.
6. Carefully dry off the excess KCl solution with a Kimwipe. Weigh the liver and submerge the minced liver in ice-cold NEHB buffer.

3.2. Tissue Homogenization and Isolation of Nuclei (Scale: Pool of Three to Four Mouse Livers) (See Note 1)

1. Homogenize the liver in a 30 ml Potter-Elvehjem Tissue Grinder with three to four strokes. Use a buffer-to-tissue ratio of ~6 (v/w) (see Note 3).
2. Carefully overlay 25 ml of homogenate on top of 10 ml NEHB in a SW28 tube.
3. Centrifuge at 25,000 rpm (90,000 × *g*) for 45 min at 2 °C. The nuclei should be found in a clear pellet at the bottom of the tube (see Note 4).
4. Remove tissue debris floating on the top and decant the supernatant. Turn the tube upside down in a cold room and wipe the sides of each tube with a Kimwipe soaked with PBS.
5. Resuspend the pellet in a minimal volume of nuclear storage buffer by applying a few strokes of a Dounce homogenizer (see Note 5).
6. Dilute a sample of the resuspended nuclei 100-fold with PBS and count the nuclei using a hemocytometer. Trypan Blue can be added to a final concentration of 0.04% to help visualize the nuclei. Nuclei should be counted at least three times to ensure that a reproducible count is obtained.
7. Snap-freeze the suspended nuclei in small aliquots using liquid nitrogen, for example 300 μl aliquots in 1.5 ml micro-centrifugation tubes, and store at −80 °C (see Note 6).

4. Notes

1. Nuclei may be isolated from a single adult mouse liver if DNase-seq data are to be collected for individual animals, otherwise it may be more convenient to pool livers from three to four mice and prepare a single preparation of nuclei, using a 30 ml or 55 ml tissue grinder for liver homogenization. If the tissue weight is ~2 g or less (e.g., when isolating nuclei from a single mouse liver), a 10 or 15 ml tissue grinder can be used and the buffer volumes and sizes of the ultracentrifuge tubes and rotor should be scaled down accordingly.
2. NEHB is prepared and stored at 4 °C as 10 mM HEPES pH 7.9, 25 mM KCl, 1 mM EDTA, 2 M sucrose, and 10% glycerol. The other buffer components: spermine, spermidine, NaF, orthovanadate, PMSF, DTT, and protease inhibitor cocktail are prepared separately as concentrated stock solutions (100× for NaF, orthovanadate (both in water), PMSF (in isopropanol), and protease inhibitor cocktail (in DMSO); 2,000× for spermine, spermidine, and DTT (all in water)), and stored in aliquots at −20 °C. The concentrated components are then added to NEHB just before tissue dissection and homogenization.

3. A buffer-to-tissue ratio of 6:1 (v/w) is recommended for mammalian tissues. Insufficient buffer during tissue homogenization can lead to impure nuclear preparations. Note that the presence of 2 M sucrose makes it difficult to carry out the homogenization step. As such, gloves and goggles should be worn during homogenization for personal protection in the event of glassware failure. Three to four strokes of the homogenizer are required to disrupt >90% of the liver tissue. Other tissues may require more vigorous conditions for effective homogenization.
4. A nuclear pellet with a red tinge indicates that the nuclei are contaminated by red blood cells and should be discarded if replalcement material is readily available. Increasing the volume of buffer used for homogenization may help avoid this problem. It may be possible to wash the nuclei by resuspending the nuclear pellet in NEHB and then repeating the centrifugation step.
5. The nuclei should be resuspended in a minimal volume of nuclear storage buffer prior to counting. For example, nuclei from three to four livers should initially be resuspended in approx. 0.5 ml storage buffer. Additional buffer can be added as required once the count is known to adjust the final preparation to 50–100 million nuclei per ml.
6. Although fresh nuclei may in some cases be preferrable for DHS assays, we have not noticed any difference in results when frozen mouse liver nuclei are used after storage at −80°C for several months.

Acknowledgments

Supported in part by NIH grant DK33765 (to DJW).

References

1. Bell O, Tiwari VK, Thomä NH, Schübeler D (2011) Determinants and dynamics of genome accessibility. Nat Rev Genet 12:554–564
2. Gross DS, Garrard WT (1988) Nuclease hypersensitive sites in chromatin. Annu Rev Biochem 57:159–197
3. Boyle AP, Davis S, Shulha HP, Meltzer P, Margulies EH, Weng Z, Furey TS, Crawford GE (2008) High-resolution mapping and characterization of open chromatin across the genome. Cell 132:311–322
4. Crawford GE, Davis S, Scacheri PC, Renaud G, Halawi MJ, Erdos MR, Green R, Meltzer PS, Wolfsberg TG, Collins FS (2006) DNase-chip: a high-resolution method to identify DNase I hypersensitive sites using tiled microarrays. Nat Methods 3:503–509
5. Sabo PJ, Kuehn MS, Thurman R, Johnson BE, Johnson EM, Cao H, Yu M, Rosenzweig E, Goldy J, Haydock A, Weaver M, Shafer A, Lee K, Neri F, Humbert R, Singer MA, Richmond TA, Dorschner MO, McArthur M, Hawrylycz M, Green RD, Navas PA, Noble WS, Stamatoyannopoulos JA (2006) Genome-scale mapping of DNase I sensitivity in vivo using tiling DNA microarrays. Nat Methods 3:511–518

6. Song L, Crawford GE (2010) DNase-seq: a high-resolution technique for mapping active gene regulatory elements across the genome from mammalian cells. Cold Spring Harb Protoc 2010:pdb.prot5384
7. Ling G, Sugathan A, Mazor T, Fraenkel E, Waxman DJ (2010) Unbiased, genome-wide in vivo mapping of transcriptional regulatory elements reveals sex differences in chromatin structure associated with sex-specific liver gene expression. Mol Cell Biol 30(23): 5531–5544
8. Lichtsteiner S, Wuarin J, Schibler U (1987) The interplay of DNA-binding proteins on the promoter of the mouse albumin gene. Cell 51:963–973

Chapter 3

DNase I Digestion of Isolated Nulcei for Genome-Wide Mapping of DNase Hypersensitivity Sites in Chromatin

Guoyu Ling and David J. Waxman

Abstract

DNase I hypersensitivity (DHS) analysis is a powerful method to analyze chromatin structure and identify genomic regulatory elements. Integration of a high-throughput detection method into DHS analysis makes genome-wide mapping of DHS sites possible at a reasonable cost. Here we describe methods for DHS analysis carried out with mouse liver nuclei, involving DNase I digestion followed by isolation of DNase I-released DNA fragments suitable for high-throughput, next generation DNA sequencing (DNase-seq). A real-time PCR-based assay used to optimize DNase I digestion conditions is also described.

Key words: DNase I hypersensitivity assay, Next generation DNA sequencing, DNase-seq, Chromatin structure

1. Introduction

Open chromatin regions are associated with active chromatin and can be identified by their hypersensitivity to nuclease digestion (1). Hypersensitive sites can be identified by DNase hypersensitivity (DHS) assays, in which intact, isolated nuclei are subjected to limited digestion with the enzyme DNase I. Using classical methods, DNase I cleavage sites are identified by restriction digestion followed by Southern blotting using radiolabeled genomic fragments to probe a specific genomic region of interest (2). Although widely used for many years, this method is slow and labor intensive and provides only a low resolution map of DHS sites, typically for a limited number of regulatory elements in close proximity to a single gene. Thus, classical DHS assay methods suffer from limited capacity to elucidate DHS sites that are distant (>100 kb) from their target genes (3). Classical DHS assays cannot readily be used to identify such distant DHS sites; moreover, the classical methods do not provide a global view of the overall chromatin landscape.

Minou Bina (ed.), *Gene Regulation: Methods and Protocols*, Methods in Molecular Biology, vol. 977,
DOI 10.1007/978-1-62703-284-1_3, © Springer Science+Business Media, LLC 2013

These limitations can be overcome using high-throughput methods to identify DNA fragments generated by DHS assays, notably tiling arrays (DNase-chip) and next generation DNA sequencing (DNase-seq) (4–7). Notably, DNase-seq generates genome-wide DHS maps that identify functional DNA elements with single nucleotide resolution, as exemplified by several studies, including the ENCODE project (8) and by work from this laboratory studying global DHS sites in mouse liver (9).

Two experimental methods have been developed for high-throughput DHS assays (5, 6), the main difference being how samples are processed after DNase I digestion. One method detects single cuts and requires several steps, including blunt-ending followed by ligation of a biotinylated linker, DNA shearing, and streptavidin bead affinity capture (5). The second method detects two nearby cuts and is much simpler, requiring only purification of short DNA fragments released by DNase I digestion (6). By selecting for and sequencing released DNA fragments that are short (typically 200–400 nt), the second method (DNase fragment release method) minimizes the impact of the nonspecific, random DNA cuts that inevitably occur during isolation of nulcei, but which are rarely in close proximity to each other. In contrast, increased background may result when random DNA cut sites are end labeled with biotinylated linkers using the single cut method of (5). Thus, when using the DNase fragment release method, the impact of nonspecific DNA shearing during sample handling is appreciably reduced.

As originally described, both methods used tiling arrays for detection of DNase I-released fragments (5, 6). However, characterization of the released fragments can presently be carried out at substantially reduced cost using high-throughput, next generation DNA sequencing (DNase-seq). Instrumentation such as the Illumina, Inc. HighSeq 2000 presently yields in excess of 150 million, 40 nt-long sequence tags from a single sequencing lane. This is sufficient to obtain high resolution genome-wide DHS maps of mouse liver for up to five individual DNase-seq samples, which can be multiplexed and sequenced together on a single sequencing lane using sample-specific sequence tags (bar codes).

The following protocol for DNase I digestion and fragment isolation is based on the DNA fragment release method of Sabo et al. (6). We have found this method to be very reliable, and have successfully implemented it to generate high quality, unbiased genome-wide DHS maps of mouse liver chromatin under a number of different biological states (9). Short fragments released by DNase-I digestion of isolated nuclei are separated from the bulk of genomic DNA by sucrose gradient ultracentrifugation (Fig. 1). The size distribution of DNA fragments across the sucrose gradient is then determined by DNA electrophoresis and fluorescence imaging. DNA fragments of the desired size are then isolated, typically from a single fraction of the sucrose gradient (fraction 7 under

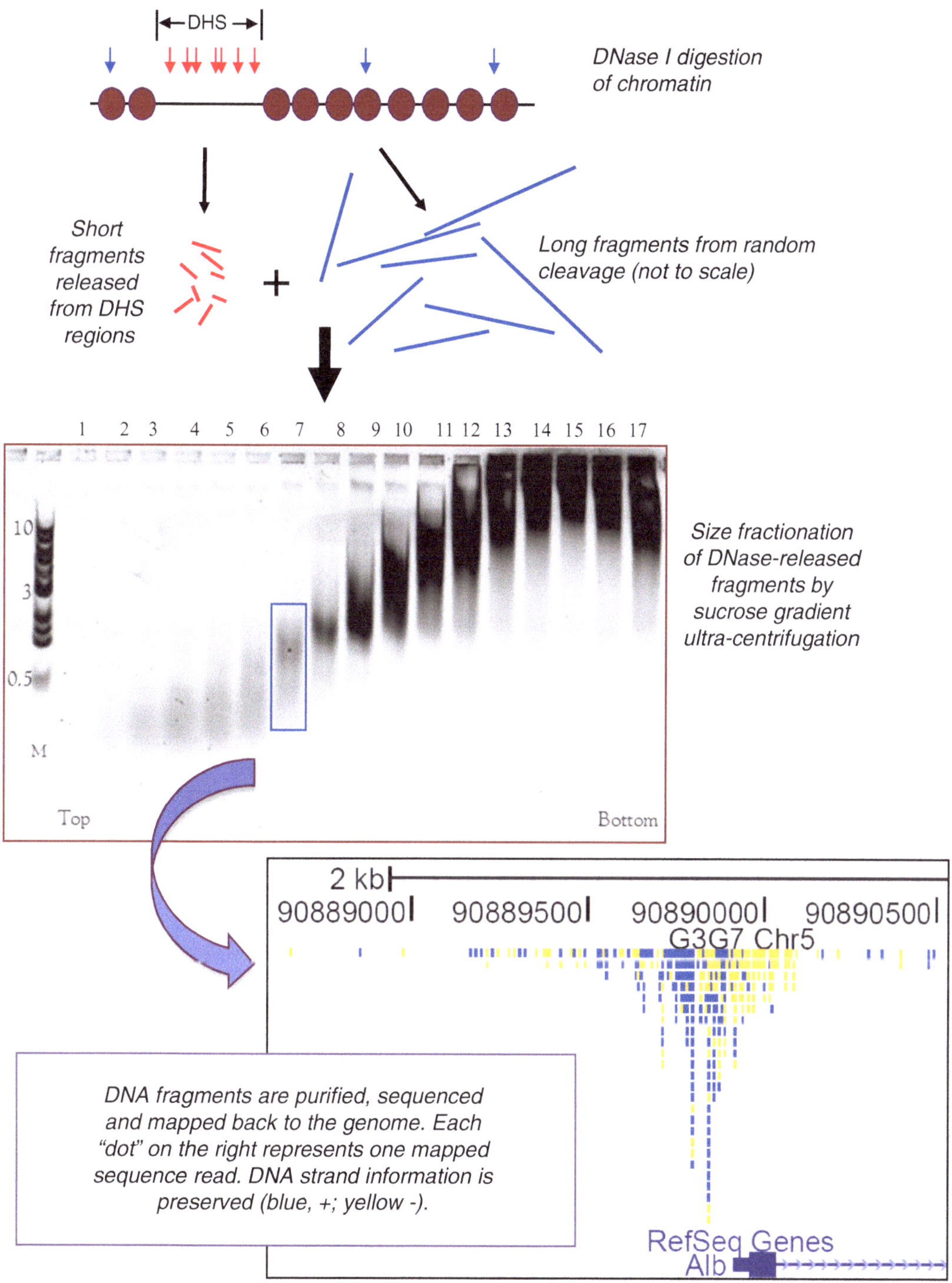

Fig. 1. Schematic diagram of DNase-seq method. A single DHS site surrounded by heterochromatin is shown at the *top*. The *ovals* represent highly packed nucleosomes, and the *arrows* are DNase I cleavage sites, which occur much more frequently in DHS regions. Limited DNase I digestion releases short DNA fragments from DHS regions (*red arrows*, and *red fragments*), as well as larger fragments that result from random cutting of genomic DNA (*blue arrows*, and *blue fragments*). The short DNA fragments are then separated from larger fragments by sucrose gradient ultracentrifugation, as shown in the gel at the center. High-throughput sequencing (typically using DNA purified from fraction 7) is then carried out to obtain 40 nt sequence tags, which are then mapped back to the genome. Mapped DNase-seq data can then be uploaded to the UCSC genome browser to visualize clusters of sequence tags, which indicate a DHS site. DHS peaks such as the one shown here at the *Alb* promoter on mouse chromosome 5 typically show an asymmetric distribution of positive and negative sequence reads, with (+) strand sequences marked in *blue* and (–) strand sequences marked in *yellow*. Quantitative analysis of the full DHS dataset is then carried out using peak-calling software (e.g., (9)).

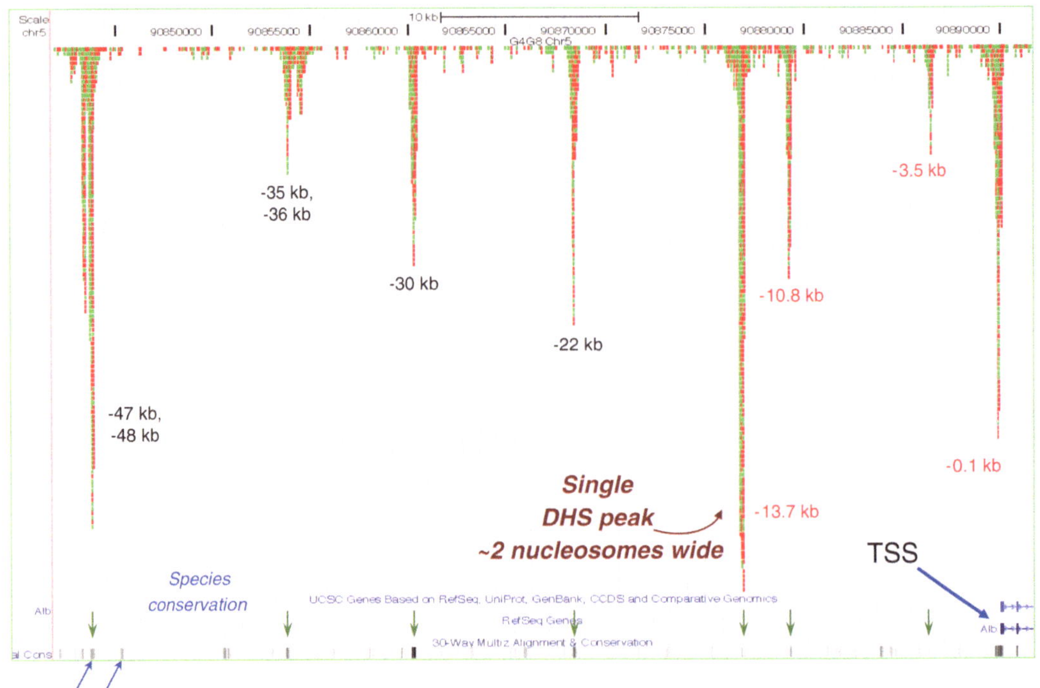

Fig. 2. DHS sites upstream of mouse *Alb* gene. Shown is a UCSC browser screen view of a 50 kb region upstream of the *Alb* promoter on mouse chromosome 5. Individual sequence tags are (*green*, (+) strand sequences; *red*, (–) strand sequences). The locations of the ten DHS sites in view are marked relative to the *Alb* transcription start site (TSS). The DNase-seq peaks shown (i.e., DHS sites) are very sharp, generally 350–500 nt wide and corresponding to an average of two nucleosomes. The DHS sites shown coincide with regions exhibiting high cross-species conservation, as shown along the *bottom* track (*green arrows*).

the experimental conditions described below) (Fig. 1). Downstream steps required to generate libraries of fragments for high-throughput sequencing (e.g., using standard Illumina sample preparation protocols) include a size-selection step, which further improves the signal to noise in the final DNase-seq dataset by limiting sequencing to DNase-released DNA fragments ~200–400 nt in length.

Quantitative PCR (qPCR) is used to optimize the conditions required to achieve limited DNase I digestion. qPCR can also be used to validate the final size-selected fragments prior to DNA sequencing. If one or more specific DHS sites has already been identified for the tissue or cell line under study, those sites serve as positive controls for DNase-released fragments in the qPCR validation assay. For example, for DHS analysis of mouse liver chromatin, qPCR primer pairs that map to a strong DHS site associated with the *Alb* gene (i.e., an open genomic region) can be used as a positive control (Fig. 2). qPCR primer pairs that map to an intergenic region that is closed and, therefore, insensitive to DNase I digestion, are used as a negative control. Samples validated by qPCR to show strong enrichment of DNase I-released fragments in the open, hypersensitive region compared to the closed, intergenic region are deemed suitable for library preparation and high-throughput DNA sequencing. DNase digestion followed by qPCR analysis can thus be used to optimize conditions for DHS cleavage and fragment isolation. If no DHS sites are known for the tissue or cell line under study, then a preliminary DNase-seq experiment can be carried out using digestion conditions recommended below. DHS sites identified in the preliminary experiment can then be used to design qPCR primers to optimize DNase digestion conditions.

2. Materials

2.1. DHS Assay

1. 37 and 55°C water baths.
2. 2 ml Safe-Lock Eppendorf tubes or equivalents.
3. 15-ml and 50-ml conical tubes.
4. Microcentrifuge at 4°C.
5. Buffer A: 15 mM Tris-Cl, pH 8.0, 15 mM NaCl, 60 mM KCl, 1 mM EDTA, 0.5 mM EGTA, 0.5 mM spermidine, and 0.3 mM spermine. Store Buffer A without spermine and spermidine at 4°C. Add concentrated stock solutions of spermidine and spermine to the buffer just before use.
6. 10× DNase I Digestion Buffer: 60 mM $CaCl_2$, 750 mM NaCl.
7. Buffer D (Digestion buffer): mix 1 volume of 10× DNase I digestion buffer with 9 volume of Buffer A.
8. Stop Buffer: 50 mM Tris-Cl, pH 8.0, 100 mM NaCl, 0.1% SDS, 100 mM EDTA, 20 μg/ml RNase A, 0.5 mM spermidine, and 0.3 mM spermine. Store the Stop Buffer without RNase A, spermidine, and spermine at 4°C. Warm up in a 37°C water bath, and add RNase A, spermidine, and spermine just before use.
9. RQ1 DNase I (RNase-free; 1 U/μl, Promega, Madison, WI; cat. # M6101).
10. 20 mg/ml proteinase K (Sigma, St. Louis, MO; cat. # P4850).
11. 25:24:1 (v/v/v) phenol–chloroform–isoamyl alcohol saturated with 10 mM Tris, pH 8.0, 1 mM EDTA (Sigma, St. Louis, MO; cat. # P3803).

2.2. Isolation of DNase I-Released Fragments

1. Agarose gel electrophoresis apparatus.
2. DNA ladder (100 bp DNA ladder from New England Biolabs, Ipswich, MA; cat. # N3231).
3. SYBR Green I nucleic acid staining solution, 10,000× in DMSO (Invitrogen, Carlsbad CA; cat. # S7563).

4. Typhoon Trio laser scanner and image system (GE Healthcare Life Sciences, Piscataway, NJ) or similar instrument.
5. Qiagen PCR Purification kit (Qiagen, Valencia, CA; cat. # 28104) and vacuum manifold for DNA purification (Qiagen, Valencia, CA; cat. # 19413). Qiagen kits that have been stored for long periods of time give low yields of recovered DNA and should not be used.
6. Quant-IT high sensitivity DNA assay kit (Invitrogen, Carlsbad CA; cat. # Q33120), multi-well plates, and fluorescent plate reader for DNA concentration measurement.
7. 2× Centrifugation buffer: 40 mM Tris-Cl, pH 8.0, 10 mM EDTA, 2 M NaCl.
8. Sucrose solutions, prepared at concentrations (w/v) of 40, 35, 30, 25, 20, 17.5, 15, 12.5, and 10% in 1× centrifugation buffer. These solutions are prepared by adding sucrose to 2× centrifugation buffer; water is then added to adjust to 1×. For example, to prepare 40% sucrose solution in 1× centrifugation buffer, dissolve 400 g sucrose in 500 ml of 2× centrifugation buffer, and then use water to bring up to 1 L. The other sucrose solutions can be prepared in the same way, or can be prepared by mixing 40% sucrose with 1× centrifugation buffer without sucrose as required.
9. PB buffer for DNA purification (Qiagen, Valencia, CA).
10. 1× TAE buffer: 40 mM Tris-acetate, pH 8.0, 1 mM EDTA.
11. NaCl, 5 M.

2.3. Validation of DHS Sample Quality by qPCR

1. qPCR primer sets targeting known DHS sites and as well as genomic regions that are insensitive to DNase I.
2. Power SYBR Green PCR Master Mix (Applied Biosystems, Foster City, CA; cat. # 4368702, or equivalent).
3. qPCR instrument, ABI 7500 (Applied Biosystems, Foster City, CA) or equivalent.
4. Genomic DNA template and DNase-released DNA fragments.

3. Methods

Subheadings 3.1–3.3 describe DNase I digestion carried out on a scale of 30 million mouse liver nuclei, divided into six 2-ml microcentrifugation tubes, each containing a 1 ml reaction volume (see Note 1). DNase I treatment of purified genomic DNA (in place of isolated nuclei) is used to generate control samples, which are used to define the background level of intrinsic hypersensitivity of genomic DNA; DNase fragmentation of these control samples requires a much lower concentration of DNase I than the DNase-digesed nuclei (see Note 2). The quality of the DHS samples obtained can be assessed by qPCR analysis using one or more known DHS sites as positive controls (see Note 3).

The use of an optimal DNase I concentration and digestion time is critical for the success of DHS-seq analysis. Subheading 3.4 describes a protocol that can be used to optimize DNase I digestion conditions using a DHS site cleavage assay, modified from McArthur et al. (12). This protocol is based on the concept that DNase digestion at a DHS site destroys the ability to PCR amplify a genomic region that spans the DHS site. The qPCR amplicon should therefore be several hundred bp in length, and must span the DHS site used to optimize the DNase digestion conditions with respect to DNase concentration and/or time of digestion. Several DNase I concentrations are typically tested in parallel using digestion time and reaction conditions that are identical to the full-scale digestion protocol described in Subheadings 3.1 and 3.2. Alternatively, the DNase I concentration may be fixed and the incubation time may be varied. This optimization protocol requires fewer nuclei and is less time-consuming than the full-scale DNase fragment release assay. Once an optimal DNase I concentration (and/or digestion time) is identified by qPCR validation, a larger scale DNase digestion can be performed to purify released DNA fragments for sequencing. As an alternative to the above DHS site cleavage assay for optimization of the DNase digestion conditions, optimization can be carried out using the qPCR-based fragment release assay described in Subheading 3.3. This involves comparison of the qPCR signal for release of DNA fragments from a known hypersensitive region to the qPCR signal for a non-hypersensitive control (background) genomic region.

3.1. DNase I Digestion

1. Add spermine, spermidine, and RNase A fresh to Buffer A and to Stop Buffer (see previous chapter). Prepare Buffer D from freshly made Buffer A and 10× DNase I buffer. Warm up the required amount of Buffer D and Stop Buffer (for example 10 ml each for a 6 ml digestion vol) in 15 ml conical tubes to 37°C at least 10 min before digestion.
2. Thaw on ice an aliquot of frozen nuclei. Aliquot total amount of nuclei required (five million nuclei/ml digestion reaction; 30 million nuclei in total for six 1-ml reactions run in parallel) into a 1.5-ml tube pre-aliquoted with 500 μl ice cold Buffer A. Mix by inverting the tube several times and spin at 500 rcf for 10 min at 4°C. Decant the supernatant and tap the tubes to loosen the nuclear pellet. Wash with Buffer A once more, remove most of the supernatant, loosen the pellet, and place the tubes on ice (see Note 4).

3. Aliquot the desired amount DNase I enzyme needed for each reaction, for example, 40 μl of 1 U/μl RQ1 DNase I, into each of six 2-ml tubes pre-aliquoted with 60 μl Buffer D (see Note 5).
4. Carry out DNase I digestion in each of six tubes, each containing 1-ml reaction volume and the same batch of nuclei (see Note 6):
   (a) Resuspend nuclei in 5.5 ml of Buffer D (prewarmed to 37°C for at least 10 min) in a 15-ml tube (master tube of the nuclei). Some clumping of nuclei may occur but should not affect the assay.
   (b) Place one of the reaction tubes in a 37°C water bath every 20 s, thereby staggering the start of DNase digestion for each sample. Incubate each tube for 3.0 min before adding the nuclei, as described in the next step.
   (c) Transfer 850 μl of suspended nuclei from the master tube of nuclei to a 2-ml prewarmed reaction tube at the precise time specified for each tube. Stop each reaction by adding 950 μl of Stop Buffer after precisely 2.0 min of digestion with DNase I.
5. Immediately transfer each reaction tube to a 55°C water bath for 15 min, then add 10 μl of 20 mg/ml proteinase K.
6. Continue incubation at 55°C overnight.
7. Perform a phenol-chloroform extraction using an equal volume of buffered phenol–chloroform–isoamyl alcohol (i.e., 6 ml for a set of six 1-ml reactions). Store the liquid phase at 4°C or proceed to fragment isolation directly (see Note 7).

3.2. Isolation of DNase I-Released Fragments

1. Add 5 M NaCl to the extracted sample to give a final concentration of 0.8 M NaCl (e.g., 1.76 ml 5 M NaCl per 11.0 ml of supernatant recovered after phenol–chloroform–isoamyl alcholol extraction).
2. Prepare sucrose step gradient by carefully overlaying 3-ml cushions of each sucrose solution in each of two SW28 ultracentrifuge tubes, placing the denser gradient fractions at the bottom. Sequentially layer the following final sucrose concentrations of 40, 35, 30, 25, 20, 17.5, 15, 12.5, and 10% in 1× centrifugation buffer.
3. Load 5.7 ml of the sample from step 1 on top of each SW28 tube filled with the sucrose step gradient. Centrifuge for 24 h at 25,000 rpm (90,000 × *g*) and 25°C in an ultracentrifuge.
4. The next day, carefully collect sequential 1.9 ml fractions into a set of clean numbered tubes, beginning at the top of the gradient. Store fractions at 4°C.
5. Electrophorese a 45 μl aliquot of each fraction on a 1.2% TAE agarose gel at 120 v for 1 h, running 2 ng of a 100 bp DNA

ladder in a separate lane. Stain the gel with 1× SYBR green solution (Invitrogen, Carlsbad CA) in 1× TAE for 30 min. Capture the gel images using a Typhoon Trio laser scanner according to the manufacturer's instructions.

6. Purify DNA from the sucrose fraction(s) that contain the desired size of DNA fragments using a Qiagen PCR Purification Kit according to the manufacturer's instructions. We routinely recover DNA from fraction 7 (Fig. 1; also see Note 8). Use one Qiagen column per sample. A vacuum device is preferred over centrifugation as the Qiagen column has to be loaded multiple times.
7. Determine the DNA concentration with a Quant-IT DNA high sensitivity assay kit. Store the purified DNA samples at −20°C, or at −80°C if long term storage is required.

3.3. Validation of DHS Sample Quality by qPCR

1. Identify genomic sequences corresponding to known DNase I-hypersensitive regions (positive controls) and insensitive regions (negative controls).
2. Use a suitable qPCR primer design program, such as Primer Express (Applied Biosystems, Foster City, CA), to design qPCR primers targeting DNase I hypersensitive and insensitive regions. The amplicon length should be kept short relative to the lengths of the released fragments (ideally about 50–60 bp) to increase the chance of detecting short fragments released by DNase digestion.
3. Use NCBI's Blat or Blast search tools to ensure the specificity of the PCR primers.
4. Perform qPCR assays based on SYBR Green-based methodology using 10 ng of genomic DNA (denoted as *g*) and 1% of the purified DHS sample (denoted as *s*).
5. Calculate the fold-enrichment adjusted for differences in primer efficiency, as follows: Fold enrichment of DHS site (denoted as DH) over DNase I-insensitive regions (denoted as IN) = 2 to the power of $\{(DHg - DHs) - (INg - INs)\}$, where DH*g* is the qPCR threshold cycle number (Ct number) obtained for the DHS qPCR primer pair using genomic DNA as template, and IN*s* is the qPCR Ct number obtained for the DNase I insensitive region qPCR primer pair using the purified, released DNA fragments as template.
6. For DNase-seq studies in mouse liver, the following qPCR primers can be used as a positive control for validation of DNase digestion. *Alb* gene DHS primers: 5′-CAATGAAAT GCGAGGTAAGTATGG-3′ and 5′-TCTTTAACCAATAACT GTAGATCATTAACCA-3′, corresponding to a 52 bp amplicon length. DNase I insensitive region primers located on Chr 3: 71,026,628–71,026,685 (mouse genome build mm9 and

amplicon length 58 bp) can be used as a negative control (background region): 5′-GCAGCAGATGGCAAGTAATACT AAGAT-3′ and 5′-CCCTTATTCTCTGAGCATTAGACAG TTATA-3′.

3.4. Method for Optimization of DNase I Digestion Conditions Using DHS Site Cleavage Assay

1. Carry out DHS digestion as described in Subheading 3.1, steps 1–7, but with the following changes: (a) use 25% of the reaction volume, i.e., about 1.25 million nuclei/0.25 ml digestion; and (b) vary the DNase I concentration in each tube, for example, 0, 10, 20, 40, 80, and 120 U of DNase I per ml. Alternatively, the concentration of DNase I may be fixed, and a variable incubation time may be employed (e.g., 1–5 min incubation with DNase I).
2. After phenol-chloroform extraction, precipitate the DNA by adding 1/10th volume of 3 M sodium acetate, pH 5.2 (Sigma, St. Louis, MO, S7899) and 2 vol of 100% ethanol.
3. Incubate the mixture at −80°C for 20 min; pellet the DNA by centrifugation using a microcentrifuge at full speed.
4. Wash the pellet with 1 ml of 70% ethanol. Spin down the pellet. Remove the ethanol and air dry for 5 min. Resuspend the pellet in 100 μl EB buffer provided in the Qiagen PCR purification kit (see Note 9).
5. Measure the DNA concentration in each dissolved DNA sample using a Quant-IT high sensitivity DNA assay kit, as described above.
6. Perform a SYBR-green based qPCR assay using 5 ng of DNA from each sample and two pairs of qPCR primers. One qPCR primer pair flanks a DNase I hypersensitive region, while the second primer pair flanks a DNase I insensitive region (see Note 10). At the optimal DNase I concentration, the hypersensitive region is selectively destroyed; as a consequence, the qPCR threshold cycle (Ct) number for the primer pair that flanks the hypersensitive region increases but there is little or no change in the qPCR Ct number for the primer pair that flanks the insensitive region (12).

4. Notes

1. In general, one batch of DNase-digested samples containing a total of ~30 million mouse liver nuclei will generate 10–30 ng of purified, DNase I-released fragments, which is more than sufficient for high-throughput sequencing on the Illumina sequencing platform (e.g., HiSeq 2000 instrument).

2. Control DHS samples (i.e., digestion of purified, chromatin-free genomic DNA) are required for downstream data analysis. The amount of DNase I required for control sample DNase digestion is much lower than that required for digestion of chromatin, typically requiring only 0.3–1.0 U of DNase I per ml. It is important that the extent of digestion of control genomic DNA samples vs. chromatin samples should be similar in terms of the DNA size distribution obtained after sucrose step gradient fractionation. Multiple DNase I concentrations should therefore be tested with genomic DNA samples to determine the optimal digestion conditions. In our experience, sucrose step gradient fraction 7 is used (Fig. 1), as this fraction primarily contains released DNA fragments <1 kb in size. After digestion, the control DHS assay continues in exactly the same way as in the experimental DHS assay.
3. High quality DHS sample show >10-fold enrichment of known DNase-releasable fragments (positive control region) compared to DNase I insensitive regions (negative control region) when assayed by qPCR. qPCR using genomic DNA as template is also carried out to quantify any differences in qPCR primer efficiency, which must be taken into account when calculating the enrichment of positive control region sequences in the preparation of released fragments.
4. Keep the nuclei on ice until DNase I digestion is initiated. Do not use extra g force to spin down the nuclear pellet. Dislodge the pellet gently by tapping after each spin, which helps prevent clumping of the nuclei.
5. Pre-dilution of DNase I in Buffer D will decrease the glycerol concentration and make it easier to mix the samples properly. Optimal and reproducible reaction conditions are required for a successful DHS assay, especially when DHS profiles are to be compared between two or more biological samples. If the nuclei are under-digested, the final DNA yield will be low and a greater portion of the released DNA will be derived from nonspecific DNA fragmentation that occurs during preparation of the nuclei. Excessive digestion with DNase I will result in extensive release of DNA fragments from genomic regions that show little or no true hypersensitivity. We have found that at a concentration of five million nuclei per ml, digested for 2 min at 37°C with 40–60 U of DNase I per ml, works well and gives reproducible DHS maps. These reaction conditions can serve as a useful starting point when optimizing DNase I digestion conditions.
6. The multi-tube format presented here is also useful when optimizing DNase I digestion conditions by testing a range of DNase concentrations and/or digestion times. DNase I treatment can also be carried out using a single tube format,

for example, by setting up a 6-ml reaction in a 15-ml test tube. Care must be taken to allow sufficient time to fully equilibrate the samples to 37°C before DNase I digestion is initiated, making sure that the samples are mixed thoroughly.

7. Avoid violent vortex during extraction. If a significant amount of jelly-like material remains in the liquid phase, or if large chunks of denatured protein are present at the interface after phenol–chloroform extraction and centrifugation, it is likely that too many nuclei were digested, which can lead to poor results. In the original protocol (6), DNase I digestion is carried out using up to ten million nuclei per ml. However, it is preferable to use five to seven million nuclei per ml to avoid overloading.

8. The rate of migration of DNA through the sucrose step gradient is DNA size-dependent, and provides a useful way to separate small DNase I-released fragments from genomic DNA that is largely uncut (10). We routinely analyze on an agarose gel a full set of sucrose gradient fractions collected from each batch of six parallel DNase I digestion reactions to insure that the desired size separation has been achieved. Because of the low DNA abundance in gradient fractions containing short DNA fragments, fluorescence staining by SYBR Green and use of high sensitivity laser scan imaging is required to visualize the DNA fragment size distribution (Fig. 1). Using the methods described here, sucrose gradient fraction 7 typically contains DNA fragments less than 1 kb in size and gives excellent DNase-seq results. However, other fractions, for example, pooled fractions 6–8, may also be used (11). Further steps required for sample processing prior to high-throughput sequencing, including adaptor ligation and library generation for Illumina sample preparation are not discussed here. Of note, there is an additional size selection step during the Illumina sample preparation protocol, where fragments 200–400 bp in length are selected for sequencing. This size selection can substantially enhance the signal to noise ratio in the final DNA sequencing data.

9. It is critical for downstream steps that the DNA pellet be completely dissolved. Over-drying the DNA pellet makes it very difficult to dissolve. Incubation at 60°C for 1 h followed by incubation at 4°C overnight may be required to completely dissolve the DNA sample.

10. Ideally, qPCR amplicons of the DNase I hypersensitive and the DNase I insensitive genomic regions should be of similar length and extend for several hundred bp. The amplicon should cover the entire hypersensitive region in order to encompass as many DNase cutting events as possible. However, excessively long DNA amplicons will decrease the

qPCR amplification efficiency and affect the assay quality. Multiple primer pairs should be tested with genomic DNA to identify primer pairs that give the best amplification efficiency and the highest specificity.

Acknowledgments

Supported in part by NIH grant DK33765 (to DJW).

References

1. Cockerill PN (2011) Structure and function of active chromatin and DNase I hypersensitive sites. FEBS J 278(13):2182–2210, 3060
2. Lu Q, Richardson B (2004) DNaseI hypersensitivity analysis of chromatin structure. Methods Mol Biol 287:77–86
3. West AG, Fraser P (2005) Remote control of gene transcription. Hum Mol Genet 14(Spec No 1):R101–R111
4. Boyle AP, Davis S, Shulha HP, Meltzer P, Margulies EH, Weng Z, Furey TS, Crawford GE (2008) High-resolution mapping and characterization of open chromatin across the genome. Cell 132:311–322
5. Crawford GE, Davis S, Scacheri PC, Renaud G, Halawi MJ, Erdos MR, Green R, Meltzer PS, Wolfsberg TG, Collins FS (2006) DNase-chip: a high-resolution method to identify DNase I hypersensitive sites using tiled microarrays. Nat Methods 3:503–509
6. Sabo PJ, Kuehn MS, Thurman R, Johnson BE, Johnson EM, Cao H, Yu M, Rosenzweig E, Goldy J, Haydock A, Weaver M, Shafer A, Lee K, Neri F, Humbert R, Singer MA, Richmond TA, Dorschner MO, McArthur M, Hawrylycz M, Green RD, Navas PA, Noble WS, Stamatoyannopoulos JA (2006) Genome-scale mapping of DNase I sensitivity in vivo using tiling DNA microarrays. Nat Methods 3: 511–518
7. Song L, Crawford GE (2010) DNase-seq: a high-resolution technique for mapping active gene regulatory elements across the genome from mammalian cells. Cold Spring Harb Protoc 2010:pdb.prot5384
8. ENCODE Project Consortium, Myers RM, Stamatoyannopoulos J, Snyder M, Dunham I, Hardison RC, Bernstein BE, Gingeras TR, Kent WJ, Birney E, Wold B, Crawford GE (2011) A user's guide to the encyclopedia of DNA elements (ENCODE). PLoS Biol 9(4):e1001046
9. Ling G, Sugathan A, Mazor T, Fraenkel E, Waxman DJ (2010) Unbiased, genome-wide in vivo mapping of transcriptional regulatory elements reveals sex differences in chromatin structure associated with sex-specific liver gene expression. Mol Cell Biol 30(23):5531–5544
10. Weis JH, Quertermous T (2001) Size fractionation using sucrose gradients. Curr Protoc Mol Biol, Chapter 5:Unit 5.3
11. Siersbæk R, Nielsen R, John S, Sung MH, Baek S, Loft A, Hager GL, Mandrup S (2011) Extensive chromatin remodelling and establishment of transcription factor 'hotspots' during early adipogenesis. EMBO J 30(8):1459–1472
12. McArthur M, Gerum S, Stamatoyannopoulos G (2001) Quantification of DNaseI-sensitivity by real-time PCR: quantitative analysis of DNaseI-hypersensitivity of the mouse beta-globin LCR. J Mol Biol 313(1):27–34

Chapter 4

Isolation and Analysis of DNA Derived from Nucleosome-Free Regions

Matthew Murtha, Yatong Wang, Claudio Basilico, and Lisa Dailey

Abstract

Precise regulation of the levels and timing of gene expression is fundamental to all biological processes and is largely determined by the activity of *cis*-regulatory modules, containing the binding sites for transcription factors, within promoters and enhancers. The global identification of these transcriptional regulatory elements within mammalian genomes, and understanding when and where they are active, is an important effort that will require the development and implementation of several different technologies. Here we detail a means for the identification of transcriptional regulatory elements using functional assays. The success of this approach relies on focusing the functional assay on DNA derived from nucleosome-free regions (NFRs), i.e., the 2% of the genome within a given cell in which regulatory elements reside. Accordingly, we present a simple method for isolating NFR DNA, and a functional assay that can be used for the identification of promoter and enhancers components within this population.

Key words: Transcription, Chromatin, *cis* Regulatory modules, Functional assay, Promoters, Enhancers, Nucleosome-free regions

1. Introduction

The specialized features that distinguish one cell type from another reflect the differential expression of distinct repertoires of gene subsets. While many determinants impact gene expression, the regulation of gene transcription remains key among them. Accordingly, much effort has gone into identifying and mapping the genomic loci of the regulatory elements, i.e., *cis*-regulatory modules, within promoters and enhancers, which serve as binding sites for transcriptional activators, and determine the level and timing of gene expression. Among the approaches that have been applied towards this challenge, ChIP-chip or ChIP-seq analyses have become a popular means for determining the genomic loci bound by specific transcription factors (TFs) or enriched for specific

Minou Bina (ed.), *Gene Regulation: Methods and Protocols*, Methods in Molecular Biology, vol. 977,
DOI 10.1007/978-1-62703-284-1_4, © Springer Science+Business Media, LLC 2013

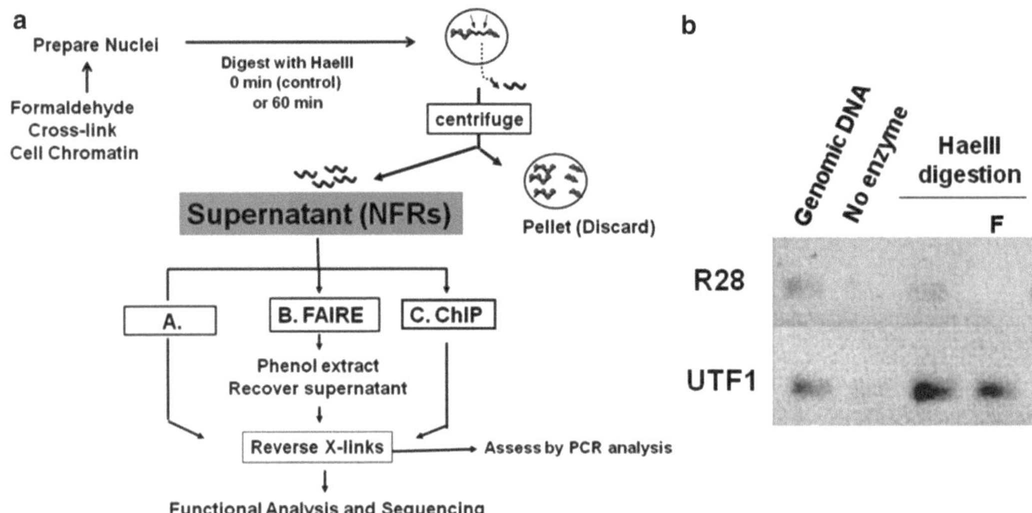

Fig. 1. (**a**) Summary of NFR isolation and preparation for functional analysis. Permeabilized nuclei prepared from formaldehyde-treated cells are incubated with a restriction enzyme (e.g., HaeIII as depicted here) to digest and release DNA fragments from open chromatin regions. After centrifugation, the released NFRs are recovered from the supernantant and can then be processed in one or more of the following ways prior to cloning and functional analysis, as described in the text: A. Reverse the formaldehyde cross-links, B. perform FAIRE treatment prior to cross-link reversal, or C. Perform ChIP using specific antibodies to isolate NFR DNAs targeted by a specific TF, followed by cross-link reversal. Semiquantitative PCR can be used to assess the quality of the NFR preparation prior to cloning and functional analysis. (**b**) PCR analysis to assess the selective enrichment for open chromatin regions. Two sets of nuclei were prepared from E14 embryonic stem cells (ES cells) and treated with HaeIII. A third nuclei preparation was incubated without enzyme for 1 h as depicted in a. Formaldehyde cross-links were reversed in the Untreated sample and one of the Hae-treated samples, while FAIRE was performed in the other Hae sample prior to cross-link reversal. The relative enrichment for regions of open chromatin was assessed using PCR and primers complementary to the UTF1 enhancer, which is active in ES cells. The presence of closed chromatin was assessed using primers complementary to the region R28 (3). The sample analyzed is indicated at the top of each lane. Genomic DNA = 10 ng of purified total mouse DNA; No enyme = 2 μl of supernatant from nuclei incubated in the absence of HaeIII; HaeIII digestion = 2 μl of supernatant from nuclei that had been incubated with HaeIII. The sample labled "F" = 2 μl of supernatant from nuclei that had been incubated with HaeIII and had also been treated with FAIRE. Approximately 28–31 cycles of PCR are sufficient. The enrichment of UTF sequences over R28 is clearly evident in the supernatant of nuclei treated with HaeIII, and somewhat further enhanced after FAIRE treatment.

modified histones. However, complementary approaches that are based on functional assays have been largely lacking. Functional assays not only provide essential independent validation of ChIP data, but also can lead to the de novo discovery of regulatory element- or transcription factor activities, and can detect dynamic changes in the transcriptional machinery during differentiation that static, ChIP-based studies may not.

In this chapter we detail a method for a mid-throughput functional assay that can be implemented for the de novo identification of endogenous transcriptional regulatory modules from cultured mammalian cells (see method summary in Fig. 1a). The key aspect of this assay is that the analysis is focused on DNA derived from "open chromatin" corresponding to so-called Nucleosome-depleted- or Nucleosome-Free Regions (NFRs), the 2% of the genome in which the important regulatory elements for a given cell reside (1, 2). To this end, a simple, restriction enzyme-mediated method is used for isolating NFR-DNAs in order to enrich for transcriptional regulatory elements (3). The resulting population of NFR DNAs is so specific and devoid of background DNA that it is amenable for direct functional interrogation. Accordingly, the DNAs can be assembled into reporter plasmids that are used in functional assays to identify those NFR DNAs that can act as promoter or enhancer elements. As reported previously, the functional modules identified in this manner represent elements that are likely to regulate transcription in situ, and include an array of novel cell type-specific enhancers (3). The functional analysis of DNA from NFRs can also be adapted for the identification of other types of regulatory elements associated with open chromatin regions. Thus, this approach provides an additional and unique tool that can be applied towards the annotation of noncoding elements within mammalian genomes.

2. Materials

2.1. Formaldehyde Cross-linking

1. 37% Formaldehyde, Molecular Biology grade (Fisher).
2. Dulbecco's Modified Eagle Medium (DMEM), stored at room temperature.
3. Phosphate Buffered Saline, pH 7.4, no $CaCl_2$, no $MgCl_2$ (PBS), filtered and stored at 4°C.
4. 2.5 M Glycine.
5. Sterile polyethylene Cell Lifters.
6. 17 × 120 mm 15 ml conical polypropylene tubes.

2.2. Preparation of Permeabilized Nuclei and Extraction of NFR DNAs

1. Buffer 1: 50 mM Hepes, pH 7.5, 140 mM NaCl, 1 mM EDTA, pH 8, 10% Glycerol, 0.5% NP-40, 0.25% Triton X-100. Filtered and stored at 4°C.
2. Buffer 2: 10 mM Tris–HCl, pH 8, 0.2 M NaCl, 1 mM EDTA, pH 8, 0.5 mM EGTA, filtered and stored at room temperature.
3. RE Buffer: 50 mM NaCl, 10 mM Tris–HCl, pH 7.9, 10 mM $MgCl_2$, 1 mM DTT (This is identical to New England Biolab's Buffer NEB 2 (1×)).
4. 0.4% Trypan Blue solution in PBS.
5. HaeIII Restriction enzyme, 50 units/μl, New England Biolabs.
6. 0.5 M EDTA, pH 8.
7. Tris-saturated Phenol–Chloroform–Isoamyl Alcohol (25:24:1), (Ambion).

8. Sodium Acetate, 3 M, pH 5.2.
9. 100% Ethanol, molecular biology grade.
10. 10 mg/ml RnaseA stock, aliquoted and stored at −20°C.
11. 20 mg/ml Proteinase K stock in 100 mM Tris–HCl, pH 8, aliquoted and stored at −20°C.
12. 5 mg/ml Glycogen (Ambion).
13. Qiaquick PCR purification kit (Qiagen).

2.3. Chromatin Immunoprecipitation of NFR-Derived DNAs

1. Chromatin immunoprecipitation (ChIP) dilution Buffer: 167 mM NaCl, 1.2 mM $MgCl_2$, 0.01% SDS, 20 mM Tris–HCl pH 8, 1.1% Triton X-100, filtered and stored at 4°C. Add protease inhibitors fresh.
2. Protein G sepharose (Amersham).
3. ChIP-grade Transcription factor-Specific antibody and IgG control antibody.
4. Elution buffer: 10 mM Tris–HCl, pH 8, 1% SDS, 1 mM EDTA, pH 8.

2.4. LMPCR Amplification of NFR DNAs

1. Oligonucleotides: Linker 1A: 5′-GCGGTGACCCGGGAGAT CTGAATTGG-3′; Linker 1B: 5′-CCAATTCAGATC-3′; HAE AMP: 5′-GCGGTGACCCGGGAGATCTGAATTGGCC-3′

   Resuspend Linkers 1A and 1B to 0.1 mM in H_2O. Resuspend HaeAMP linker to 20 μM in H_2O.
2. STE buffer: 10 mM Tris–HCl, pH 8, 100 mM NaCl, 1 mM EDTA, pH 8.
3. High Concentration (HC) T4 DNA Ligase and 10× Ligation buffer (Invitrogen).
4. 10 mM dNTPs.
5. 5 M Betaine (Sigma).
6. Taq polymerase (5 units/μl) and 10× buffer A (Fisher).

2.5. Cloning and Functional Assays

1. Alkaline phosphatase and ligation reagents, e.g., Roche Rapid DNA Ligation kit.
2. BamHI restriction enzyme.
3. Transformation-Competent bacteria.
4. LB agar plates for antibiotic selection of plasmid-transformed bacteria.
5. Qiagen miniprep kit or equivalent.
6. 96-well black, flat- and clear-bottom black Viewplates (Perkin Elmer) or equivalent.
7. Standard sterile 96-well plates.

8. Adjustable 20–200 μl 8- or 12-channel multipipettor.
9. Sterile multipipettor trays.
10. Opti-MEM or serum- and Antibiotic-free DMEM.
11. Lipofectamine 2000 (Invitrogen).
12. Microplate reader, fitted for FITC detection (488 nm/520 nm).

3. Methods

3.1. Formaldehyde Cross-linking of Cultured Cell Chromatin

1. Culture adherent cells in four 15 cm tissue culture plates, or as required, so that approximately $40–80 \times 10^6$ cells will be generated upon the day of collection (see Note 1).
2. Prepare fresh 1% Formaldehyde solution by mixing 2.7 ml 37% Formaldehyde with DMEM (no serum) to a final volume of 100 ml. Prepare fresh Glycine STOP solution by mixing 2.5 ml of 2.5 M Glycine with 47.5 ml of COLD PBS to make a 0.125 M final solution. Keep the Glycine STOP solution on ice.
3. Aspirate the culture medium from the plates. Add 20 ml of 1% Formaldehyde/DMEM solution per 15 cm plate. In this and all subsequent steps, take care not to dislodge cells from the plate. Place the plates on a rocking platform and cross-link the cells for 10 min at room temperature.
4. Discard the formaldehyde solution. Wash the cells once with cold PBS. Add 10 ml of cold Gycine STOP solution to each plate. Place the plates on a rocking platform and treat the cells for 5 min at room temperature.
5. Discard the Glycine STOP solution. Wash the cells once with cold PBS. Aspirate excess PBS and then add 2 ml PBS per plate. Use a polyethylene Cell Lifter to detach the cells from the culture plate. Tilt the plate, scrape all of the PBS/cells toward the lower edge, and use a pipetteman to transfer and combine the cell suspensions from all plates to a single 15 ml conical polypropylene tube on ice.
6. Pellet the cells at 2.5K rpm for 10 min at 4°C using a tabletop centrifuge. Remove the PBS and flash freeze the cell pellet. Store at −80°C (see Note 2).

3.2. Preparation of Permeabilized Nuclei

1. Allow the frozen cell pellet to thaw on ice. Using your fingers, flick the bottom of the tube sharply to evenly disperse the cells in the pellet (see Note 3).
2. Resuspend the cells in 8 ml of cold Buffer 1 and incubate for 10 min on ice. Mix intermittently by gently inverting the tube.

3. Transfer the cell suspension to a glass dounce (that had been kept cold on ice), and dounce the cells ten times using a B pestle.
4. Transfer equal portions of the dounced cell suspension into two 15 ml conical polypropylene tubes and pellet the cells at 2K rpm, 7 min at 4°C using a tabletop centrifuge.
5. Withdraw the liquid. Using your fingers, flick the bottom of the tube sharply to evenly disperse the pellet. Add 3 ml of Buffer 2 to each tube and mix by gently pipetting up and down using a 10 ml pipette. Rock the suspension for 10 min at room temperature.
6. Pellet the nuclei at 2K rpm for 10 min at 4°C using a tabletop centrifuge, and discard the supernatant. Using your fingers, flick the bottom of the tube sharply to evenly disperse the nuclei. At this point the pellet may be somewhat "sticky" and harder to disperse. Add 1.3 ml of RE Buffer to each tube and continue to evenly resuspend the nuclei by gently pipetting up and down using a 1 ml pipetteman. After resuspension, combine the two samples for a final volume of approximately 2.6 ml.
7. To check the integrity and homogeneity of the nuclei suspension, withdraw a 5 μl aliquot and transfer onto a glass slide. Add 5 μl of a 0.4% Trypan Blue solution to the slide, and use the pipetteman tip to mix with the nuclei suspension. Place a coverslip over the Trypan blue-stained nuclei and observe under a light microscope. If the preparation has been successful, this treatment will result in the appearance of nearly 100% blue spheres (i.e., stained nuclei). Also ascertain that there is minimal clumping of the nuclei.

3.3. Isolation of DNA from Nucleosome-Free Regions (NFRs)

1. Invert the tube containing the sample of nuclei from Subheading 3.2, step 6 to mix, and distribute 500 μl aliquots into five 1.7 ml polypropylene microfuge tubes. Add 100 units of HaeIII restriction enzyme (see Note 4) to four of the tubes. One tube should not receive any enzyme and will serve as the untreated control (see Note 5). Transfer all tubes to a 30°C water bath and incubate the samples for 1 h. Invert the tubes approximately every 15 min to mix.
2. Remove the tubes from the water bath, place on ice, and add 20 μl of 0.5 M EDTA to stop the digestion.
3. Centrifuge at maximum speed for 15–20 s using a microfuge to pellet the nuclei. Carefully withdraw the supernatants with a 1 ml pipetteman, taking care not to disturb or touch the pellet, and transfer and combine the supernatants of the HaeIII-treated nuclei to a single 15 ml conical tube on ice. This sample contains the preparation of "Total HaeIII NFR DNAs". Similarly transfer the supernantant from the untreated control sample into a separate tube 1.7 ml microfuge tube.

4. At this point the NFR DNAs can either be aliquoted and frozen at −20°C before continuing, or processed as follows:
   (a) If the Total HaeIII NFR DNAs prepared in Subheading 3.3, step 3 will be used for functional analysis or sequencing, proceed to Subheading 3.4. If desired, the samples may be stored as 500 μl aliquots at −20°C before proceeding to Subheading 3.4.
   (b) If the Total HaeIII NFR DNAs prepared in Subheading 3.3, step 3 will be used for Chromatin Immunoprecipitaion prior to functional analysis, distribute the sample into 800 μl aliquots and either store them at −20°C for later processing, or proceed directly to Subheading 3.5.

3.4. FAIRE Treatment and Cross-link Reversal

FAIRE (Formaldehyde Assisted Isolation of Regulatory Elements) consists of a phenol extraction step. It has been reported that when formaldehyde-cross-linked chromatin is mixed with phenol–chloroform and then centrifuged, nucleosome-bound chromatin partitions into the interphase whereas DNA that is free of histones partitions into the upper, aqueous phase (4). The NFR DNA samples of Subheading 3.3, step 3 will be highly enriched for NFR DNAs, and can be processed directly for cloning and functional analysis after cross-link reversal by proceeding to Subheadings 3.6 and 3.7. However, in some cases it may be desirable to first perform an additional step of FAIRE treatment, as outlined below, to further ensure the exclusion of any remaining nucleosome-bound DNA (see Note 6). DO NOT PERFORM FAIRE IF ISOLATED NFR FRAGMENTS ARE TO BE USED FOR ChIP EXPERIMENTS.

1. Distribute the preparation of Total HaeIII NFR DNAs and the untreated control from Subheading 3.3, step 3 into 500 μl aliquots in 1.7 ml microfuge tubes.
2. Add 500 μl (i.e., 1 volume) of phenol–chloroform–IAA to each tube, shake vigorously (by hand) for 10 s, and then separate the aqueous and organic phases by centrifugation in a microfuge for 10 min at maximum speed, room temperature.
3. Transfer the upper phase to a new 1.7 ml microfuge tube, taking care to avoid touching or withdrawing material from the interphase.
4. Repeat steps 2 and 3.
5. Add 5 μl of a 10% SDS solution (0.1% SDS final) to each 500 μl sample and incubate the tubes at 65°C overnight to reverse the formaldehyde cross-links.
6. Let the samples cool to room temperature. Precipitate the NFR DNA by adding 35 μl of 3 M Sodium Acetate (0.2 M final), 1 μl of 5 mg/ml Glycogen, and 1 ml (2 volumes) of 100% Ethanol. Precipitate the DNAs overnight at −20°C.
7. Pellet the NFR DNAs by centrifugation for 5 min at maximum speed, 4°C using a microfuge. Withdraw the supernantant.

8. Wash the pellet with 200 µl of cold 80% Ethanol and centrifuge again. When loading the samples into the microfuge rotor, position the hinge of the microfuge cap to point upwards so that the DNA pellet, which may not be clearly visible, will be easily located after centrifugation (i.e., on the hinge "side").
9. Withdraw as much of the ethanol as possible while taking care not to disturb the (position of) pellet. Air-dry for 15 min.
10. Resuspend each pellet in 30 µl water by scraping the DNA from the wall of the tube, vortexing, and pippeting up and down with a pipetteman. Add 1 µl of RNase stock and incubate for 1 h at 37°C.
11. Add 1 µl of Proteinase K stock and incubate for 1 h at 37°C.
12. Purify the NFR DNAs using a spin column (e.g., Qiaquick PCR kit, Qiagen) as instructed by the manufacturer. Elute the NFR DNAs using 35 µl water.
13. Proceed to Subheading 3.6

3.5. Chromatin Immunoprecipitation of NFR-Derived DNAs

Although the NFR DNAs are relatively depleted of bound nucleosomes, those that correspond to active promoters or enhancers will be bound by their cognate Transcription Factors (TFs). Thus, ChIP may be utilized to analyze the subset of genomic NFR DNAs that is targeted by a specific TF. A number of ChIP protocols can be used to this end and the reagents and conditions that are most appropriate for the ChIP of a particular TF, e.g., ChIP-grade Antibody or wash buffers etc, will need to be determined independently. Basic parameters that were used for the ChIP of NFR-derived DNAs bound by Sox2 are outlined below and can be used as a guide towards the IP of NFR DNAs bound by other TFs.

1. Place two 2 ml polypropylene tubes on ice. For each tube, combine 800 µl of the NFR preparation from Subheading 3.3, step 3 with 800 µl of ChIP dilution buffer, 3 µl of 10% NP-40, and fresh protease inhibitors.
2. To the first tube, add Protein G sepharose beads that had been prebound using 4 µg of ChIP-grade antibody recognizing the TF of interest. To the second tube, add Protein G sepharose beads that had been prebound to 4 µg of IgG control antibody. Incubate the samples overnight on a nutator, 4°C.
3. The following day, wash the Immunoprecipitated complexes extensively.
4. To recover the IP'd target NFR DNAs, resuspend the pelleted complexes in 150 µl of Elution Buffer, vortex briefly, and incubate at 65°C for 10 min.
5. Centrifuge the samples for 30 s at maximum speed in a microfuge and transfer the supernantant to a clean 1.7 ml microfuge tube.

6. Add 20 μl more Elution Buffer plus 9 μl of 5 M NaCl. Incubate overnight at 65°C to reverse the cross-links.
7. Let the samples cool to 37°C. Add 1 μl of RNase and incubate for 1 h at 37°C.
8. Add 1 μl of Proteinase K stock and incubate for 1 h at 37°C.
9. Purify the NFR DNAs using a spin column (e.g., Qiaquick PCR kit, Qiagen) as instructed by the manufacturer. Elute the NFR DNAs using 35 μl water.

3.6. Quality Control to Assess the Successful Isolation of NFR DNAs

1. The methodology described above has been successfully implemented for the isolation of NFR-derived DNAs from several different cell lines including F9 teratocarcinoma cells, E14 Embryonic stem cells, Neural Stem cells, and the OB1 osteoblast cell line ((3) and Murtha et al., in preparation), and should be similarly applicable for the isolation of NFRs from additional cell types. However, one may choose to assess the quality of the NFR preparation at this point in the procedure using semiquantitative PCR to ascertain the selective enrichment for "open chromatin" genomic regions. If ChIP was performed, this procedure can additionally be used to assess its success.
2. Oligonucleotide primers need to be designed that are complementary to target genomic regions representing either open or closed chromatin in the cell type from which the NFRs were prepared. Since the restriction enzyme HaeIII was used to isolate the NFRs, it is essential that no HaeIII site be present in the genomic DNA segment between the Forward and Reverse PCR primer sequences. A well characterized promoter or enhancer element that is known to be active in the cells would be ideal representations of "open" chromatin. If the exact location of an active element is not known, however, this approach could prove challenging. The same is true for the targeting of a negative control, "closed" region. We have identified a region within the mouse genome, designated R28 ((3), see Note 7), which resides within "closed" chromatin in all of the cell lines we tested so far, and therefore may be more broadly useful as a negative control for additional cell types. If necessary, the chromatin status of a candidate genomic region can first be assessed by testing the primers in a classic DNaseI accessibility experiment such as that performed in (3).
3. Semiquantitative PCR analysis can then be used to determine the relative enrichment for open and closed genomic regions in the NFR preparation compared to total genomic DNA. 10 ng of total genomic DNA, 2 μl of the reversed HaeIII-NFR DNA sample, or 2 μl of the reversed, Untreated, negative control sample is used as template for PCR amplification using the positive or negative primer sets designed above. A detectable difference in the ratio of positive and negative PCR products

should be observed in the HaeIII-NFR sample compared to that using total genomic DNA. Furthermore, the amounts of both putative "positive" and "negative" PCR products observed after amplification of the Untreated control should be negligible and equivalent. Figure 1b shows a representative assessment of NFRs isolated from mouse Embryonic stem cells, and analyzed for the relative enrichment of the ES cell-specific, UTF1 enhancer DNA compared to the negative control region R28.

3.7. Ligation-Mediated PCR Amplification of NFR-Derived DNAs

1. The primary advantage provided by isolating DNA from NFRs using the protocol described in this chapter is that the specificity of the method is high enough, and the background low enough, to make mid- or high throughput functional analyses of the DNA elements feasible. Several types of functional elements have been reported to reside within nucleosome-depleted regions including transcriptional enhancers, promoters, silencers, insulators, and replication origins (1, 2). Thus, the subset of isolated NFR DNAs that can direct any of these activities may be identified by cloning the NFR DNAs into an appropriate reporter plasmid and performing functional assays. By way of example, this and the following sections describe the preparation of the NFR DNAs for analyses using a plasmid-based approach for the functional identification of promoters and enhancers within the NFR DNA population.
2. Prior to cloning of the NFR DNAs, they are ligated to an adaptor that contains a restriction enzyme site that is used for subsequent cloning into a reporter plasmid. The adaptor used in this protocol is designed for ligation to DNA fragments possessing blunt 5′ and 3′ ends, such as those produced by HaeIII cleavage. The adaptor-ligated NFR DNA population is then amplified using PCR.
3. To prepare the adaptor, combine 6.7 μl of 0.1 mM Linker 1A and 6.7 μl of 0.1 mM Linker 1B with 86.6 μl of H_2O. Incubate in a heat block set to 100°C for 5 min. Prepare a glass beaker filled with boiled H_2O covered by a layer of foil, and insert the tube containing the denatured oligonucleotides through a small hole in the foil such that the bottom of the tube is submerged in the hot water. Place the beaker on a benchtop and allow the sample to cool for several hours to facilitate adaptor annealing.
4. Ligate the adaptor to the NFR DNAs by combining 30 μl of the reversed NFR sample from Subheading 3.4, step 12 or Subheading 3.5, step 9 with 10 μl of 5× Ligation Buffer (Invitrogen), 6.7 μl annealed adaptor, and 1 μl (5 units) of HC T4 DNA Ligase (Invitrogen) in a total volume of 50 μl.
5. Allow the ligation to proceed for 1 h at room temperature, and continue incubation overnight at 16°C.

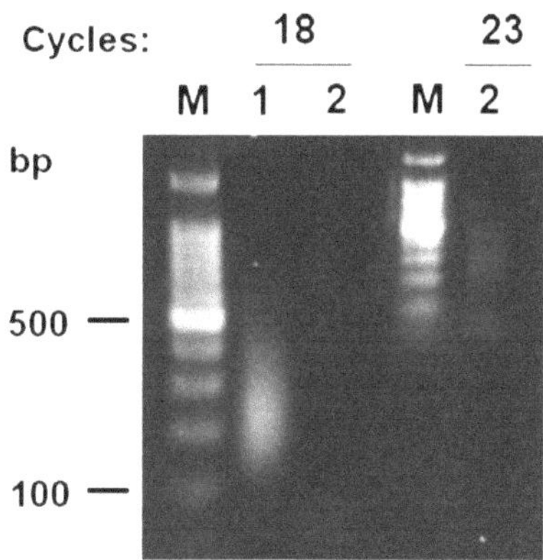

Fig. 2. LMPCR Amplification of NFR DNAs. Isolated Total HaeIII NFR DNAs or αSox2 ChIP-NFR DNAs were subjected to LMPCR as described in Subheading 3.7. The reactions were paused, and 5 μl aliquots were analyzed by electrophoresis in a 2% agarose gel in TAE after cycles 18 and 23. A smear of PCR products is seen for the LMPCR amplified Total Hae-NFRs after 18 cycles of amplification (sample 1), whereas 23 cycles were required to visualize products within the ChIP-NFR sample (*lane 2*). M = 100 bp ladder DNA marker.

6. Purify the adaptor-ligated NFR DNAs using a Qiagen PCR purification spin column. Elute the DNAs using 30 μl H_2O.
7. Assemble the PCR reaction with 5 μl of 10× Taq amplification Buffer A (Fisher), 1.4 mM dNTPs, 6.5 μl 5 M Betaine, 2.5 μl 20 μM HaeAMP oligonucleotide primer, 1 μl (5 units) of Taq polymerase (Fisher), and 8.6 μl H_2O. Add 25 μl of adaptor-ligated NFR DNAs from Subheading 3.7, step 5 (see Note 8).
8. Mix, quick spin the sample and then PCR amplify using the following program: Pre-amplify with 55°C, 2 min, 72°C, 5 min, and 95°C, 2 min to allow release of Linker 1B and extension of the ligated adaptor sequences. Then amplify the reaction using 15–25 cycles of 95°C 30 s, 55°C 30 s, 72°C 1 min (see Note 9 and Fig. 2), and end with a final extension step for 1 min at 72°C.
9. Test 5 μl of the PCR reaction using electrophoresis in a 2% agarose gel containing Ethidium Bromide, in TAE buffer. The average size of NFR DNAs is approximately 150 bp, but the total PCR-amplified population will appear as a heterogeneous smear, ranging in size from approximately 100–500 bp, as shown in Fig. 2.
10. Purify the LMPCR-amplified NFR DNAs using a spin column (PCR purification kit) and elute using 52 μl H_2O.

3.8. Cloning of the NFR DNAs

1. An appropriate reporter plasmid for cloning and detecting transcriptional activation (or other activity) by an inserted NFR in the cell type under analysis must be designed. We constructed the –64promGFP reporter plasmid for this purpose ((3), Fig. 3a). –64promGFP contains a minimal promoter region (TATA box

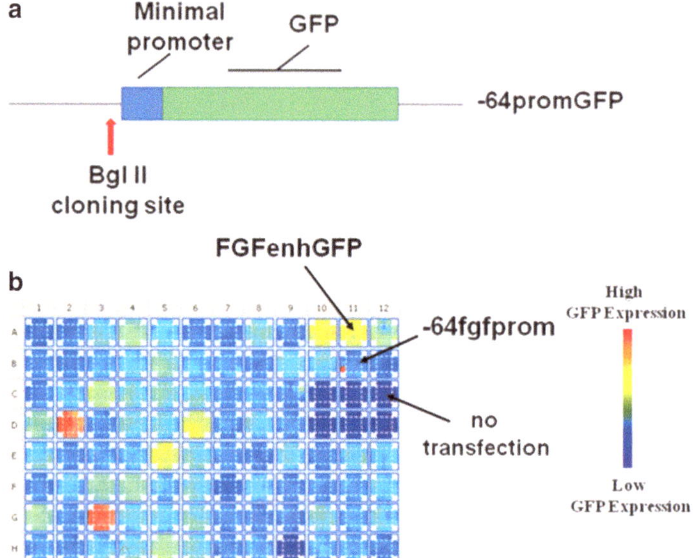

Fig. 3. (**a**) Reporter plasmids used for the functional identification of NFRs that can activate transcription. As described in the text, the –64promGFP reporter plasmid consists of a minimal promoter placed upstream of GFP coding sequences, and has very little activity on its own. A BglII site located immediately upstream of the promoter is used as a cloning site for the insertion of NFR DNAs. (**b**) Representative microplate reader-generated image of functional assay results. Wells of a 96-well plate containing cells that had been transfected with the basal –64promGFP plasmid, the positive control FGFenhGFP plasmid, no plasmid, or each of the test plasmids containing a different cloned NFR DNA are indicated. After transfection, a microplate reader was used to detect GFP expression (FITC, 488 nm). The distict levels of GFP expression generated from each of the transfected plasmids is shown here colorimetrically, and correspond to numerical entities that can be normalized and quantified as "fold induction" relative to the basal –64promGFP plasmid. Generally, approximately 20% of the cloned NFR DNAs will activate GFP expression at least twofold in this assay.

and transcriptional start site) placed upstream of GFP gene coding sequences. –64promGFP also contains a BglII site for cloning NFR DNAs immediately upstream of the promoter sequence. A positive control plasmid, FGFenhGFP, contains the FGF4 enhancer at this position.

2. Digest 1 μg of –64promGFP plasmid DNA with BglII and remove the 5′ phosphate groups using any number of commercially available phosphatases. Inactivate the phosphatase after treatment.
3. Digest 25 μl (i.e., half) of the NFR DNA sample from Subheading 3.7, step 9 overnight with 20 units of BglII (or with another enzyme if a different recognition site has been engineered into the adaptor).
4. Purify the BglII-digested NFR DNAs using a Qiagen spin column (PCR purification kit) and elute the DNAs using 50 μl H_2O. Quantify the eluted DNA sample using NanoDrop or a comparable device.

5. Ligate the NFR DNAs overnight at 16°C with 100 ng of BglII-digested vector at a molar ratio of 5:1. (Assume that the average size of the NFR fragments is in the 150–200 bp range.)
6. Use 1/10 volume of the ligation reaction to transform competent bacteria and spread onto LB-agar plates with selective antibiotic.
7. The following day, use PCR and primers complementary to vector DNA sequences flanking the BglII cloning site to screen the transformed colonies for those containing plasmids harboring an NFR DNA insert. As per standard protocol, touch the tip of the pipetteman on a colony, mix up and down in a tube containing the PCR reaction components, and then transfer 1 μl of this to an LB-agar plate with antibiotic that had been marked with a numbered grid allowing correspondence between the PCR reactions and specific bacterial colonies. We routinely screen 60 colonies at a time, and usually observe that at least 70–80% of the plasmids contain an inserted NFR.
8. Isolate plasmid DNA from those colonies containing an NFR DNA insert using a mini prep kit (e.g., Qiagen). We previously determined that approximately 20% of the NFR DNAs act as promoters or enhancers in the functional assay (i.e., have the capacity to activate GFP expression in transfected cells). Thus, for mid-throughput functional analyses, we processed several hundred minipreps per NFR or ChIP-NFR sample.
9. Determine the concentration of each miniprep DNA. Store at −20°C.

3.9. Functional Analysis (See Note 10)

1. All steps in this section should be performed in a Laminar Flow Hood for Tissue culture using standard sterile technique. It is important that the function of the cloned NFR DNAs be assessed after transfection of the same cell type from which the NFRs were derived, as many of these activities may be cell-type-specific (3).
2. Plate the cells 1 day before transfection at a density of 40×10^3 cells/100 μl/well, in 96-well clear-bottom, black Viewplates.
3. The following day, replace the medium with 100 μl of Opti-MEM, plus serum as required. Importantly, antibiotics should never be added as they interfere with the Lipofectamine.
4. Return the plates to the incubator while the DNA/Lipofectamine samples are prepared.
5. For each Viewplate of cells, 93 NFR-GFP miniprep DNAs will be assayed. In addition, one well each should be reserved for the positive-, negative, and untransfected controls. In our case, the positive control plasmid consists of FGFenhGFP described in Subheading 3.8, step 2. The negative control plasmid is the basic reporter construct lacking NFR insert DNA.

6. To assemble the DNA/Lipofectamine samples, use a second, sterile standard 96-well plate (i.e., not a Viewplate). In a tissue culture hood, transfer 3 ml Opti-MEM into a sterile Multipipette tray. Fill each of the 8 or 12 channels of the Multipipettor with 28 μl of Opti-MEM and transfer to each row of the sterile 96-well plate. Add 0.6 μg of each miniprep DNA to one well containing the Opti-MEM. Add 0.6 μg each of the positive- and negative control plasmids, one well each. One well will not receive any DNA (Untransfected control) (see Note 11).
7. Assemble a Transfection master mix by combining 100 μl of Lipofectamine 2000 and 2.9 ml Opti-MEM in a sterile tube, and let it stand at room temperature for 5 min.
8. Pour the Lipofectamine–Opti-MEM mixture into a sterile Multipipette tray and add 30 μl to each well of the DNA/Opti-MEM plate using the Multipipettor. Change the pipettor tips after each transfer to ensure that there is no cross-contamination among the DNAs. Cover the 96-well assembly plate and mix the samples in the wells by moving the plate vigorously, but carefully, back and forth. Let the plate stand for 20 min at room temperature to allow the DNA–Lipofectamine complexes to form.
9. Bring the Viewplates containing the cells from the incubator. Use the Multipipettor to transfer 50 μl of each DNA–Lipofectin mixture to the corresponding row of wells in the Viewplate. The 50 μl DNA–Lipofectamine solution is *added* to the 100 μl Opti-MEM already in the well, and therefore, each well of the Viewplate will now contain 150 μl of solution. Remember to change pipette tips after each transfer.
10. Place the Viewplates in the incubator and allow the transfection to proceed for at least 4 h.
11. Replace the medium with that used for standard culture, and continue incubation of the cells overnight.
12. Approximately 24 h after transfection, wash the cells at least twice with PBS (see Note 12). The levels of GFP expression can then be determined using a microplate reader such as that provided by EnVision (see Fig. 3b). It is important to take multiple readings per well to minimize variations arising from uneven cell densities. (Save all values in Excel). We typically record ten readings per well and use an average of the top three readings.
13. Subtract the background reading of untransfected cells from all values, and determine the level of GFP expression relative to that of the negative control plasmid (i.e., Fold Induction).
14. Plasmids displaying at least twofold greater GFP expression than the negative control plasmid are analyzed in duplicate in a second round of transfection and microplate reading.

15. NFR inserts of those plasmids confirmed to activate GFP transcription after the secondary screen represent active Transcriptional Regulatory Modules and should be sequenced using oligonucleotide primers complementary to vector sequences near the cloning site. The genomic colocalization of sequenced NFRs with respect to annotated genes, ChIP data for modified histones or transcription factors, and other genomic features can be visualized using Genome browsers such as that available from UCSC (http://genome.ucsc.edu/). While active NFRs localizing near TSSs may activate transcription in a cell-specific manner, we and others have found that cell type-specific regulatory elements are most often located at distal positions (i.e., >2 kb) from the TSS and can act as enhancers (2, 3, 5). Thus, the identification of these elements using the approach outlined in this chapter may facilitate the further identification of components of cell-specific transcriptional circuitries operative in the cells under analysis.

4. Notes

1. This protocol will produce a 2.6 ml preparation of NFR DNAs which can be used for several downstream analyses, for example, 1 negative control sample (500 μl) and 1 HaeIII-NFR sample (500 μl) for FAIRE treatment and cloning, and, if ChIP is to be performed before cloning, 1 ChIP-NFR sample (800 μl), and 1 IgG control ChIP-NFR sample (800 μl). However, the distribution and use of the NFR preparation, as well as the number of starting cells required, can be adjusted accordingly. Cells grown in suspension can also be used.
2. Freezing the cell pellet at this point appears to be helpful, and may facilitate disruption to the cell membrane in subsequent steps. Be sure to use polypropylene tubes to prevent the development of cracks during the freeze–thaw process.
3. It is important to gently, but completely, dislodge and disperse the cell or nuclear pellet *following each centrifugation step* to ensure homogeneity, and to avoid excessive pipeting up and down, and potential rupture of the nuclei, during resuspension.
4. HaeIII was chosen as the enzyme to digest DNA within cellular Nucleosome-Free Regions for the following reasons: (a) It generates fragments with a blunt-end; (b) its 4-base recognition sequence GGCC appears frequently in the genome; and (c) its digestion buffer is compatible with maintaining the integrity of the permeabilized nuclei. However, several other enzymes (e.g., RsaI) also fulfill these criteria and

can be used in place of, or in addition to, HaeIII if desired. In this case however, the oligonucleotide primer used for NFR amplification in the LMPCR Subheading 3.7, step 6 will need to be redesigned (see Note 9).

5. We find that the successful preparation of NFR DNAs, as outlined in this chapter, is very consistent for each of the cell lines tested. Thus, once successful isolation of NFR DNAs for a given cell line has been established, the use of cells for preparation of the negative control, untreated sample need not be included in subsequent NFR isolations.
6. Although FAIRE is included here as further ensurance for the selection of NFR DNAs, our previous characterisation showed that FAIRE treatment of NFR preparations did not result in any further enrichment for functional elements (3), and therefore may be considered optional.
7. Oligonucleotide primers for amplification of the 143 bp negative control genomic region R28 are: Forward primer 5′-TGAGTCACAACTCCGGTC-3′; Reverse primer 5′-GAGG GGGGGTAACTAAGG-3′.
8. The sequence of the Hae AMP primer is the same as that of the LINK 1A adaptor oligonucleotide, plus an extension of two CC residues at the 3′ end. This is to enhance the specificity for amplification of only Hae-digested DNAs ligated to the adaptor (the HaeIII site is GG/CC). Thus, if an enzyme other than HaeIII is used to generate the NFR DNAs, the amplification primer needs to be redesigned accordingly.
9. Use the minimum number of amplification cycles possible. Adequate amplification is determined by pausing the PCR reaction after completion of the 72°C step of a cycle, and analyzing a 5 μl aliquot of the reaction on a 2% agarose gel as described in Subheading 3.7, step 9. The PCR reaction is terminated when amplification products are visible (see Fig. 2). Generally, approximately 16–18 cycles are adequate for amplification of the Total Hae NFRs, whereas approximately 22–25 cycles are needed to visualize ChIP-NFRs.
10. This section details the mid-throughput functional assay described in Yaragatti et al. (3) and can be used to identify many ubiquitous cell-specific transcriptional regulatory modules. Approximately 20% of the NFRs analyzed displayed transcriptional activation in our previous analyses. However, functional analysis of the NFRs may be adapted into a high-throughput format by using lentiviral- or retroviral-based vectors. This adaptation, currently under development in our group, will harness the full potential of this technique for regulatory element discovery.

11. The Lipofectamine–DNA ratio and duration of exposure of the cells to the transfection mixture described here works well for F9 cells. Different cell types may require different conditions.
12. It is essential to rinse the wells adequately with PBS since residual phenol red from the DMEM will interfere with FITC readings of GFP.

Acknowledgements

This work was supported by an Empire State Stem Cell Board grant through the New York State Department of Health (NYSTEM Contract #CO24322) to L.D.

References

1. Cockerill PN (2011) Structure and function of active chromatin and DNase I hypersensitive sites. FEBS J 278(13):2182–2210
2. Xi H, Shulha HP, Lin JM et al (2007) Identification and characterization of cell type-specific and ubiquitous chromatin regulatory structures in the human genome. PLoS Genet 3(8):1377–1388
3. Yaragatti M, Basilico C, Dailey L (2008) Identification of active transcriptional regulatory modules by the functional assay of DNA from nucleosome-free regions. Genome Res 18(6):930–938
4. Giresi PG, Kim J, McDanielle RM et al (2007) FAIRE (Formaldehyde-Assisted Isolation of Regulatory Elements) isolates active regulatory elements from human chromatin. Genome Res 17(6):877–885
5. Song L, Zhang Z, Grasfeder LL et al (2011) Open chromatin defined by DNaseI and FAIRE identifies regulatory elements that shape cell-type identity. Genome Res 21(10):1757–1767

Chapter 5

Acquisition of High Quality DNA for Massive Parallel Sequencing by In Vivo Chromatin Immunoprecipitation

M. van den Boogaard, L.Y.E. Wong, V.M. Christoffels, and P. Barnett

Abstract

ChIP-seq is rapidly becoming a routine technique for the determination of the genome wide association of DNA binding proteins and histone modifications. Here we provide a protocol for the isolation, purification, and immunoprecipitation of DNA fragments associated with a target transcription factor of interest. Although the method makes use of adult mouse hearts, it can, with relative ease, be adapted for the in vivo ChIP isolation of DNA from other cell and tissue sources with the intention of massive parallel sequencing.

Key words: Chromatin immunoprecipitation, ChIP-seq, Cross-linking, Heart, Transcription factor

1. Introduction

Chromatin Immunoprecipitation (ChIP) is a method commonly used to determine the location of DNA binding sites on the genome for a particular protein of interest (1, 2). This technique provides a temporal view of the protein–DNA interactions that occur inside the nucleus of living cells or tissues. Insights at this level of protein–DNA interaction can provide crucial information about the position of regulatory elements, such as enhancers, and their occupation and function during development and disease progression.

ChIP-Sequencing, also known as ChIP-seq, combines chromatin immunoprecipitation (ChIP) with massive parallel DNA sequencing to identify the cistrome of DNA-associated proteins (3). In this way, protein–DNA interactions can be studied on a genome-wide level. ChIP-qPCR provides a confirmation of the presence of a given protein at a known, fixed position within the genome. Prior knowledge of a confirmed or suspected binding site is required to assess the quality of the ChIP. This binding site can either be

Minou Bina (ed.), *Gene Regulation: Methods and Protocols*, Methods in Molecular Biology, vol. 977,
DOI 10.1007/978-1-62703-284-1_5, © Springer Science+Business Media, LLC 2013

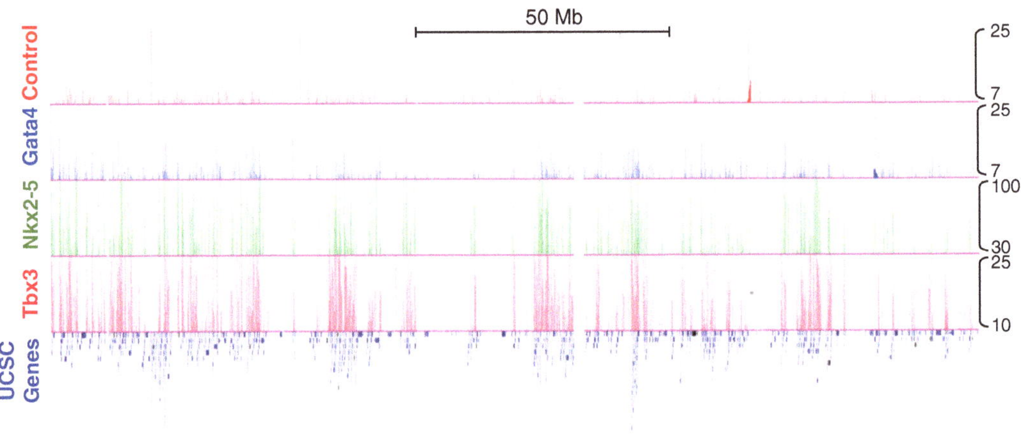

Fig. 1. ChIP-seq of three transcription factors in the adult mouse heart. The figure shows the ChIP-seq genome peak alignment to chromosome 1 of the mouse. The three transcription factors are key proteins with key roles in heart development: Tbx3, Nkx2-5, and Gata4. The control sample was carried using the Tbx3 antibody in hearts lacking induced Tbx3 expression. Nkx2-5 and Gata4 represent normal endogenous expression experiments.

compared to a site known not to be bound by the protein or to a reference ChIP carried out using a control antibody.

Since it is an acceptable hypothesis that two spatiotemporally co-expressed interacting transcription factors may share a set of gene targets and that the function of such an interaction may be the coordinated regulation of shared targets (4–6), we recently made use of chromatin immunoprecipitation (ChIP) coupled to genome wide sequencing technology to assist in the identification of regulatory elements across the genome bound by the transcription factor Tbx3 (7) (see Fig. 1). Although our initial study made use of an over expression system in mouse hearts, we have since applied the same approach to endogenous transcription factors of interest in the heart, with similar success. With the increasing application of this technology (ChIP-seq, (8)) to different transcription factors, available data resources can be mined and compared to help shed light on not only gene regulation in general, but provide new insights into transcription factor complexes and their in vivo function. The protocol we present here was designed to perform ChIP on adult mouse hearts. However, with minimal adjustments it can be applied to most tissues and cells, delivering DNA fragments which can either be submitted for sequencing on a suitable platform or used in focused single/multi-target quantitative PCR (qPCR) experiments. The latter is particularly applicable if the tissue resources are limited. Following this approach we were able to verify our results with qPCR during early development using relatively small numbers of embryos.

2. Materials

Ultrapure water is defined as purified deionized water with a resistivity of approximately 18 MΩ at 25°C. Sterilization, unless otherwise stated, is by autoclaving at 15 psi (121°C) for 15 min.

2.1. Materials and Chemicals

1. Protein G plus agarose beads (Pierce).
2. Specific antibody to target protein of interest (see Note 1).
3. Complete Protease Inhibitor Cocktail (Roche).
4. 37% Formaldehyde (P.A. grade) (Merck). Bottle should be as fresh as possible (unopened <6 months).
5. PBS (Phosphate buffered saline) tablets (Invitrogen). Solubilize and sterilize by autoclaving.
6. RNase A (Sigma).
7. Proteinase K (Invitrogen). Stock 10 mg/mL in ultrapure sterile water.
8. Glycine.
9. NaCl.
10. HEPES.
11. Tris–HCl.
12. EDTA.
13. EGTA.
14. LiCl.
15. $NaHCO_3$.
16. Sodium dodecyl sulfate (SDS).
17. Triton X100.
18. NP-40.
19. Deoxycholic acid.
20. Phenol (Tris-buffered biophenol pH 8.0 (Biosolve)).
21. Chloroform.
22. Glycogen (Roche).
23. Absolute Ethanol (100 and 70% solution made with sterile ultrapure water).
24. Sterile ultrapure water.

2.2. General Equipment

5.0 mL Stopper Tubes (Sterile).

2.0 mL Eppendorf tubes (Sterile).

1.5 mL LoBind Eppendorf tubes (Sterile).

Lab-bench Rotator.

Centrifuge with cooling.

Dounce homogenizer (2 mL).

IKA Ultra Turrax T5 FU (Or equivalent macerator).

Sonics VCX-130 (Or equivalent ultra-sonication device).

2.3. Working Solutions

Working from stock solutions (see Note 2).

1. X-link buffer (freshly prepared): 100 mM NaCl, 50 mM HEPES (pH 7.9), 0.5 mM EGTA, 1 mM EDTA (see Note 3). Add glycine to a final concentration of 125 mM.
2. Lysis Buffer: 50 mM Tris–HCl (pH 8.1), 10 mM EDTA (pH 8.0), 1% SDS. Add 1× Complete Protease Inhibitor Cocktail tablet just before use.
3. Dilution Buffer: 20 mM Tris–HCl (pH 8.1), 150 mM NaCl, 2 mM EDTA (pH 8.0), 1% Triton X-100.
4. TSE I (Low Salt Washing Buffer): 20 mM Tris–HCl (pH 8.1), 150 mM NaCl, 2 mM EDTA (pH 8.0), 0.1% SDS, 1% Triton X-100.
5. TSE II (High Salt Washing Buffer): 20 mM Tris–HCl (pH 8.1), 500 mM NaCl, 2 mM EDTA (pH 8.0), 0.1% SDS, 1% Triton X-100.
6. LiCl/detergent: 10 mM Tris–HCl (pH 8.1), 1 mM EDTA (pH 8.0), 250 mM LiCl, 0.5% NP-40, 0.5% (w/v) Deoxycholic acid.
7. TE buffer: 10 mM Tris–HCl (pH 8.1), 1 mM EDTA (pH 8.0).
8. Elution Buffer: 100 mM $NaHCO_3$, 1% SDS (This solution should be made fresh just before use and kept at room temperature).
9. Proteinase K. Make a stock solution of 20 mg/mL in ultrapure sterile water. Store at −20°C.

3. Methods

The following protocol is based on a single adult mouse heart sample. A single adult mouse heart contains approximately 1×10^7 cells. Although sufficient chromatin can be obtained from smaller sample sizes, we recommend starting with an equivalent of this number before optimizing for smaller numbers of cells.

3.1. Tissue Fixation

1. Isolate the tissue/cells in sterile 1×PBS. Good fixation will require that the fixative can enter the material (e.g., for adult mouse heart, cut it into small pieces). Once the sample has been sliced in to small pieces, transfer to a 5 mL stoppered tube.

2. Fix the material with 1% formaldehyde in X-link buffer (see Note 4) for 30 min (this has been optimized for adult mouse heart (see Note 5)) at room temperature (RT) with constant steep rocking/tilting.
3. Remove formaldehyde by pipetting off the supernatant
4. Add fresh X-link buffer containing 125 mM glycine
5. Remove the X-link buffer and rinse 2× with sterile, ice-cold PBS.
6. Transfer tissue to a 2 mL Eppendorf Tube.

Note: At this stage the sample may be snap frozen and stored at −80°C.

3.2. Lysis and Sample Shearing

1. Add 1 mL of sterile PBS to the material. The ratio of PBS to sample is approximately 10:1.
2. Make an even suspension of the material. If the sample is a tissue this can best be done by physically chopping via maceration (see Note 6).
3. Centrifuge at 2300×g for 5 min at 4°C and discard the supernatant
4. Resuspend in 1 mL of Lysis Buffer containing 1× Protease Inhibitor Cocktail and incubate for 1 h at 4°C with rotation (see Note 7).
5. Physically grind the cells with a pre-cooled Dounce homogenizer 20× (see Note 8).

At this stage the sample may be snap frozen and stored at −80°C.

3.3. Chromatin Isolation and Shearing

The use of LoBind (Eppendorf) or similar reaction tubes with a low attractive surface is strongly recommended in all subsequent steps.

1. Transfer material to a 1.5 mL tube (see comment above) prior to sonication (see Note 9).
2. Adjust the settings of the sonicator to medium power. For the sonicator used here we use an amplitude of 40%, though for other instruments this stage may need optimization. Pulse using successive rounds of 30s ON/45s OFF Depending on the type of tissue being used, the number of rounds may need optimization (see Note 10).
3. Centrifuge at full speed for 15 min at 4°C and collect the supernatant (see Note 11).
4. Save 2.5% (=25 μL) supernatant sample to determine shearing efficiency, continuing with the protocol for this test sample listed under Subheading 3.5 (see Note 12).
5. Add 1 mL PBS and 1 mL Dilution Buffer, both containing 1× Protease Inhibitor Cocktail and mix well. (see Note 7).

6. Aliquot in 1.5 mL tubes. We recommend using 1 mL of sample per experiment; this can be adjusted depending on the estimated expression of the protein of interest.

At this stage the sample may be snap frozen in liquid nitrogen and stored at −80°C before continuing.

3.4. Immunoprecipitation

1. Gently resuspend Protein A or G beads (see Note 13) and then wash 30 μL of beads in an eppendorf tube 3× with TSE I (without protease inhibitor cocktail). Washing definition: add 500 μL of TSE I to beads, gently resuspend (see Note 14), and centrifuge at low speed (2,800 rpm), 2 min, 4°C. Carefully pipette supernatant away (see Note 15), but avoid disturbing the beads. Repeat.
2. Add 1 mL of isolated chromatin to washed beads for preclearing.
3. Incubate for 1 h with rotation at 4°C.
4. Centrifuge at low speed (730×*g*) for 2 min at 4°C.
5. Remove a 2.5% (=25 μL) supernatant sample. This sample will function as a future input reference (see Note 16).
6. Transfer the remaining supernatant to new tube and add 2 μg of antibody against the protein of interest.
7. Incubate from a minimum of 2 h to overnight with rotation at 4°C (see Note 17).
8. Wash 30 μL protein A or G beads and wash them 3× with TSE I (without protease inhibitor cocktail) as in step 1 above.
9. Add antibody/chromatin mix from steps 6–7 to washed beads.
10. Incubate for 1 h with rotation at 4°C.
11. Centrifuge at low speed (730×*g*) for 2 min at 4°C. Carefully pipette away the supernatant.
12. Wash the beads 2× with 1 mL TSE I (containing protease inhibitor cocktail). Centrifuge at low speed (2,800 rpm) for 2 min at 4°C. Carefully pipette away the supernatant.
13. Wash the beads 1× with 1 mL TSE II (containing protease inhibitor cocktail). Centrifuge at low speed (2,800 rpm) for 2 min at 4°C. Carefully pipette away the supernatant.
14. Wash the beads 1× with 1 mL LiCl/detergent. Centrifuge at low speed (730×*g*) for 2 min at 4°C. Carefully pipette away the supernatant.
15. Wash the beads 1× with 1 mL TE buffer. This is the last wash. Remove as much buffer as possible (see Note 18) without disturbing the beads.
16. Add 200 μL freshly prepared Elution Buffer to the beads.
17. Incubate for 15 min with rotation at room temperature (see Note 19).

18. Centrifuge, at room temperature, at low speed (730×*g*) for 2 min.
19. Collect the supernatant in a new tube.
20. Perform elution again, repeat steps 16–18. Pool eluates in one tube (total volume will be approximately 400 μL).

3.5. Cross-link Reversal and proteinase K Treatment

The input reference sample should also be processed in the following manner.

1. Add 16 μL 5 M NaCl (end concentration will be 200 mM) to the 400 μL sample.

   For input: add 75 μL sterile ultrapure water to the 25 μL input sample and then add 4 μL of 5 M NaCl.
2. Incubate from a minimum of 4 h to overnight at 65°C.
3. Add 2 μL RNase A to ChIP sample (add 1 μL to input sample) and incubate for 30 min at 42°C.
4. Add 8 μL 0.5 M EDTA, 8 μL 1 M Tris–HCl (pH 6.5) and 8 μL Proteinase K solution (20 μg/μL).

   For input use 2 μL 0.5 M EDTA, 2 μL 1 M Tris–HCl (pH 6.5), and 2 μL Proteinase K solution (20 μg/μL). (See Note 20).
5. Incubate for 2 h at 42°C.

3.6. DNA Purification

(See Note 21)

1. Add 1 volume of biophenol, mix by vortexing, and centrifuge at 4°C for 10 min at full speed in a microfuge.
2. Collect top aqueous phase in a sterile microfuge tube.
3. Add 1 volume of freshly prepared, settled, 50:50 phenol–chloroform, mix by vortexing, and centrifuge at 4°C for 10 min at full speed in a microfuge.
4. Collect top aqueous phase in a sterile microfuge tube.
5. Add 1 volume of chloroform, mix by vortexing, and centrifuge at 4°C for 10 min at full speed in a microfuge.
6. Collect top aqueous phase in a sterile microfuge tube and add 0.5 μL Glycogen, 0.1× volumes of 3 M NaAc pH 5.2, mix and then add 3 volumes of ice-cold absolute ethanol.
7. Leave in −20°C freezer for 30 min (see Note 22).
8. Centrifuge at 4°C for 30 min at full speed.
9. Wash with 500 μL of cold 70% ethanol.
10. Centrifuge at 4°C for 5 min at full speed.
11. Remove all traces of ethanol from the pelleted DNA and air-dry until all moisture droplets have gone.
12. Resuspend pellet in 30–50 μL sterile utlrapure water. A typical ChIP-experiment yields about 5–10 ng DNA. This is sufficient for quantitative PCR and sequencing.

3.7. Sample Concentration Determination and Further Applications

The sample concentration can be accurately measured by fluorometric quantitation (e.g., Qubit Quant-iT). This is especially important for downstream applications such as ChIP-sequencing, where it is vital to know that the protocol is entered with sufficient DNA for ligation of adapters and amplification of the material.

Another application for which the material can be used is quantitative PCR. The enrichment of a DNA-fragment can be measured compared to the input sample, a sample treated with a control antibody or a control DNA-fragment that is not bound by the protein of interest. ChIP-qPCR can be used as a final result or to determine the quality of a ChIP experiment for other downstream applications. The quality can best be assessed by checking DNA fragments that are known to bind the protein of interest over DNA fragments that are not binding. We find an enrichment of at least tenfold over a control region to be a reasonable standard for ChIP-sequencing quality.

4. Notes

1. Our experience with choosing an antibody for ChIP-seq is that polyclonal antibodies raised against goat work very well with this protocol. However, antibodies raised against rabbit and other species have been commonly used for ChIP-seq throughout literature as well. When using an IgG control antibody, consider matching this to your antibody against the protein of interest (i.e., both raised against the same species).
2. We usually make 5–10× stock solutions of all components for the working solutions. These are then autoclaved for longer term storage (typically for up to 1 month). We typically then make solutions in either sterile glassware or pre-sterilized tube (e.g., Greiner). Adequate sterile ultrapure water must be available for diluting the components to make up the final working concentrations. Additionally, all working solutions used after the immunoprecipitation should be filter-sterilized.
3. The amount of X-link buffer required will depend on the number samples. We recommend making at least 5 mL per sample. We also recommend using a stock solution of 0.5 M EDTA pH 8.0 for the addition of EDTA to the X-link buffer.
4. We typically use 5 mL of X-link buffer per extraction when using adult mouse heart.
5. The time required for fixation will vary between samples and will probably require optimization. As a general rule, the younger or softer the tissue, the less time required. For example, early embryonic samples (ED10.5) use a reduced fixing time (15 min at RT). Optimization of this step can be done

in two ways; fixation can be varied by changing the percentage of the fixative or by changing the time of fixation. We recommend choosing one of the two, make a range around the standard protocol and compare the shearing efficiency and qRT-PCR outcome to the standard protocol (for adult or embryo) in one experiment.

6. We make use of an IKA Ultra Turrax T5 FU with the power set to 40,000/min, keeping the sample on ice during the maceration process.
7. For some tissue types cell lysis can be a problem. It is recommended to check the lysis of the cells under a microscope when optimizing the protocol for a new tissue type. When lysis is insufficient, the volume of lysis buffer can be increased. The volume of other buffers throughout the protocol should then also be equally adjusted. When doing this, it should be noted that the immunoprecipitation in this protocol is performed in 1/3 of the total sample, based on the amount of chromatin that can be isolated from one sample.
8. The dounce tube should have a well-fitted pestle.
9. Shearing is more efficient in a conical shaped tube compared to a round-bottomed tube.
10. An adult mouse heart requires 25 rounds of sonication. For embryonic tissue 15 rounds are sufficient. The objective is to shear the chromatin (DNA) into pieces of approximately 150–500 bp in length. During optimization, samples may be withdrawn, reverse cross-linked (see Subheading 3.5) and the DNA purified for gel analysis see Fig. 2. During sonication it is advisable to keep the sample on ice to prevent over-heating. However, SDS will precipitate if left on ice reducing the efficiency of DNA shearing. During sonication, the heat generated keeps the SDS in solution. Therefore care should be taken to ensure that just prior to sonication the sample is warmed between finger and thumb to ensure SDS is in solution.
11. At this stage only a small pellet should be visible, possessing a white/grey color. A black pellet may be a sign of "burnt" cellular debris and as a rule we do not continue with these samples. A large pellet may mean incomplete lysis which can result in reduced shearing efficiency. A small sample can be checked on gel (see Notes 7, 10 and 12). In this case the sample may be resuspended in fresh lysis buffer and sheared again before proceeding.
12. We recommend checking the shearing efficiency before continuing with the immunoprecipitation. The remaining sample can be frozen at −80°C. When shearing is insufficient, the sample can be subjected to an additional round of sonication. After this, the step needs to be repeated after which the chromatin sample may be refrozen at −80°C.

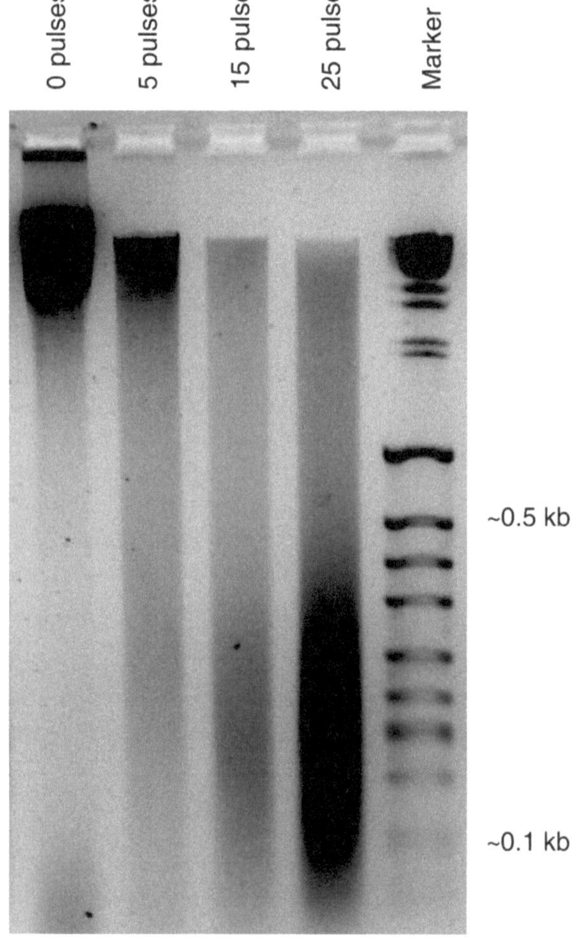

Fig. 2. Chromatin shearing using sonication. Optimization of the number of rounds of sonication may be required. A typical average fragment bp length for efficient ChIP sequencing is around 200–300 bp.

13. Choice of antigen bead (A or G) may depend on the primary antibody. See Table 1.
14. Resuspend Protein A or G beads by gently flicking the tube until there is no visible pellet anymore.
15. Pipet using a tip with a large opening or cut the end of a 200 μL tip off.
16. This sample may be snap frozen and stored at −80°C.
17. A 2 h incubation period works well for most proteins.
18. Use of a needle and syringe is an option.
19. The buffer will precipitate at 4°C.
20. Making a mastermix is an option when treating a large number of samples.

Table 1
IgG Affinity table

Ig origin	Protein-A binding	Protein-G binding
Goat IgG1	Weak	Strong
Goat IgG2	Strong	Strong
Human IgG1, 2, and 4	Strong	Strong
Human IgG3	No binding	Strong
Mouse IgG1	Weak	Strong
Mouse IgG2a, 2b, and 3	Strong	Strong
Rat IgG1	Weak	Weak
Rat IgG2a	No binding	Strong
Rat IgG2b	No binding	Weak
Rabbit IgG	Strong	Strong
Sheep IgG1	Weak	Strong
Sheep IgG2	Strong	Strong

21. DNA purification may also be carried out using a commercial DNA purification kit and following the manufacturer's instructions.
22. Incubations at −20°C will increase DNA yield. The period of incubation may be extended to overnight.

Acknowledgments

This work was supported by European Community's Sixth Framework Programme contract HeartRepair LSHM-CT-2005-018630.

References

1. Kuo MH, Allis CD (1999) *In vivo* cross-linking and immunoprecipitation for studying dynamic protein:DNA associations in a chromatin environment. Methods 19: 425–433
2. Orlando V (2000) Mapping chromosomal proteins *in vivo* by formalehyde-crosslinked-chromatin immunoprecipitation. Trends Biochem Sci 25:99–104
3. Johnson DS, Mortazavi A, Myers RM, Wold B (2007) Genome wide mapping of *in vivo* protein-DNA interactions. Science 316: 1497–1502
4. Farnham PJ (2009) Insights from genomic profiling of transcription factors. Nat Rev Genet 10:605–616
5. He A, Kong SW, Ma Q, Pu WT (2010) Co-occupancy by multiple cardiac transcription factors identifies transcriptional enhancers active in heart. Proc Natl Acad Sci 108:5632–5637
6. MacQuarrie KL, Fong AP, Morse RH, Tapscott SJ (2011) Genome-wide transcription factor binding: beyond direct target regulation. Trends Genet 27:141–148
7. Boogerd CJ, Wong LY, van den Boogaard M, Bakker ML, Tessadori F, Bakkers J, 't Hoen PA,

Moorman AF, Christoffels VM, Barnett P (2011) Sox4 mediates Tbx3 transcriptional regulation of the gap junction protein Cx43. Cell Mol Life Sci 68:3949–3961

8. Robertson G, Hirst M, Bainbridge M, Bilenky M, Zhao Y, Zeng T, Euskirchen G, Bernier B, Varhol R, Delaney A, Thiessen N, Griffith OL, He A, Marra M, Snyder M, Jones S (2007) Genome-wide profiles of STAT1 DNA association using chromatin immunoprecipitation and massively parallel sequencing. Nat Methods 4:651–657

Chapter 6

Luciferase Assay to Study the Activity of a Cloned Promoter DNA Fragment

Nina Solberg and Stefan Krauss

Abstract

Luciferase based assays have become an invaluable tool for the analysis of cloned promoter DNA fragments, both for verifying the ability of a potential promoter fragment to drive the expression of a luciferase reporter gene in various cellular contexts, and for dissecting binding elements in the promoter. Here, we describe the use of the Dual-Luciferase® Reporter Assay System created by Promega (Promega Corporation, Wisconsin, USA) to study the cloned 6.7 kilobases (kb) mouse (m) Tcf3 promoter DNA fragment in mouse embryonic derived neural stem cells (NSC). In this system, the expression of the firefly luciferase driven by the cloned mTcf3 promoter DNA fragment (including transcription initiation sites) is correlated with a co-transfected control reporter expressing *Renilla* luciferase from the herpes simplex virus (HSV) thymidine kinase promoter. Using an internal control reporter allows to normalize the activity of the experimental reporter to the internal control, which minimizes experimental variability.

Key words: Promoter, Transfection, Dual-Luciferase, Firefly, *Renilla*, NSC

1. Introduction

Luciferase assay is a widely used technique for studying many aspects of the biology of cells (1–4), and it has also become an invaluable tool when studying the activity of cloned promoter DNA fragments in vitro in eukaryotic cell lines. The luciferase protein was first cloned from the firefly (*Photinus pyralis*) in 1985 (5), and is one of the most commonly used reporter genes for several reasons, including its high sensitivity, and the tight coupling of the luciferase protein concentration with the luminescence output. Additionally, the protein is available immediately upon translation without the requirement of posttranslational processing (6, 7). Thus, the luciferase assay provides a rapid, sensitive, and quantitative

Minou Bina (ed.), *Gene Regulation: Methods and Protocols*, Methods in Molecular Biology, vol. 977, DOI 10.1007/978-1-62703-284-1_6, © Springer Science+Business Media, LLC 2013

basis in the analysis of factors that potentially regulate mammalian gene expression in vitro. In a standard setting, a promoter DNA fragment of interest is cloned in front of the luciferase cDNA in a luciferase reporter vector lacking eukaryotic promoter elements, and is further assessed for its ability to drive the expression of luciferase after introduction into a proper cell-line, either by transient transfection, electroporation or by viral transduction. If a cloned promoter DNA fragment identified through in silico analysis contains promoter activity, it will be able to drive the expression of the luciferase reporter to a statistically relevant level above basal luciferase level created by the promoter-less vector control. In principle there are two predominant versions of reporter constructs; one which is devoid of both promoter and transcription initiation sites where the cloned fragment needs to contain both elements in order to drive expression of the reporter gene, and the second version which contain a minimal promoter for the investigation of enhancer sequences. The empty vector control used as a negative control in the experiment is the same reporter vector used to clone the promoter fragment of interest without any promoter fragment added. In principle this vector should yield no luciferase activity (or very low if the enhancer version is used), only reflecting the background illuminesence in the sample. Due to cellular and environmental factors, an in vitro analysis of transfected cells may only partially reflect the in vivo activity of the cloned promoter DNA fragment, in particular if the promoter regulates the expression of a gene with tissue specific activity. After the initial in vitro analysis to verify the potential of a cloned DNA fragment to act as a promoter, it might be desirable to analyze its activity in vivo in transgenic mice by using the Xenogen IVIS optical imaging system (Xenogen Corporation) to monitor live activity of the promoter driven expression of luciferase in the entire organism (see refs. 8 and 9 for example).

Here we describe as an example the use of the Dual-Luciferase® Reporter Assay System produced by Promega for a deletion analysis of a 6.7 kb cloned mTcf3 promoter DNA fragment in embryonically derived NSC cultured as neurospheres (10). The mTcf3 promoter fragments (including the transcription initiation sites) were cloned into the pGL3-basic reporter vector (Promega) containing cDNA for the firefly luciferase. The pGL3-basic vector does not contain any promoters or enhancers, but it contains several features aiding in the structural characterization of the putative regulatory sequences under investigation, including a modified coding region for firefly luciferase that has been optimized for monitoring transcriptional activity in transfected eukaryotic cells (Promega). The luciferase activity generated by pGL3-basic-mTcf3 in transfected NSC was further normalized against a co-transfected internal pRL-TK control plasmid expressing *Renilla* (*Renilla reniformis*) luciferase driven by the constitutively active (11)

HSV-thymidine kinase promoter. Taking advantage of the evolutionary distinct origin of the firefly and the *Renilla* luciferase, and their requirements for different substrates to illuminate, makes it possible to discriminate between their respective bioluminescent reactions in the same sample (12). After lysing the cells, the firefly luciferase reporter is measured first by adding sample to the Luciferase Assay Reagent II (LARII). After quantifying the firefly luminescence in a luminometer this reaction is quenched, and the *Renilla* luciferase reaction is initiated by simultaneously adding Stop & Glo® Reagent to the same tube. The sequential measurement of these two luciferases from a single sample minimizes experimental variability, such as differences in cell viability, transfection efficiency, pipetting errors, cell lysis efficiency, and assay efficiency.

Sequential 5′ or 3′ truncations of the cloned promoter fragment allow narrowing down the functional core elements involved in promoter activity and specificity. The truncations are often designed based upon in silico analysis showing the presence and location of putative regulatory elements. Cloned promoter fragments are created in a way that they initially include elements of predicted importance and subsequently exclude some of these elements through 5′, 3′ or internal truncations. Regulatory elements could be involved in both activating and silencing of the controlled gene and may vary between different cellular contexts, including culture conditions and cellular heterogeneity of the cell-line used. These variables may influence the presence and absence of transcription factors, cofactors, or other factors that may regulate the activity or cellular localization of transcription factors, and furthermore the accessibility of the promoter through methylation and chromatin condensation. Potential regulatory elements discovered in silico, and confirmed by promoter/enhancer truncations, are further analyzed by introducing point-mutations. Likewise, putative transcription factor binding sites may be verified with chromatin immunoprecipitation (ChIP), using an antibody against the transcription factor of interest to immunoprecipitate cross-linked, sonicated DNA, where the presence of the transcription factor binding site is verified by PCR. In addition, involvement of a transcription factor in promoter activity can be tested by knocking it down with siRNA, or by altering its presence or activity through drugs.

Our study showed that in neurospheres, both the 6.7 kb mTcf3 promoter fragment, as well as a 4.5 kb fragment of it, provided a moderate 2.5-fold activation of firefly/*Renilla* luciferase activity above background level from the vector control. However, subsequent additional 5′ truncations of the mTcf3 promoter DNA fragment (Fig. 1) led to a 40-fold increase in firefly/*Renilla* luciferase activity above background level, which suggests presence of repressive regulatory elements within the

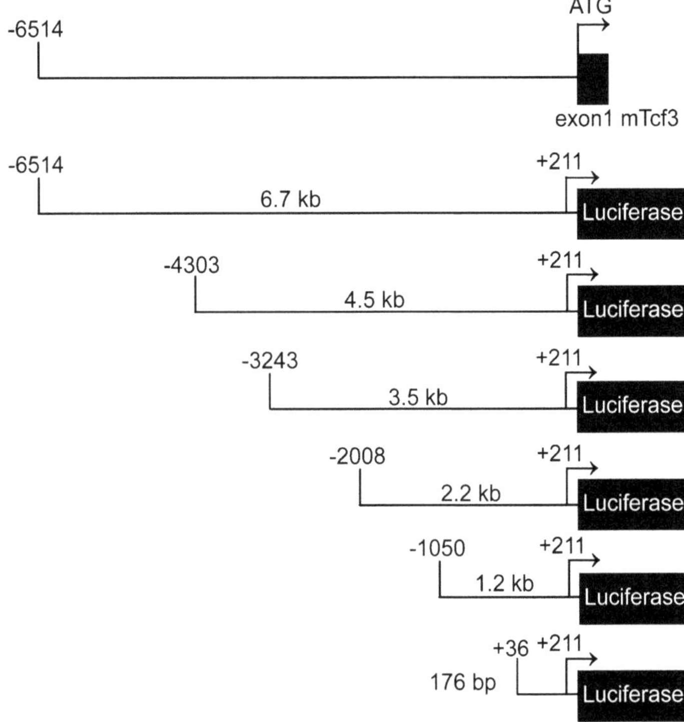

Fig. 1. The cloned mTcf3 promoter DNA fragments. The upper cartoon shows the genomic DNA, including the translation start codon ATG and exon1 of the mTcf3 gene. The second line shows the 6.7 kb cloned mTcf3 promoter fragment containing transcription initiation sites for the mTcf3 gene (but not the translation start site ATG). The various promoter fragments containing 5′ truncations are indicated below. (Reproduced from (10) with kind permission from Springer Science + Business Media).

truncated fragment between the 4.5 kb and the 3.5 kb promoter fragment(Fig. 2). In silico analysis predicted binding sites for a number of known regulatory transcription factors within this fragment, one (Oct1) which was confirmed to bind to the predicted site by further ChIP analysis (10). Functional binding sites for NF-κB were also confirmed by activity measurements of various deletion fragments after treatment with tumor necrosis factor alpha (TNFα).

2. Materials

All steps regarding cultivation and transfection of cells, as well as coating of 48-well plates, should be performed in the cell culture hood. Store reagents according to the manufacturer's instructions, and avoid repeated freeze thawing of premade LARII reagent by storing it as aliquots.

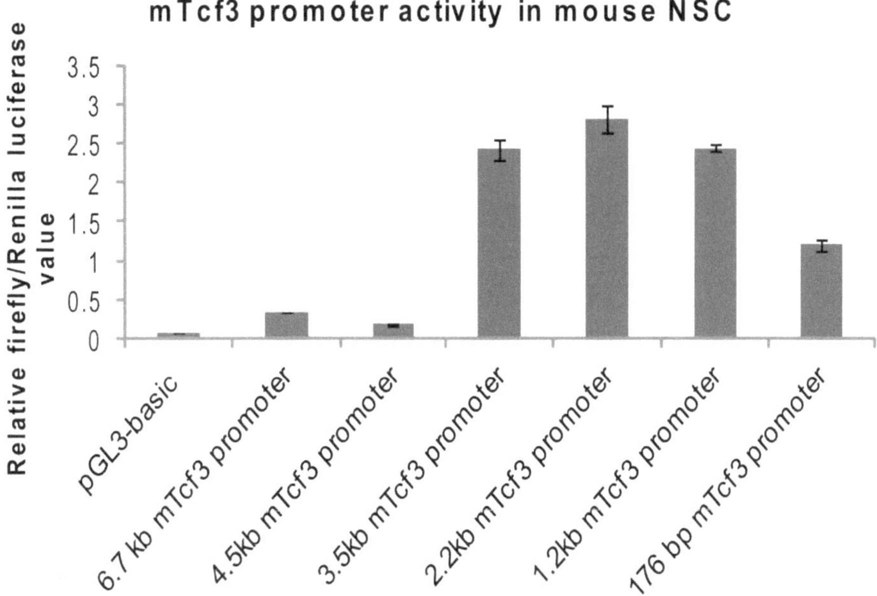

Fig. 2. The figure show the relative firefly/Renilla luciferase values of both the 6.7 kb mTcf3 promoter fragment, as well as 5′ truncations, after transient transfection into mouse NSC. The empty vector control is included to the left. Bars represent relative luciferase values from one set of triplicate measurements with standard deviation.

2.1. Culture of Neurospheres

1. Neurospheres in culture.
2. Media for cultivation of neurospheres: To one bottle of 500 ml Neurobasal-A medium (Gibco Products International, Oklahoma, USA), add 10 ml 1×B27 supplement (Gibco Products International), epidermal growth factor (EGF) to 20 ng/ml (R&D Systems, Minnesota, USA), basic fibroblast growth factor (bFGF) to 8 ng/ml (R&D Systems), 5 ml Penicillin/streptomycin (Pen-Strep) (100×) (BioWhitaker, Maryland, USA).
3. Trypsin-EDTA (BioWhitaker).
4. 1× sterile Phosphate buffered saline (PBS).
5. Tissue culture flasks.
6. 50 ml tubes.
7. Centrifuge to spin 50 ml tubes.
8. Incubator with 5% CO_2.

2.2. Coating of Tissue Culture Plates

1. Preparation of poly-L-lysine stock and working solution: Resuspend one vial of 25 mg poly-L-lysine (Sigma) in 5 ml of tissue culture grade water in a 15 ml tube to achieve a concentration of 5 mg/ml, followed by filtration with a 0.2 μM filter. In a second 15 ml tube, dilute 140 μl of the poly-L-lysine stock in 9.86 ml tissue culture grade water to achieve a concentration of 70 μg/ml.

2. 48-well Nunc tissue culture plates (VWR, Pennsylvania, USA).
3. Autoclaved distilled (d) H_2O.

2.3. Plating of Cells

1. Media for cultivation of neurospheres (see Subheading 2.1).
2. Trypsin-EDTA (BioWhitaker).
3. 1× sterile PBS.
4. 50 ml tubes.
5. Centrifuge for 50 ml tubes.
6. Cell counter of choice.
7. Incubator with 5% CO_2.

2.4. Transfection of Cells

1. pRL-TK plasmid (Promega, Wisconsin, USA) expressing *Renilla* luciferase from the HSV-thymidine kinase promoter, diluted to 10 ng/μl in TE buffer, pH 8.0.
2. Promoter fragment of interest (6.7 kb mTcf3 promoter fragment and 5′ truncated versions) cloned into pGL3-Basic reporter vector (Promega) containing firefly luciferase cDNA, diluted to 0.1 μg/μl in TE buffer, pH 8.0.
3. FuGENE 6 Transfection Reagent (Roche holding Ltd, Basel, Switzerland).
4. Opti-MEM® I Reduced Serum Medium, Gluta MAX™ (Invitrogen Corporation, California, USA).
5. 1.5 ml tubes.
6. 15 ml tube.
7. Vortexer.
8. Table centrifuge for 1.5 ml tubes.
9. Cell culture media.

2.5. Cell Lysis

1. 1× Passive Lysis Buffer (PLB): Add 4 volumes of dH_2O to 1 volume of 5× PLB from the Dual-Luciferase®Reporter Assay System (Promega) in a 15 ml tube. Mix well.
2. 1× PBS.
3. Orbital shaker.

2.6. Luciferase Assay

1. 1× Stop & Glo Reagent: Add 50 volumes of Stop & Glo buffer to 1 volume of 50× Stop & Glo substrate from the "Dual-Luciferase® Reporter Assay System" kit (Promega) in a 15 ml tube.
2. Luciferase Assay Buffer II (LARII): Resuspend the lyophilized Luciferase Assay Substrate in Luciferase Assay buffer II provided with the Dual-Luciferase® Reporter Assay System kit. Store in aliquots at −70°C.

3. 20/20^{n} Luminometer (Turner Biosystems, California, USA) (or luminometer of choice).
4. Vortexer.
5. 1.5 ml tubes.

3. Methods

3.1. Coating of Cell Culture Plates

1. Add 500 μl of the 70 μg/ml poly-L-lysine working solution to as many wells of a 48-well Nunc plate as needed (see Notes 1 and 2). Incubate at room temperature (RT) for 2 h.
2. Remove poly-L-lysine (see Note 3).
3. Wash each well with 1 ml autoclaved dH_2O, and aspirate.
4. Leave the plate open to air-dry in the cell culture hood for 15–20 min (see Note 4).

3.2. Plating Cells

1. Spin down cultured neurospheres (see Note 5) at 693 × *g* for 3 min in a 50 ml tube. See ref. 13 for how to obtain neurospheres.
2. Carefully remove media from the cell pellet.
3. Add 1.5 ml Trypsin-EDTA. Mix well and incubate at 37°C for 5–10 min to obtain a single cell suspension. Mix every now and then to help dissociating the spheres.
4. Add 10 ml 1× PBS to dilute Trypsin-EDTA and spin down at 693 × *g* for 3 min.
5. Remove trypsin-EDTA/PBS from the cell pellet.
6. Dissolve the cell pellet in 2 ml medium and count cells in a cell counter.
7. Dilute cells in media to obtain 350,000 cells/ml and aliquot 500 μl of the cell suspension into each poly-L-lysine coated well (175,000 cells per well) (see Note 6).
8. Leave the plated cells in the incubator with 5% CO_2 overnight (ON).

3.3. Transfection of Plasmids into Cells

1. Mix 2.3 μl of plasmid midiprep (0.1 μg/μl) to be analyzed (6.7 kb-mTcf3-pGL3-basic and 5′ truncations) (see Note 7) with 2 μl *Renilla* luciferase plasmid (pRL-TK, 10 ng/μl) in separate 1.5 ml tubes to obtain a total amount of 0.25 μg DNA per well (see Note 8). Vortex slightly and spin down to obtain an even mixture of the two plasmids. Include control plasmids in the setup (see Note 9).

2. Mix the FuGENE 6 by light vortexing and let it come to room temperature prior to use. Prepare a master-mix of 0.75 μl FuGENE 6 and 19.95 μl Opti-MEM (see Note 10) per well to be transfected in a 15 ml tube (see Notes 11–13). Vortex and spin down. Incubate at RT for 5 min.
3. Aliquot 20.7 μl FuGENE 6/Opti-MEM master-mix per biological replicate into each tube of premade sample to be transfected (see Note 14). Mix by brief vortexing and spin down. Incubate at RT for 15–30 min such that a complex between the FuGENE 6 reagent and DNA (plasmids) can be formed.
4. Change media on the plated cells by adding 500 μl of new prewarmed culture media (see Note 15).
5. Add 25 μl of DNA/FuGENE 6 master-mix to each well in a drop-wise manner, and distribute the mix evenly to cover all cells (see Note 16)
6. Put cells in the incubator for 48 h (see Note 17).

3.4. Lysis of Cells

1. Prepare 65 μl of 1× PLB per well to be analyzed (see Notes 18 and 19).
2. Remove all media from the cells, rinse once with PBS, and add 65 μl of 1×PLB.
3. Incubate with gentle shaking on an orbital shaker with rough shaking for 15 min at RT (see Note 20).

3.5. Luciferase Assay

1. Bring all reagents stored at –20°C to room temperature prior to analysis, including the samples if they are frozen. Prepare a master-mix of 1× Stop & Glo substrate, enough to use 25 μl per sample to be analyzed (see Notes 21 and 22).
2. Aliquot 25 μl of LARII into 1.5 ml tubes, one tube per sample to be analyzed (see Note 23).
3. Prepare 1× Stop & Glo reagent, enough to use 25 μl per sample to be analyzed (see Note 24).
4. Transfer 5 μl of cell lysate from the well to be analyzed (see Note 25) into 25 μl of LARII solution in a 1.5 ml tube. Pipette up and down three times (see Notes 26 and 27), and immediately measure firefly luciferase in a luminometer.

Remove the tube from the luminometer and add 25 μl premixed 1× Stop & Glo reagent to quench the firefly luciferase luminescence and concomitant activation of *Renilla* luciferase. Immediately vortex for 3 s to make sure all firefly luciferase is quenched, and measure *Renilla* luciferase in the luminometer. Discard the tube and record the readings of both luciferases.

3.6. Data Analysis

All luminescence data for both firefly and *Renilla* luciferases for each individual sample should be recorded in a spreadsheet (see Note 28), and a relative firefly/*Renilla* luciferase value should be calculated for each sample. The mean firefly–*Renilla* luciferase ratio of biological replicates is subsequently calculated. To obtain a fold enrichment of the relative firefly/*Renilla* luciferase activity of the experimental samples compared to the vector control, the mean value of the experimental sample is divided on the mean value of the vector control. Every experiment should be performed with at least three biological repetitions and a minimum of three biological parallels, generating at least three sets of data. The mean value of fold activation from several repetitive experiments is calculated for each experimental plasmid together with a standard deviation, which is further used to visualize the results in a graph (see Note 29). Statistical analysis should be performed using a One Way Analysis of Variance, and a Student *t*-test to identify sources of difference in order to validate the statistical relevance of the data.

4. Notes

1. The culture plate used for the assay needs to contain a negatively charged surface in order for the positively charged poly-L-lysine to adhere to the plastic and subsequently anchor the cells to the plastic surface. We used Nunc plates from Thermo Scientific (Thermo Fisher Scientific, Illinois, USA), and the assay can be scaled both up and down as desired.
2. NSCs are cultured as floating cells and need coating for attachment to tissue culture dishes. Different cell types may require different coating.
3. The poly-L-lysine coating solution can be reused several times after removal from the plate, and stored at +4°C.
4. Coated plates can be stored at +4°C and used at a later time point.
5. It may be useful to choose a cell line where the gene of interest shows a clear endogenous expression, such that the promoter DNA fragment is analyzed in a natural environment with the presence of its most common regulatory factors. However, human HEK 293 cells are also a widely used cell line. When using primary cells one make sure to use a low passage number to avoid differentiation or adaption of cells that may occur in more mature cell cultures.

6. The number of cells to be plated in each well should be optimized such that the wells are 50–80% confluent upon transfection. Cells are more efficiently transfected when they are in the exponential log-phase. The most accurate plating of cells between replicate wells is obtained by preparing an excess of master mix of cells in media. Make sure that cells are evenly distributed in the media by inverting the tube gently several times. Immediately aliquot into wells since cells rapidly tend to fall down to the bottom of the tube, diluting the upper part of the cell–media mix and subsequently concentrating the lower part.
7. One should take into consideration that transfection of larger plasmids is less efficient than transfection of smaller plasmids, which may result in an inverse correlation between the construct size and the luciferase activity (14).
8. Transfect at least three biological replicates with each plasmid to be analyzed in order to obtain enough data to create good statistics. Prepare a master-mix for each experimental plasmid according to number of replicates. This ensures that each biological parallel is treated as equal as possible. If using a cell line that is difficult to transfect, the number of replicates should be increased. In this way, outliers showing very divergent luciferase values may be omitted without disturbing the statistical relevance of the data. Similarly, the control plasmids should also be transfected with more replicates to ensure the robustness of the data if something happens with one of the control replicates (fungus, cell death, low transfection efficiency etc.), and those data must be omitted.
9. In our example we used an empty pGL3-basic vector as a negative control to measure the background level of luciferase. As a positive control, a vector containing a constitutively active promoter to drive the expression of luciferase could have been used. Promega provides a pGL3-Control vector containing a SV40 promoter and enhancer sequences which should yield strong expression of luciferase in many types of mammalian cells.
10. Opti-MEM is used because it is important that the complex between the FuGENE 6 reagent and DNA is formed in a serum-free media, even if the cells are transfected in the presence of serum. However, any serum free media will work. Always mix the FuGENE 6 Reagent, and let it come to RT, prior to use.
11. A master mix of Opti-MEM and FuGENE 6 makes sure that all plasmids are given the same amount of transfection reagent, which otherwise could be a source of variability between samples in the assay. As with all master mixes, prepare an excess

volume since multiple pipetting has a tendency to exhaust the volume prior to the last aliquot if only accurate volume is prepared.

12. The FuGENE 6 Reagent:DNA ratio needs to be optimized for each cell line to be transfected. However, three times the amount of μl FuGENE 6 to the amount of μg DNA (3:1) worked well for both mouse NSC and human HEK 293 cells.
13. Always pipette the FuGENE 6 reagent directly into the serum-free medium, and avoid contact with the plastic surface. This because the biological activity of the FuGENE 6 Reagent can be significantly decreased from chemical residues in the plastic vials. Contact with the plastic surface may result in no, or very low transfection efficiency.
14. Roche recommends adding DNA into the FuGENE 6 Reagent/Opti-MEM mix, but in our hands it also worked adding the FuGENE 6 Reagent/Opti-MEM mix into pre-mixed DNA in 1.5 ml tubes. This makes it easier to handle larger amount of samples to be analyzed since they can be pre-mixed into separate tubes prior to the addition of the FuGENE 6 Reagent/Opti-MEM mix.
15. Changing media is not required, and may be omitted. However, changing the media removes dead or unattached cells.
16. Pipette carefully up and down a few times in the upper layer of the media to make sure that the entire volume of DNA/FuGENE 6 complexes are aliquoted into the well. Carefully shake the plate in several directions to make sure that the complex mix is evenly distributed to all cells in each well. By swirling the plate in circles one may end up with all complexes in the middle of the well.
17. Investigation of a desired drug or treatment on its role on the promoter activity is usually performed by adding treatment 24 h post-transfection. Alternatively, cells may be co-transfected with expression plasmids with a transcription factor of choice together with the initial DNA mix to be transfected. In such cases, the amount of experimental promoter plasmid is reduced to a level where an equal amount of promoter plasmid and the third expression plasmid is transfected, and the total amount of DNA transfected into each well is still kept at 0.25 μg (48-well plate). It is also recommended to include a positive control for the treatment in the experimental setup.
18. Look at the cells prior to lysing. In this way one may discover wells where cells have died, or other changes that may affect the obtained luciferase values, which may be kept in mind when obtaining luciferase data.

19. Prepare master mix of 1× PLB in excess to cover all wells to be analyzed. Equal treatment of all wells reduces the risk of experimental variability. 65 μl is enough to cover the cell monolayer in a well of a 48 well plate. Using other lysis buffers in this assay is not recommended since the PLB is specifically designed to promote rapid lysis of cells, to provide optimum performance and stability of the firefly and *Renilla* luciferases, and to minimize background autoluminescence. Leftover of 1× PLB may be stored at +4°C for up to a month, but is preferred to be prepared fresh just before use.
20. Using an orbital shaker is not necessary, but may enhance lysis of the cells. The firefly and *Renilla* luciferases are stable for at least 6 h in RT, but the lysed cells may also be frozen at either −20°C (up to a month), or at −80°C (several months) if analysis needs to be performed at a later time point.
21. Prepare excessive Stop & Glo master-mix (substrate and buffer). Sometimes it is necessary to measure a sample twice to validate the measurements if the obtained values from one sample are very different from those obtained from the biological parallels. The Stop & Glo master-mix is best when prepared just before use, but might also be stored at −20°C for some weeks without decrease in activity. If stored frozen, it should be thawed in a waterbath at RT, and mixed well before use. Leftovers might be mixed with freshly prepared Stop & Glo reagent. However, make sure that all samples are measured using the same mix.
22. If using a luminometer with an auto-injector system one should take great care to properly clean the tubes if the system is to be used for LARII. This is because the Stop & Glo Reagent has a moderate affinity for plastic materials, and exhibit a reversible association with the interior surfaces of plastic tubings used in the auto-injection systems. If not properly cleaned, the Stop & Glo Reagent may therefore leach trace quantities of quench reagent into the solutions, which may further cause significant inhibition of firefly luciferase reporter activity.
23. Avoid air bubbles when pipetting the various solutions in the tubes to be measured since these may affect the readings of the luciferase luminescence.
24. Prepare master mix in excess in case some samples needs to be measured twice. Leftovers may be stored at −20°C for weeks without losing activity.
25. Avoid the white precipitate of cell membranes, etc. The reading is more repeatable when only the clear lysate is measured.
26. It is not important whether you pipette up and down three or five times, but by following the exact same procedure for every sample to be measured, both with firefly and *Renilla* luciferase

measurements, one eliminates a source of error. The signal created from the reaction between the luciferase enzyme and the substrate is rapidly degrading, and readings may therefore be affected with major differences in time spend on each sample.

27. It is important that the cell lysate is mixed with LAR II by pipetting and not by vortexing. Vortexing may coat the sides of the tube with luminescent solution which can escape mixing with the subsequently added Stop & Glo Reagent.

28. The *Renilla* luciferase value functions as an internal control for transfection efficiency between triplicates, and between different experimental conditions. Drop in *Renilla* luciferase value may be a result of toxicity from the treatment (if added), inefficient transfection, or changes in the expression of the vector DNA containing *Renilla* luciferase. Usually both the experimental plasmid and the *Renilla* luciferase are affected equally such that the relative ratio between the two may not affect the median relative ratio between replicates. However, in these cases the internal control can be used to normalize the triplicate wells to each other. This is done by averaging the *Renilla* luciferase values for the three wells, dividing the individual *Renilla* luciferase values for each well on the average *Renilla* luciferase values, and lastly; multiplying the firefly luciferase values for each well to the normalized *Renilla* luciferase value. Including enough replicates, at least for the controls, may strengthen the experimental data and making it less vulnerable to outliers.

29. One may also correct for background luminescence by subtracting readings from non-transfected cells from each of the luciferase readings. However, this is rarely necessary since most cells show very little background luminescence, if any at all.

Acknowledgment

The research was granted by the Norwegian Research Council.

References

1. Fan F, Wood KV (2007) Bioluminescent assays for high-throughput screening. Assay Drug Dev Technol 5:127–136
2. Massoud TF, Paulmurugan R, De A, Ray P, Gambhir SS (2007) Reporter gene imaging of protein-protein interactions in living subjects. Curr Opin Biotechnol 18:31–37
3. de Almeida PE, van R Jr, Wu JC (2011) In vivo bioluminescence for tracking cell fate and function. Am J Physiol Heart Circ Physiol 301:H663–H671
4. Greer LF III, Szalay AA (2002) Imaging of light emission from the expression of luciferases in living cells and organisms: a review. Luminescence 17:43–74
5. De W Jr, Wood KV, Helinski DR, Deluca M (1985) Cloning of firefly luciferase cDNA and the expression of active luciferase in

Escherichia coli. Proc Natl Acad Sci USA 82:7870–7873

6. Ow DW, Wit DE Jr, Helinski DR, Howell SH, Wood KV, Deluca M (1986) Transient and stable expression of the firefly luciferase gene in plant cells and transgenic plants. Science 234:856–859
7. De W Jr, Wood KV, Deluca M, Helinski DR, Subramani S (1987) Firefly luciferase gene: structure and expression in mammalian cells. Mol Cell Biol 7:725–737
8. Zhang W, Purchio AF, Chen K, Wu J, Lu L, Coffee R, Contag PR, West DB (2003) A transgenic mouse model with a luciferase reporter for studying in vivo transcriptional regulation of the human CYP3A4 gene. Drug Metab Dispos 31:1054–1064
9. Padmanabhan P, Otero J, Ray P, Paulmurugan R, Hoffman AR, Gambhir SS, Biswal S, Ulaner GA (2006) Visualization of telomerase reverse transcriptase (hTERT) promoter activity using a trimodality fusion reporter construct. J Nucl Med 47:270–277
10. Solberg N, Machon O, Krauss S (2012) Characterization and functional analysis of the 5 -flanking promoter region of the mouse Tcf3 gene. Mol Cell Biochem 360:289–299
11. Zipser D, Lipsich L, Kwoh J (1981) Mapping functional domains in the promoter region of the herpes thymidine kinase gene. Proc Natl Acad Sci USA 78:6276–6280
12. Lorenz WW, McCann RO, Longiaru M, Cormier MJ (1991) Isolation and expression of a cDNA encoding Renilla reniformis luciferase. Proc Natl Acad Sci USA 88: 4438–4442
13. van den Bout CJ, Machon O, Rosok O, Backman M, Krauss S (2002) The mouse enhancer element D6 directs Cre recombinase activity in the neocortex and the hippocampus. Mech Dev 110:179–182
14. Yin W, Xiang P, Li Q (2005) Investigations of the effect of DNA size in transient transfection assay using dual luciferase system. Anal Biochem 346:289–294

Chapter 7

Promoter Deletion Analysis Using a Dual-Luciferase Reporter System

Yong Zhong Xu, Cynthia Kanagaratham, Sylwia Jancik, and Danuta Radzioch

Abstract

Promoter deletion analysis is a useful tool for identifying important regulatory regions involved in transcriptional control of gene expression. In this approach, a series of promoter deletion fragments are fused to a reporter gene, such as chloramphenicol acetyltransferase or luciferase gene in a vector, and then transfected into cells for induction. Screening the expression level of the reporter gene using either a qualitative or a quantitative assay, allows to identify the regulatory regions of interest (e.g., *cis*-acting elements or enhancer) in the promoter.

Luciferase genes have been widely used as reporter genes for their sensitivity and efficiency. Firefly and Renilla luciferases are two commonly used reporters, which oxidize different substrates to generate quantifiable luminescence. Therefore, the enzymatic activities of firefly and Renilla luciferases can be sequentially measured in a single sample by controlling reaction conditions. Here, we describe a dual-luciferase reporter assay, where the promoter of interest is fused to a firefly luciferase reporter and is co-transfected into cells with an internal control vector (pRL-CMV) to express Renilla luciferase. Both the Firefly and Renilla luciferases are measured using a dual-luciferase reporter assay system which improves experimental accuracy.

Key words: Promoter deletion, Construction of plasmid, Cell transfection, Firefly luciferase, Renilla luciferase, Dual-luciferase reporter assay

1. Introduction

In eukaryotic cells, genes are surrounded by DNA regulatory elements such as promoter regions and enhancer regions that are involved in the regulation of transcriptional activity (1). A promoter sequence contains a cluster of functional DNA sequences that interact with regulatory proteins such as transcription factors and gene-specific DNA-binding proteins (2–4). By binding to a gene promoter, transcriptional factors are typically assembled to form

Minou Bina (ed.), *Gene Regulation: Methods and Protocols*, Methods in Molecular Biology, vol. 977,
DOI 10.1007/978-1-62703-284-1_7, © Springer Science+Business Media, LLC 2013

"transcriptional switches" capable of controlling gene expression. In eukaryotic cells, RNA polymerase II requires transcription factors for polymerase II (TFIIs), such as TFIIA and TFIIB, to form a transcription initiation complex on a DNA template (5). Promoters have been classified into three types based on the regulation of gene transactivation: constitutive promoters, tissue-specific promoters activated in a specific cell type or a specific developmental stage; and inducible promoters activated by internal or external stimuli.

Promoter deletion analysis represents one of the primary and widely used techniques to determine the *cis*-acting elements or specific transcription factor binding sites within a promoter that are responsible for transcriptional regulation of a gene. In principle, a promoter sequence needs to be identified, isolated and then fused to a reporter gene such as chloramphenicol acetyltransferase (6), luciferase (7) or green fluorescence protein (8). The generated reporter construct can then be transfected into cells to express reporter gene. Prior to promoter deletion analysis, the reporter construct need to be evaluated to ensure that the promoter remains the responsive to stimuli known either to induce or repress transactivation. Once the construct is positively evaluated, small portions of the promoter are deleted until the region controlling transcription is identified (Fig. 1a) by observing the expression of the reporter gene.

The luciferase reporter system has been widely used for the study of gene expression and function because of its sensitivity and efficiency (9, 10). Firefly and Renilla luciferase genes are two commonly used reporter genes. Firefly luciferase gene encodes a monomeric 61-kDa enzyme protein that does not require posttranslational modification for its activity (11, 12). The enzyme catalyzes the oxidation of D-luciferin into oxyluciferin in the presence of ATP, Mg^{2+}, and oxygen. The catalysis releases light and allows for the quantification of the luciferase activity (Fig. 2a). Addition of coenzyme A to the reaction enhances the light intensity and extends the light emission lifetime (13). Renilla luciferase, a 36 kDa monomeric protein, catalyzes coelenterate-luciferin (coelenterazine) oxidation to generate luminescence (Fig. 2b). Similarly, although Renilla luciferase is glycosylated in its natural host, the posttranslational modifications are not required for its activity to be monitored in the surrogate host (14). Coelenterazine also emits light via enzyme-independent oxidation, a process known as autoluminescence (15). The autoluminescence can be reduced to a level that is undetectable with a specific coelenterazine derivative, cell lysis buffer and luciferase assay reagent.

The dissimilarities in the enzyme structures and substrates between Firefly and Renilla luciferases enable to sequentially measure the activity of each of the two luciferases by manipulating the reaction conditions. Using the dual luciferase reporter reagent, the luminescence of firefly luciferase can be measured by addition of

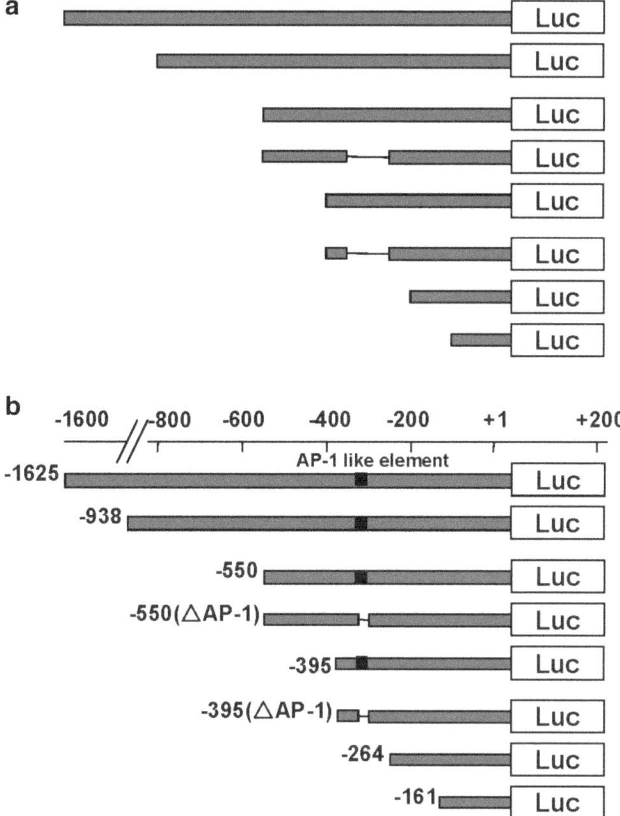

Fig. 1. Deletion analysis of promoter activity. (**a**) A schematic representation of a typical strategy to analyze a promoter. A promoter deletion series are cloned into a luciferase reporter vector. (**b**) A specific example using such a strategy to identify the *cis*-acting element that is responsive to PMA stimulation in the SLC11A1 gene promoter (Reproduced from the Journal of Biological Chemistry, see ref. 18).

the luciferin reagent, and this reaction is subsequently quenched while simultaneously activating the luminescence of the Renilla luciferase. Dual reporters are commonly used to improve the experimental accuracy. In dual reporter assay system, one luciferase is selected as an experimental reporter and another one as an internal control. Normalizing the activity of an experimental reporter to the activity of an internal control minimizes or sometimes even eliminates the experimental variability.

In this chapter, we will use the human SLC11A1 gene as an example to describe both methods for promoter deletion analysis and luciferase assay. First, construction of a series of promoter-luciferase reporter vectors is described, followed by the description of transfection of promoter-luciferase reporter vectors and control luciferase reporter vector into human myeloid leukemia cell line (HL-60 cells) using the nucleofector kit. Finally, the luciferase assay using the dual-luciferase reporter assay system is described.

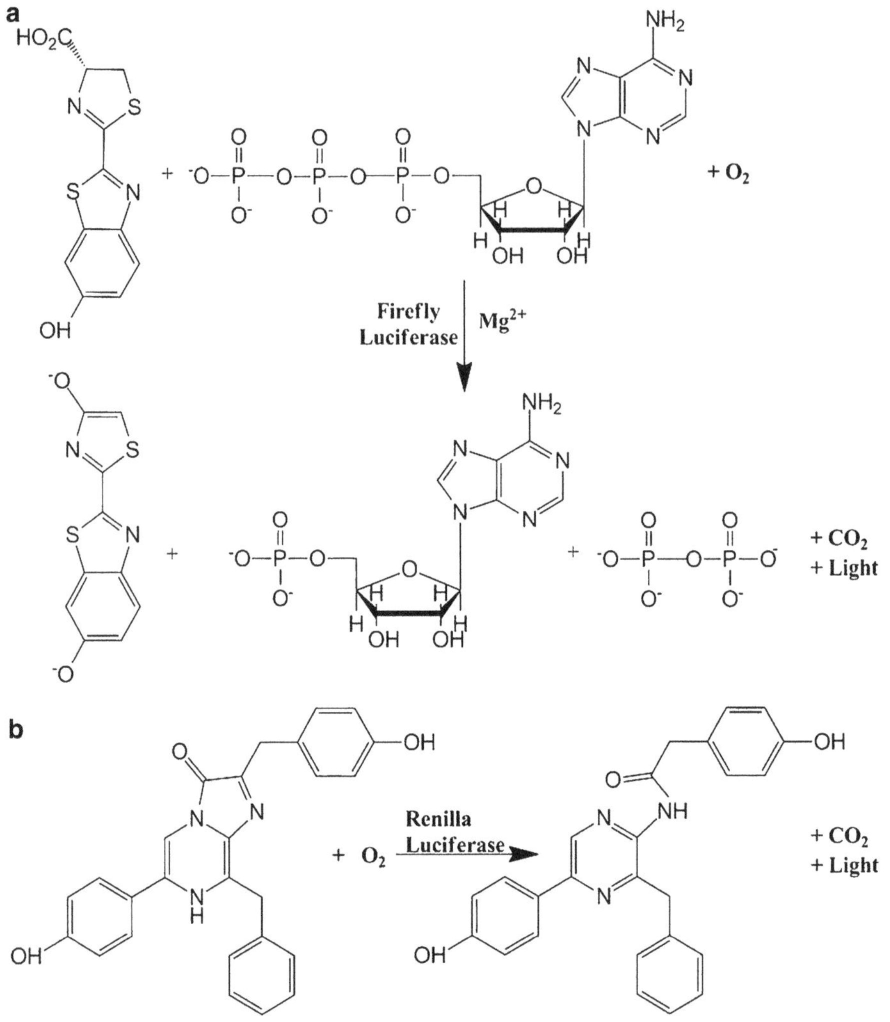

Fig. 2. Biochemical reactions catalyzed by Firefly luciferase (**a**) and Renilla luciferase (**b**) to produce luminescence.

2. Materials

2.1. Promoter Deletion Analysis and Plasmid Construction

1. Human genomic DNA (Roche, Laval, QC, Canada).
2. PfuUltra high-fidelity DNA polymerase (Stratagene, La Jolla, CA, USA).
3. 10 mM dNTP mix (Invitrogen, Burlington, ON, Canada).
4. Restriction endonuclease enzymes (see Note 1): *Nhe*I and *Hin*dIII (New England Biolabs).
5. Agarose, low melting point (Sigma).
6. DNA ladder, 100 bp and 1 kb (Invitrogen).
7. 10 mg/ml Ethidium bromide (Sigma).

8. QIAquick gel purification kit, QIAquick PCR purification kit and QIAprep spin miniprep Kit (Qiagen).
9. T4 DNA ligase (Invitrogen).
10. Plasmid pREP4-Luc (16).
11. LB medium.
12. LB plates.
13. SOC medium.
14. DH5α competent cells.
15. 100% ethanol.

2.2. Cell Culture and Transfection

1. Human myeloid leukemia cell line (HL-60 cells) from American Type Culture Collection (ATCC, Manassas, VA, USA).
2. Culture medium: Roswell Park Memorial Institute (RPMI) 1640 plus L-glutamine and 25 mM HEPES (Invitrogen) supplemented with 1% penicillin/streptomycin (Invitrogen) and 10% fetal calf serum (Hyclone). Store at 4°C (see Note 2).
3. Nucleofector II device (Lonza).
4. Amaxa cell line nucleofector kit V (Lonza) including: 2.25 ml of cell line nucleofector solution V, 0.5 ml of supplement, 30 μg of pmaxGFP vector (0.5 μg/μl in 10 mM Tris-HCl, pH 8.0), 25 certified cuvettes, and 25 plastic pipettes.
5. 12-well tissue-culture plates and 15-ml polystyrene conical tubes.

2.3. Luciferase Activity Assay

1. Dual-luciferase reporter assay system (Promega) including: 10 ml of luciferase assay buffer II, 1 vial of lyophilized luciferase assay substrate, 10 ml of stop & Glo buffer, 200 μl of stop & Glo substrate (50×), and 30 ml of passive lysis buffer (5×).
2. Dulbecco's phosphate buffered saline (DPBS).
3. TD-20/20 Luminometer (Turner biosystems, Sunnyvale, CA, USA).
4. Cell scraper (for adherent cells), 15 ml polystyrene conical tubes, and 1.5 ml microcentrifuge tubes.

3. Methods

3.1. Promoter Deletion Analysis

3.1.1. Promoter Analysis

Before performing a promoter deletion analysis, it is necessary to know as much about the promoter sequences as possible by performing a thorough literature search., Helpful online databases include PubMed, Gene Bank, tissue-specific promoter database (TiProD) (17), transcript sequence retriever (TRASER), and transcription factor database (TRANSFAC). Your search may allow you to uncover basic characteristics about your promoter of interest, e.g., the transcription start site, the quantity and type of transcription factor-binding sites, TATA box and its position, and results from other research groups that may have studied the same or similar promoter sequences. All these help you to determine which promoter regions or *cis*-acting elements may be important for your study and to construct reporter plasmids.

3.1.2. Plasmid Constructions

1. Design primers with Nhe I restriction site at 5′ end of forward primers and HindIII restriction site at 5′ end of reverse primers to amplify a series of progressive 5′-deleted promoter region and AP-1 like element of the SLC11A1 gene (Fig. 2b)(18).
2. Amplify the longest promoter fragment (spanning nucleotides −1625 to +19 of the *SLC11A1* promoter, numbers relative to the major transcription start site) using nest PCR (see Note 3). Combine the following in a PCR tube (50 μl total volume) for first PCR.

   7.5 μl 10× reaction buffer (Stratagene).

   2.5 μl dNTPs mix (10 mM for each).

   1.25 μl forward primer (20 μM stock).

   1.25 μl reverse primer (20 μM stock).

   1 μl genomic DNA (template, 200 ng/μl).

   2.0 μl PfuUltra high-fidelity DNA polymerase (2.5 U/μl).

   34.5 μl H_2O.

Set up PCR reaction conditions according to the manufacturer's instructions (Stratagene) and run first PCR for 30 cycles. Then combine the following in a PCR tube. Use 1 μl of PCR product from the first PCR as a template.

   5 μl 10× reaction buffer (Stratagene).

   1.25 μl dNTPs mix (10 mM for each).

   1 μl forward primer (20 μM stock).

   1 μl reverse primer (20 μM stock).

   1 μl PCR product from the first PCR

   1 μl PfuUltra high-fidelity DNA polymerase (2.5 U/μl).

   39.75 ml H_2O.

   Set up PCR reaction conditions according to the manufacturer's instructions (Stratagene) and run second PCR for 32 cycles.

3. Prepare 1.2% low melting agarose gel (see Note 4) containing ethidium bromide (final concentration 0.5 μg/ml) (see Note 5) and run the PCR product on the gel.
4. Cut the desired DNA fragment from the gel under a long-wavelength UV light (see Note 6). Place the gel slice in a microcentrifuge tube.

5. Purify DNA fragment from the agarose gel using gel extraction kit according to the manufacturer's instructions (Qiagen).
6. Measure the DNA concentration.
7. Digest the DNA fragment with restriction enzymes *Nhe*I and *Hin*dIII (see Note 7). Mix the following in a microcentrifuge tube (total volume 20 μl). Incubate tubes at 37°C for 2–3 h (see Note 8).

   2 μl NEBuffer 2 (10×)

   1 μl NheI (10 U/ml)

   0.5 μl HindIII (20 U/ml)

   0.2 μl BSA (100×)

   1–2 μg purified DNA fragment

   Add ddH_2O to bring the total volume to 20 μl
8. Purify the digested insert DNA using PCR purification kit according to the manufacturer's instructions (Qiagen).
9. Digest the vector REP4-luc with restriction enzymes *Nhe*I and HindIII (see Note 9). Mix the following in a microfuge tube (total volume 20 μl) and incubate at 37°C for 2–3 h (see Note 8).

   2 μl NEBuffer 2 (10×)

   1 μl NheI (10 U/μl)

   0.5 μl HindIII (20 U/μl)

   0.2 μl BSA (100×)

   2 μl pREP4-luc (1 μg/μl)

   Add 14.3 μl ddH_2O to bring the total volume to 20 μl
10. Purify the cut plasmid using gel extraction kit according to the manufacturer's instructions (Qiagen).
11. Ligate insert DNA into the plasmid using T4 DNA ligase at 14°C overnight. Combine the following in a microfuge tube (see Note 10).

    4 μl buffer (5×)

    X μl pREP4-luc (100 ng)

    Y μl insert DNA (40 ng)

    0.5 μl T4 DNA ligase

    1 μl 10 mM ATP

    Add deionized H_2O to bring the total volume to 20 μl.
12. Ethanol-precipitate the ligated plasmid in the following manner.
    (a) Add 3 M sodium acetate (pH 5.2) at 1/10 volume of the plasmid, mix by inverting tube.
    (b) Add 2.5 volume of ice-cold 100% ethanol and mix by inverting tube.

(c) Place the tube in a –20°C freezer for 1–2 h (see Note 11).

(d) Spin in a microcentrifuge at 12,000 × *g* for 15–20 min at 4°C. Carefully remove the supernatant and discard.

(e) Add 1 ml of ice-cold 70% ethanol, vortex, and centrifuge at 12,000 × *g* for 5 min (see Note 12).

(f) Aspirate the supernatant. Dry the pellet at room temperature (see Note 13).

(g) Redissolve the pellet in an appropriate volume of Qiagen's EB buffer (10 mM Tris–HCl, pH 8.5) or deionized H_2O (see Note 14).

13. Transformation of competent cells DH5α by electroporation.

(a) For each DNA sample to be transformed, take an aliquot of competent cells (25 μl) out of –80°C freezer and thaw on ice (see Note 15).

(b) Label an electroporation cuvette (0.1 cm gap width) for each sample and place on ice. Place the white chamber slide on ice.

(c) Add 1 μl of plasmid into an aliquote of competent cells, mix gently by flicking the tube, and transfer the mixture to the bottom of a cuvette on ice.

(d) Set the electroporation apparatus to 1.8 kV (volts) and 25 μFD (capacitance).

(e) Dry the outside of the cuvette and chamber slide with a Kimwipe paper. Place the cuvette into the slide and push the slide into the sample chamber. Apply the pulse immediately.

(f) Quickly remove the cuvette and immediately add 1 ml SOC medium (see Note 16) and mix gently. Transfer the mixture to a culture tube.

(g) Place the tube in a 37°C shaker for 1 h at a speed of 225 rpm (see Note 17).

(h) Spread an appropriate volume of the cells on a selective LB plate containing 100 μg/ml Ampicillin.

(i) Incubate the plate at 37°C overnight.

14. Colony screening.

(a) Label bacteria colonies and microfuge tubes. Add 25 μl deionized water to each tube.

(b) Pick a colony with a toothpick and twirl the toothpick vigorously in the tube with water. Repeat for each colony. Vortex for 10 s.

(c) Add 25 μl of cracking buffer and vortex for 10 s.

(d) Heat tubes at 65°C for 5 min and then vortex for 10 s. Spin the tubes at 14,000 rpm for 10 min at 4°C.

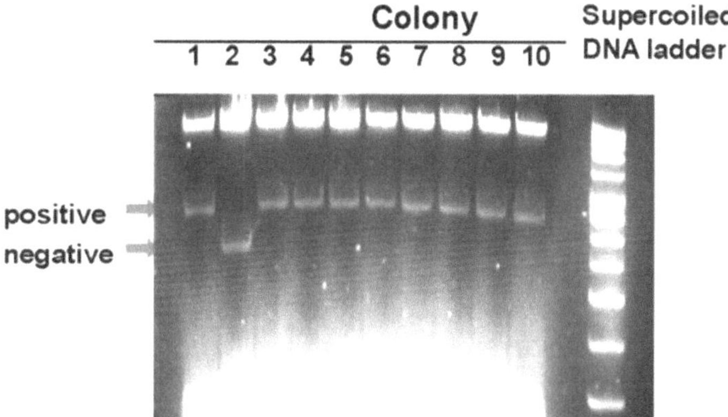

Fig. 3. Agarose electrophoresis shows different electrophoretic mobility between positive plasmid with insert DNA and negative plasmid without insert DNA. Positive plasmid moves slower than negative plasmid. Several positive colonies and one negative colony are illustrated as an example.

(e) Prepare 0.8% agarose gels containing ethidium bromide.

(f) Load 25 μl of the supernatants into wells of gel and load supercoiled DNA ladders. Run gel at 100 V.

(g) Identify the positive colonies according to the molecular weight and electrophoretic mobility difference between the plasmids with and without insert DNA. Plasmids with insert DNA move slower than plasmids without insert DNA due to the increase in molecular weight (Fig. 3).

(h) Pick several positive colonies from the labelled plate and inoculate them in 5 ml LB broth, separately. Incubate the culture overnight in a shaker at 37°C. Extract plasmids using QIAprep Miniprep Kit following the manufacturer's instructions (Qiagen).

15. Confirm the result by digestion with restriction enzymes *Nhe*I and *Hin*dIII and sequencing of the constructed plasmid. The plasmid is named −1625 Luc.

16. Plasmid −1625 Luc was used as a template for several other constructs: −938Luc, −550Luc, −395Luc, −264Luc, and −161Luc. The two deletion mutants (−550(ΔAP-1) Luc and −395(ΔAP-1) Luc), lacking the AP-1-like element TGACTCT, were constructed using QuickChange II XL site-directed mutagenesis kit following the manufacturer's instructions (18).

3.1.3. Transfection of Cells

1. Culture HL-60 cells in RPMI 1640 medium supplemented with 25 mM HEPES, 2 mM L-glutamine, 10% heat-inactivated fetal bovine serum, and 1% penicillin/streptomycin (see Note 18).

2. Three days before the transfection, plate cells at a density of 1×10^5 cells/ml. Cells will be ready for transfection when they achieve a density of $5–7 \times 10^5$ cells/ml.

3. Prepare supplemented nucleofector solution by adding entire supplement (0.5 ml) to the vial containing cell line nucleofector solution V (2.25 ml) and mix well. This solution can be stored at 4°C for 3 months.
4. Prepare 12-well plates by filling the appropriate number of wells with 1 ml of RPMI 1640 plus L-glutamine and 25 mM HEPES supplemented with 1% penicillin/streptomycin and 20% fetal calf serum. Pre-incubate the plates at 37°C in a humidified 5% CO_2 incubator.
5. Determine cell density by counting the cells using hemocytometer or an alternative cell counting method.
6. Transfer required numbers of cells (2×10^6 cells per transfection) to a 15-ml tube and spin at $100 \times g$ for 8 min. Use a separate tube for each transfection.
7. Remove the supernatant as much as possible and gently resuspend the cells in 100 μl nucleofactor solution (bring to room temperature prior to use) (see Note 19).
8. Gently mix the 100 μl of cells resuspended in nucleofactor solution with 2.0 μg plasmid DNA and 50 ng pRL-CMV as an internal control.
9. As quickly as possible, carefully transfer the cell suspension mixed with plasmid DNA into a nucleofector cuvette. Close the cuvette with the cap and place it into the cuvette holder of Nucleofactor II device. Perform transfection using program T-019.
10. Immediately take out the cuvette and add 500 μl of culture medium pre-equilibrated to room temperature. Gently transfer the cell suspension into the previously prepared 12-well plates (as described in the step 4 above; see Note 20).
11. Incubate the transfected cells in a humidified incubator with 5% CO_2 at 37°C.
12. Six hours after transfection, treat cells with PMA (10 ng/ml) or leave untreated for 48 h, following this harvest cells for luciferase activity assay.

3.2. Luciferase Reporter Assay

3.2.1. Preparation of Cell Lysates

1. Make 1× Passive lysis buffer by adding 1 volume of 5× Passive lysis buffer to 4 volumes of deionized water and mix well (see Note 21).
2. Harvest and lyse the transfected cells

   For adherent cells (HL-60 cells treated with PMA)

   (a) Remove growth media from the wells of 6-well tissue culture plate and gently wash the cells by rinsing them twice with 1× PBS. Be careful not to dislodge the cells. Aspirate as much volume of the final wash solution as possible.

(b) Add 500 μl of 1× Passive lysis buffer to each well (see Note 22), and then shake several times to ensure complete coverage of the cells.

(c) Scrape attached cells from each well using a cell scraper. Transfer the cell lysates into individual microcentrifuge tubes. Place the tubes on ice.

For suspension cells (HL-60 cells)

(a) Transfer the cells to a 15-ml tube and centrifuge for 5 min at 100×*g*. Carefully remove culture medium. Wash cells with ice-cold 1× PBS and centrifuge at 100×*g* for 5 min. Aspirate as much of the final wash solution as possible.

(b) Add 100 μl of 1× Passive lysis buffer to the 15 ml-tube per each 10^6 cells. Resuspend the cells using a micropipettor.

(c) Transfer the cell lysate to a microcentrifuge tube. Place the tube on ice.

3. Place the microcentrifuge tubes containing cell lysates on a rocking platform and gently rock at room temperature for 15 min.
4. Vortex the tubes 10–15 s, and then centrifuge at 12,000×*g* for 10 min at 4°C. Transfer the supernatant to a new tube.
5. Store the lysates (supernatant) at −80°C for later use or place them on ice for immediate luciferase activity assay.

3.2.2. Luciferase Activity Assay (See Note 23)

1. Prepare luciferase assay reagent II (LARII) by adding 10 ml of luciferase assay buffer II to the vial containing the lyophilized luciferase substrate and mixing well. Aliquot the reagent to avoid multiple freeze–thaw cycles and store the aliquots at −70°C (see Note 24).
2. Calculate the required volume of Stop & Glo reagent according to the numbers of assays (100 μl per assay). Prepare the reagent by mixing 1 volume of 50× Stop & Glo substrate with 50 volumes of Stop & Glo buffer in a polypropylene tube (see Note 25).
3. Add 100 μl of the LAR II into a luminometer tube, one tube per assay.
4. Program the luminometer with appropriate parameters: a 2-s of measurement delay followed by a 10-s of integration time for luciferase activity assay.
5. Add 20 μl of cell lysate to the tube containing the LAR II and mix gently by pipetting 2–3 times.
6. Place the tube in the luminometer and initiate reading.
7. Record the light units of firefly luciferase activity.
8. Take the tube out of the luminometer, immediately add 100 μl of Stop & Glo reagent, and mix by vortexing briefly.

9. Replace the tube in the luminometer and measure.
10. Record the light units of Renilla luciferase activity.
11. Discard the reaction tube and proceed with the next sample (repeat steps 5–11).

4. Notes

1. Restriction endonuclease enzymes are selected according to the promoter sequence and the vector sequence. Make sure that the restriction sites are not present in the promoter sequence but within the multiple cloning sites of the vector.
2. Culture medium can be quickly acidified by HL-60 cells, so adding HEPES helps to dampen changes in pH.
3. Conventional PCR can be used to amplify DNA fragment; however, one of the most encountered problems is the annealing of primers at a wrong location (or locations) to generating a wrong product (or products). It may be possible to resolve the problem by raising the annealing temperature in the reaction conditions or changing the ionic concentrations in the reaction mix. Here, we use nested-PCR to increase the specificity of amplification.
4. Prepare appropriate concentration of agarose gels for separating DNA fragment of different size (19).
5. Ethidium bromide is known carcinogen. Handle solution containing ethidium bromide with adequate precautions, such as wearing double gloves as well as covering all skin that could come in contact with the solution.
6. Use a long-wavelength UV light (360 nm) to minimize DNA damage by UV. When using a short wavelength UV, reduce the exposure time of DNA to UV as much as possible. Keep the gel on a glass or plastic plate when excising DNA fragment. Alternatively, use dye visible in ambient light to visualize DNA (20).
7. Digesting a DNA fragment with two restriction enzymes simultaneously is a common time-saving procedure. However, if a compatible buffer for both enzymes cannot be found or the two recognition sites are separated by less than ten base pairs, the digestion reaction should be performed sequentially with individual enzymes.
8. A few enzymes require special conditions for digestion, such as different incubation temperatures, therefore be sure to check manufacturer's instructions before using. In addition, to ensure complete digestion, extra enzyme and longer incubation (8–12 h) may be needed.

9. The restriction enzymes used for digesting vector and insert DNA fragment should be the same to ensure optimal ligation. Noncompatible ends produced by enzyme digestion are ideal because they prevent self-ligation. Some enzyme combination, however, produce compatible ends. To reduce the chance of self-ligation, dephosphorylate the 5′ phosphorylated ends of digested vector with alkaline phosphatase such as calf alkaline phosphatase or shrimp alkaline phosphatase following the manufacturer's instructions. The double-enzyme digestion of vector may be carried out in a single reaction or in two sequential reactions as described in Note 7.
10. The molar ratio of digested insert to vector DNA can have a significant effect on the ligation efficiency. A 3:1 ratio is a common starting point, and ratios ranging from 1:1 to 10:1 will give you the desired results. When necessary try different ratios to optimize ligation reaction. If you choose molar ratio of insert to vector DNA, use the following formula to calculate the required amount of insert DNA:

$$\text{Insert DNA}(\text{ng}) = \frac{\text{size of insert DNA}(\text{bp}) \times \text{amount of vector DNA}(\text{ng})}{\text{size of vector DNA}(\text{bp})} \times \text{molar ratio}(\text{insert to vector})$$

11. The incubation time in freezer depends on the concentration and the size of DNA sample. If the concentration of your DNA sample is low or it is very small in size (less than 15 nucleotides) you may extend the incubation time to several hours or overnight. You can also put the sample in a −80°C freezer for 20 min to a few hours. It also works well to put it on dry ice for 15–20 min.
12. After addition of 70% ethanol, the pellet should be vortexed to help dissolve salts.
13. The pellet can be dried in a speedvac; however, overdrying can make it difficult to redissolve the DNA pellet, especially for longer DNA.
14. Make sure the pH of water is neutral. We have found that water can vary from pH 5.0 to pH 8.5. Low pH may cause the depurination of DNA.
15. Cells should be thawed on ice. Thawed with hand-heat significantly decreases the transformation efficiency.
16. The time period between applying pulse and adding SOC medium is crucial for recovering *E. coli* transformants (21). Transfer the cells to medium as soon as possible. Using SOC medium gives twofold higher transformation efficiency than LB medium.
17. Shaking at 225 rpm may improve the recovery of transformants.

18. Keep a basin of water on the bottom shelf of the incubator to maintain humidity.
19. Medium containing FBS may interfere with transfection so it is important to remove the medium completely.
20. Use the supplied plastic pipette and do not aspirate the cell suspension repeatedly.
21. Prepare the required volume of 1× Passive lysis buffer just before using, although it may be stored at 4°C for 3–4 weeks.
22. Add an adequate volume of 1× lysis buffer to each type of culture plate following manufacturer's instructions. For 6-well plate, 400–500 μl 1× passive lysis buffer per well should be used; however, an increased volume of lysis buffer and/or extended treatment time is required to ensure complete lysis when cells are overgrown, because overgrown cells may become more resistant to lysis buffer.
23. The protocol described here is based on a single-tube luminometer. Microplate luminometer and correspondent computer software are also commercially available to read multi-well plates (e.g., 96-well plate) and process large numbers of samples.
24. Aliquots of LAR II should be thawed in water bath at room temperature because this reagent is unstable when heated. During the process of thawing, both density and composition gradients are formed in LAR II, therefore, it is necessary to mix the solution prior to use by inverting the vial several times or by gently vortexing.
25. It is best to prepare required volume of the Stop & Glo reagent (substrate plus buffer) just before use. However, if you prepared more, it may be stored at –20°C. The frozen reagent should be thawed in water bath at room temperature and always mix before use because of the same reason described for LAR II reagent.

References

1. Riethoven JJ (2010) Regulatory regions in DNA: promoters, enhancers, silencers, and insulators. Methods Mol Biol 674:33–42
2. Fitzgerald PC, Shlyakhtenko A, Mir AA, Vinson C (2004) Clustering of DNA sequences in human promoters. Genome Res 14: 1562–1574
3. Breathnach R, Chambon P (1981) Organization and expression of eucaryotic split genes coding for proteins. Annu Rev Biochem 50:349–383
4. Dynan WS, Tjian R (1985) Control of eukaryotic messenger RNA synthesis by sequence-specific DNA-binding proteins. Nature 316: 774–778
5. Alberts B, Johnson A, Lewis J, Raff M, Roberts K, Walter P (2002) Molecular biology of the cell. Garland Science, New York
6. Bookstein R, Rio P, Madreperla SA et al (1990) Promoter deletion and loss of retinoblastoma gene expression in human prostate carcinoma. Proc Natl Acad Sci USA 87:7762–7767
7. Heckman CA, Cao T, Somsouk L et al (2003) Critical elements of the immunoglobulin heavy chain gene enhancers for deregulated expression of Bcl-2. Cancer Res 63:6666–6673
8. Soboleski MR, Oaks J, Halford WP (2005) Green fluorescent protein is a quantitative reporter of gene expression in individual eukaryotic cells. FASEB J 19:440–442

9. Alam J, Cook JL (1990) Reporter genes: application to the study of mammalian gene transcription. Anal Biochem 188:245–254
10. Bronstein I, Fortin J, Stanley PE et al (1994) Chemiluminescent and bioluminescent reporter gene assays. Anal Biochem 219: 169–181
11. Ow DW, de Wet JR, Helinski DR et al (1986) Transient and stable expression of the firefly luciferase gene in plant cells and transgenic plants. Science 234:856–859
12. de Wet JR, Wood KV, Deluca M et al (1987) Firefly luciferase gene: structure and expression in mammalian cells. Mol Cell Biol 7:725–737
13. Fraga H, Fernandes D, Fontes R et al (2005) Coenzyme A affects firefly luciferase luminescence because it acts as a substrate and not as an allosteric effector. FEBS J 272: 5206–5216
14. Matthews JC, Hori K, Cormier MJ (1977) Purification and properities of Renilla reniformisluciferase. Biochemistry 16:85–91
15. Hannah RR, Jennens-Clough ML, Wood KV (1998) Rapid luciferase reporter assay systems for high throughput studies. Promega Notes 65:9–14
16. Liu R, Liu H, Chen X et al (2001) Regulation of CSF1 promoter by the SWI/SNF-like BAF complex. Cell 106:309–318
17. Chen X, Wu J-M, Hornischer K et al (2006) TiProD: the tissue-specific promoter database. Nucleic Acids Res 34:D104–D107
18. Xu YZ, Thuraisingam T, Marino R, Radzioch D (2011) Recruitment of SWI/SNF complex is required for transcriptional activation of SLC11A1 gene during macrophage differentiation of HL-60 cells. J Biol Chem 286:12839–12849
19. Ausubel FM, Brent R, Kingston RE et al (2001) Current protocols in molecular biology, vol 1. Wiley, New York, p 2.5.3
20. Adkins S, Burmeister M (1996) Visualization of DNA in agarose gels as migrating colored bands: applications for preparative gels and educational demonstrations. Anal Biochem 240:17–23
21. Dower WJ, Miller JF, Ragsdale CW (1988) High efficiency transformation of E. coli by high voltage electroporation. Nucleic Acids Res 16:6127–6145

Chapter 8

Application of mRNA Display for In Vitro Selection of DNA-Binding Transcription Factor Complexes

Seiji Tateyama and Hiroshi Yanagawa

Abstract

Comprehensive analysis of DNA–protein interactions is important for mapping transcriptional regulatory networks at the genome-wide level. Here, we present a new application of mRNA display, using the in vitro virus (IVV) technology, for in vitro selection of DNA-binding protein complexes. Under optimal selection conditions using bait DNAs, many kinds of DNA-binding protein complexes can be successfully selected from an mRNA display library constructed from poly A$^+$ RNA after several rounds of selection. This mRNA display selection system can identify a variety of DNA-binding protein complexes in a single experiment. Remarkably, this system can also select DNA-binding protein heterooligomeric complexes. Since almost all transcription factors form heterooligomeric complexes to bind with their target DNA, this method should be broadly useful to identify DNA-binding transcription factor complexes.

Key words: DNA–protein interaction, mRNA display, Protein complex, Transcription factor, Selection, Gene regulation, Heterooligomer

1. Introduction

The specific interactions between *cis*-regulatory DNA elements and transcription factors are critical components of transcriptional regulatory networks (1–3). The whole genome and complete cDNA sequences contain large numbers of transcription factors and their are expected to provide a deep understanding of the mechanisms of cell proliferation, developmental processes in tissue morphogenesis, disease states, and control of pluripotency and self-renewal in stem cells (1). Currently, combinations of chromatin immunoprecipitation (ChIP) assay with DNA-microarrays (ChIP-chip) (4–6) and with massively parallel DNA sequencing (ChIP-seq) (1) are the most widely used high-throughput methods

Minou Bina (ed.), *Gene Regulation: Methods and Protocols*, Methods in Molecular Biology, vol. 977,
DOI 10.1007/978-1-62703-284-1_8, © Springer Science+Business Media, LLC 2013

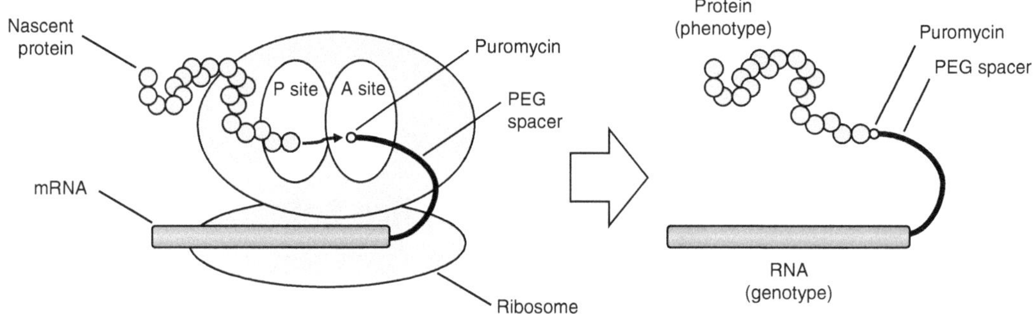

Fig. 1. Principle of IVV formation on the ribosome in a cell-free translation system. Puromycin ligated to the 3′-terminal end of mRNA through the polyethyleneglycol (PEG) spacer can enter the ribosomal A site to bind covalently to the C-terminal end of the protein that it encodes.

for discovering *cis*-regulatory DNA elements for transcription factors. In contrast, development of high-throughput methods for discovering transcription factors for *cis*-regulatory DNA elements remains at an early stage. Although current techniques using living cells are attractive candidates, these methods are not easily scalable because of the use of living cells (7, 8). In addition, as over-expression of transcription factors often affects cellular metabolism, screening can be difficult.

In order to circumvent such difficulties with cell-based systems, we developed a totally in vitro mRNA display technology (9–22), named the in vitro virus (IVV) method (9–21), for the discovery of protein–protein (11–13), DNA–protein (14), RNA–protein (15), antigen–antibody (16), and peptide–protein (17–19) interactions, and for the creation of novel functional proteins (20, 21). In mRNA display, a library of genotype (mRNA)-phenotype (protein) linked molecules (IVVs) is constructed, in which an mRNA is covalently bound to the corresponding protein through puromycin during cell-free translation (Fig. 1). After affinity selection via the protein moiety of the IVV, the mRNA moieties of the selected molecules are amplified by means of RT-PCR (Fig. 2). Therefore, even very low-copy-number proteins can be identified by iterative affinity selection from a library with high diversity and complexity, routinely in the range of 10^{13} members (9–22). The use of our IVV selection system for selection of DNA–protein interactions has several advantages over current techniques. Remarkably, for example, the IVV selection system is available for selection of DNA-binding protein heterodimeric complexes. Since almost all transcription factors form heterooligomeric complexes to bind with their target DNA, the IVV selection system is particularly suited to the search for DNA-binding transcription factor complexes. We have demonstrated that almost all members of the AP-1 family, including c-Jun, c-Fos,

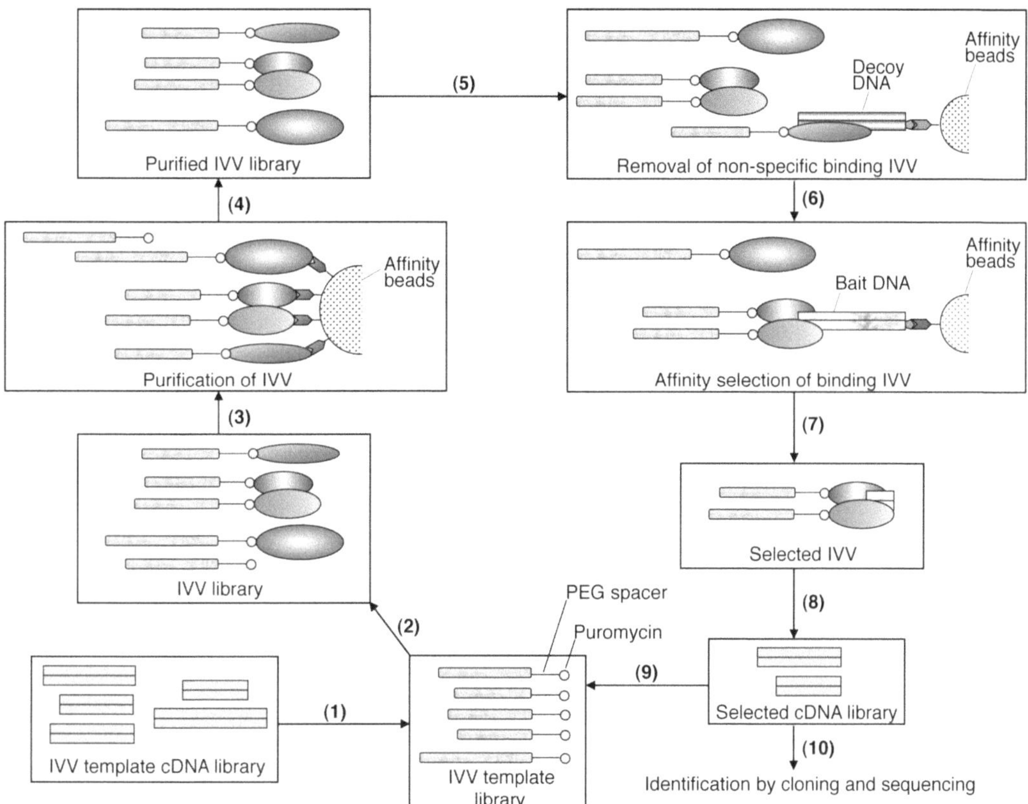

Fig. 2. Schematic representation of the IVV selection procedure. **(1)** An IVV template cDNA library is transcribed and ligated with a Fluoro-PEG Puro spacer. **(2)** The IVV template library is translated in a cell-free translation system. **(3)** The constructed IVV library is purified with anti-FLAG M2 antibody-immobilized beads to remove untranslated mRNA and impurities contained in the cell-free translation system. **(4)** The purified IVV are **(5)** subjected to affinity selection with decoy bait DNA-immobilized beads to eliminate nonspecific binders to DNA. **(6)** Unbound IVV are then subjected to affinity selection with bait DNA-immobilized beads. **(7)** After washing, DNA-binding IVV are eluted by DNase I digestion. **(8)** The mRNA portions of the selected IVV are reverse-transcribed, PCR-amplified, and **(9)** subjected to the next round of selection or **(10)** identified by cloning and sequencing.

JunD, JunB, ATF2, and b-ATF, can be successfully selected from an mRNA display library constructed from mouse brain poly A^+ RNA under improved selection conditions using a TPA-responsive element as a bait DNA, in a single experiment (14). More importantly, c-Fos and B-ATF were identified as positives in the IVV selection system. These proteins themselves do not show TRE-binding ability, but bind with TRE-DNA only in the presence of Jun or ATF (23). Therefore, the IVV selection system has unique advantages for the large-scale, comprehensive analysis of DNA-transcription factor interactions for mapping of transcriptional regulatory networks on a genome-wide level.

2. Materials

Prepare all solutions using molecular biology grade materials. Either DEPC-treated water or RNase-free water should be used to prevent degradation of RNA. Solutions may be arbitrarily divided into aliquots. Use filter tips to prevent cross-contamination of DNAs, RNAs, or RNases during pipetting tasks (see Note 1).

2.1. Primers Used in This Experiment

adaptor DNA FW: 5′-GAACAACAACAACAACAAACAACAACAAAATGGCTAGCATGACTGGTGGACAGCAAATGGCGAATTCC-3′.

adaptor DNA RV: 5′-GGAATTCG-3′.

3′random: 5′-TCATCGTCCTTGTAGTCAAGCTTN$_9$-3′.

5′ FW:

5′-GGAAGATCTATTTAGGTGACACTATAGAACAACAACAACAACAAACAACAACAAAATG-3′.

3′RV36:5′-TTTTTTTTCTTGTCGTCATCGTCCTTGTAGTCAAGC-3′.

3′RV30:5′-TTTTTTTTCTTGTCGTCATCGTCCTTGTAG-3′.

2.2. Bait DNA-Immobilized Affinity Beads

1. Streptavidin-immobilized agarose beads (Pierce) (see Note 2).
2. NeutrAvidin-immobilized agarose beads (Pierce) (see Note 2).

2.3. IVV Template Library

1. SuperScript double-strand cDNA synthesis kit (Invitrogen).
2. DNA ligase (Ligation High kit; Toyobo, etc.).
3. High-fidelity PCR kit (KOD plus; Toyobo, Phusion High-fidelity PCR kit; NEB, etc.).
4. DNA purification kit (QIAquick PCR purification kit; Qiagen, etc.).
5. CHROMA SPIN-1000 (BD Biosciences Clontech).
6. In vitro RNA production kit (RiboMAX large-scale RNA production system SP6; Promega).
7. RNA and IVV template purification kit (RNeasy RNA purification kit; Qiagen, etc.).
8. m^{7}G(5′)ppp(5′)G RNA Capping Analog (Invitrogen, etc.).
9. RNA ligase (T4 RNA ligase; Takara Shuzo).
10. Fluoro-PEG Puro spacer (p(dCp)$_2$-T(Fluor)p-PEGp-(dCp)$_2$-puromycin) (10).
11. Polyethyleneglycol (PEG) 2000 (NOF Corporation, etc.).
12. Urea-polyacrylamide gel.

2.4. IVV Library (In Vitro Translation)

1. In vitro translation system (Transdirect insect cell-free protein synthesis mixture; Shimadzu Biotech, cell-free wheat germ extract; Promega) (see Note 3).
2. RNase inhibitor (Super RaseIn RNase inhibitor; Ambion, etc.).
3. FLGA M2 Agarose beads (Sigma).
4. FLAG M2 Peptide (Sigma).
5. poly-d(I-C) (Roche Applied Science, etc.).
6. DNase I (Promega, etc.).
7. One-step RT-PCR kit (Qiagen, etc.).

2.5. Cloning and Sequencing

1. TA cloning kit (Qiagen, etc.).
2. LB medium and respective antibiotics.
3. DNA sequencing reagent kit and DNA sequence analyzer.

2.6. Buffers

1. DNA binding buffer: 10 mM Tris–HCl, pH 8.0, 1 mM EDTA, 1 M NaCl, and 0.1% Triton X-100. Make up to appropriate volume (e.g., 50 mL) and store at 4°C.
2. Selection buffer: 50 mM Tris–HCl, pH 7.6, 50 mM NaCl, 1 mM EDTA, 5 mM $MgCl_2$, 5 mM DTT, and 2 % glycerol. Low concentration of nonionic detergents such as Triton X-100 and NP-40 can be added. Make up to appropriate volume (e.g., 50 mL) and store at 4°C.
3. FLAG binding buffer; 50 mM HEPES-NaOH, pH 8.3, 150 mM NaCl, and 0.25% Triton X-100. Make up to appropriate volume (e.g., 10 mL) and store at 4°C.
4. FLAG elution buffer: 50 mM Tris–HCl, pH 7.6, 50 mM NaCl, 1 mM EDTA, 5 mM $MgCl_2$, 2% glycerol, 5 mM DTT, 20 ng/μL of poly-d(I-C), and 1 μg/μL of FLAG-M2 peptide. Make up to appropriate volume (e.g., 2 mL), arbitrarily divide into aliquots, and store at −20°C. Avoid multiple freeze–thaw cycles.
5. DNase buffer: 40 mM Tris–HCl, pH 8.0, 10 mM NaCl, 10 mM $CaCl_2$, and 6 mM $MgCl_2$. Make up to appropriate volume (e.g., 10 mL) and store at 4°C.

3. Methods

3.1. Preparation of Bait DNA-Immobilized Affinity Beads

Biotin-labeled bait DNA can be prepared by various methods, e.g., amplification of DNA by PCR in which one primer is modified with biotin, or hybridization of two complementary chemically synthesized oligonucleotide DNAs in which one side of the oligonucleotide is modified with biotin as described below. In any case, it is important to attach one biotin molecule at one terminus of the DNA fragment to attain immobilization on Streptavidin (or NeutrAvidin) beads.

1. Prepare a mixture of equal amounts (50 pmol each) of oligonucleotides dissolved in 100 μL of DNA binding buffer.
2. Heat the DNA mixture at 95°C for 10 min, and cool to room temperature for several hours. This allows double-stranded DNA to be formed.
3. Add the DNA mixture to 20 μL of either Streptavidin- or NeutrAvidin-immobilized agarose beads.
4. Incubate at 4°C for 1 h.
5. Wash the agarose beads with 1 mL of DNA binding buffer twice to remove unbound DNA.
6. Add 100 μL of selection buffer for equilibration and store at 4°C until the start of affinity selection.

3.2. Preparation of IVV Template cDNA Library

An IVV template cDNA library is prepared from a suitable mRNA library, e.g., polyA$^+$ mRNA (Fig. 3).

1. Prepare 50 μM of an adaptor DNA dissolved in RNase-free water by hybridization of adaptor DNA FW and RV (see Subheading 3.1, steps 1 and 2).
2. Prepare a double-stranded cDNA library using 1 μg (or more) of the corresponding mRNA library, a SuperScript double-strand cDNA synthesis kit, and 1–4 pmol of a 3′ random primer according to the manufacturer's instructions (see Note 4).
3. Precipitate the resulting cDNA and dissolve in 4 μL of RNase-free water.
4. Ligate 1 μL (50 pmol) of adaptor DNA to 4 μL of cDNA using an equal amount of DNA ligase at 16°C overnight.
5. Purify the ligated cDNA with a QIAquick PCR purification kit. The final volume of the purified DNA solution is 50 μL.
6. PCR-amplify using 2.5–10 μL of purified cDNA as a template, 5′FW and 3′RV36 primers, and a high-fidelity PCR kit.
7. Purify the resulting PCR product with the QIAquick PCR purification kit.
8. Confirm the pattern of the PCR products by agarose gel electrophoresis (Fig. 4, lane 1).
9. Size-fractionate with a CHROMA SPIN-1000 according to the manufacturer's instructions (see Note 5). It is recommended that total yield of the size-fractionated PCR product (IVV template cDNA library) is more than 1 μg.
10. Option: Confirm the quality of the constructed IVV template cDNA library by cloning and sequencing to see whether the library has high diversity and complexity. Check whether

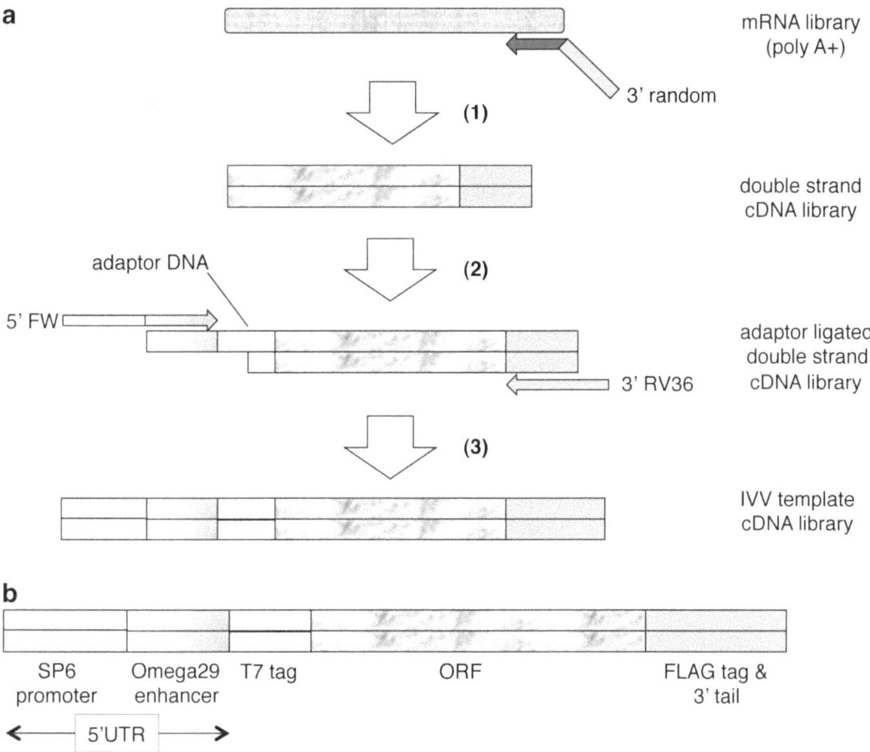

Fig. 3. Schematic representation of the IVV template cDNA library. (**a**) **(1)** An mRNA library (poly A+RNA) is reverse-transcribed and double-stranded cDNA is synthesized. **(2)** The double-stranded cDNA is ligated with an adaptor DNA. **(3)** The adaptor-ligated double-stranded cDNA is amplified. (**b**) The IVV template cDNA library is composed of 5′UTR (SP6 and Omega29 enhancer), T7 tag, ORF, FLAG tag and 3′ tail.

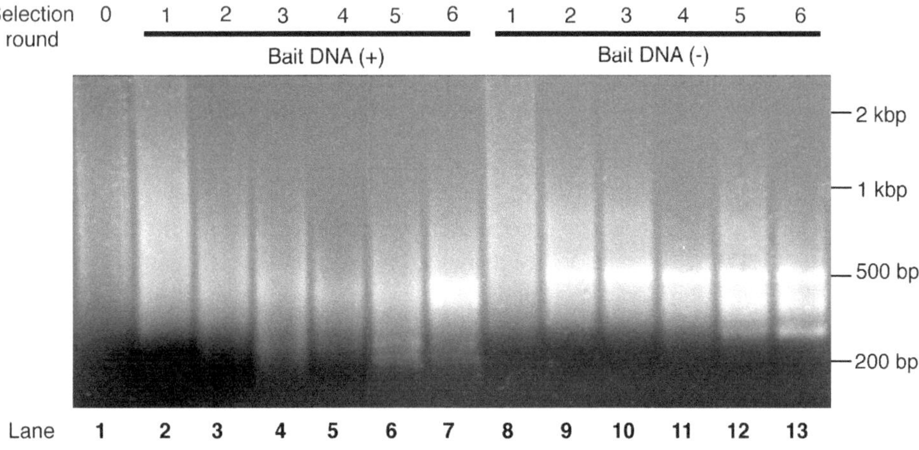

Fig. 4. Example of IVV template cDNA library profiles of initial library (*Lane 1*), libraries after one to six rounds of selection of the IVV template cDNA library in the presence (*Lane 2–7*) and absence (*Lane 8–13*) of the bait DNA. Note that chain length of the IVV template cDNA libraries is at least 200 bp.

almost all of multiple randomly chosen clones correctly contain the 5′UTR (SP6 promoter and Omega29 enhancer), T7 tag, FLAG tag, and 3′tail, and confirm that all of the randomly chosen clones are different from each other (Fig. 3).

3.3. Preparation of IVV Template Library

1. Mix 1 μg (or more) of the IVV template cDNA library as a template, 250 nmol of an capping anolog, and a RiboMAX large-scale RNA production system SP6 reaction mixture. The final volume of reaction mixture is 50 μL.
2. Incubate at 37°C for several hours.
3. Purify the resulting RNA with a RNeasy RNA purification kit (see Note 6).
4. Size-fractionate with the CHROMA SPIN-1000 (see Note 5).
5. Calculate average molecular weight of mRNA and yield (see Note 7). It is recommended that the total yield of the mRNA is more than 50 pmol.
6. Mix 10 nmol of a Fluoro-PEG Puro spacer, 10 nmol of a PEG 2000, 50 pmol of the mRNA, 3 mM DTT, 400 μM ATP, 0.006 % BSA, 20 % DMSO, 25 U of a Super RaseIn RNase inhibitor, 250 U of a T4 RNA ligase, and an appropriate amount of RNase-free water. The final volume of the reaction mixture is 50 μL.
7. Incubate at 15°C overnight.
8. Purify the resulting ligated mRNA (IVV template) with the RNeasy RNA purification kit (see Note 6).
9. Calculate the average molecular weight of IVV template (see Note 7). It is recommended that the total yield of the IVV template is more than 40 pmol.

3.4. Preparation and Purification of IVV Library

Although IVV has been confirmed to be stable under appropriate conditions, it is preferable to prepare an IVV library freshly as required.

1. First round of selection: Mix 40 pmol of an IVV template, 80 U of a Super RaseIn RNase inhibitor, and a 100 μL aliquot of a cell-free protein synthesis mixture. The final volume of the reaction mixture is 200 μL. Divide this into four (50 μL of reaction mixture per tube).

   Second and subsequent rounds of selection: scale-down the volume of the IVV formation reaction mixture as appropriate: e.g., mix 20 pmol an IVV template and 40 U of a Super RaseIn RNase inhibitor, and a 50 μL aliquot of a cell-free protein synthesis mixture. The final volume of the reaction mixture is 100 μL. Divide this into two.
2. Incubate at 24°C for 2 h.

3. Add 100 μL of constructed IVVs to 160 μL of anti-FLAG M2 antibody-immobilized agarose beads equilibrated with 40 μL of FLAG binding buffer.
4. Mix on a rotator at 4°C for 1 h.
5. Wash the beads with 400 μL of FLAG binding buffer four times.
6. Add 100 μL of FLAG elution buffer to the beads and mix at 4°C for 1 h.
7. Collect the resulting supernatant containing purified IVV library.

3.5. Affinity Selection

Perform at least two different kinds of selection, i.e., selection in the presence of bait DNA (bait DNA(+)), and selection in the absence of bait DNA (bait DNA(−)) to distinguish nonspecific binders to the agarose beads (11, 14, 24). This is done by dividing the initial IVV library into two and subjecting the two parts to the respective selection procedures. Continue selection until almost all of the selected clones are found to be positives or positive candidates. If the selection system is working well, the electrophoresis pattern of IVV cDNA templates will be different after each round (Fig. 4).

1. Preselection. Add 50 μL of a purified IVV library to either 20 μL of agarose beads or decoy DNA-immobilized agarose beads (see Note 8).
2. Incubate at 4°C for 1 h.
3. Collect the supernatant containing unbound IVVs.
4. Selection. Add the collected IVVs to bait DNA-immobilized agarose beads.
5. Incubate at 4°C for 2 h.
6. Wash the beads with 50 μL of the selection buffer several (e.g., 3–7) times.
7. Wash the beads with 50 μL of DNase buffer once.
8. Add 50 μL of the DNase buffer containing 3 U of DNase I.
9. Incubate at room temperature for 10 min.
10. Add 8 μL of 0.25 M EGTA for termination reaction and collect the supernatant containing DNA-binding IVVs (complexes).
11. Perform RT-PCR using less than 1/10 volume of the supernatant as a template according to the manufacturer's instruction using 5′FW and 3′RV30 primers. For example, 5 μL of template is added to the RT-PCR reaction mixture and the final volume is 50 μL.
12. Purify the resulting RT-PCR products with a QIAquick PCR purification kit. The purified RT-PCR products are size-fractionated (as required, Subheading 3.2, item 9) and used for the next round of selection as described above (Subheading 3.3).

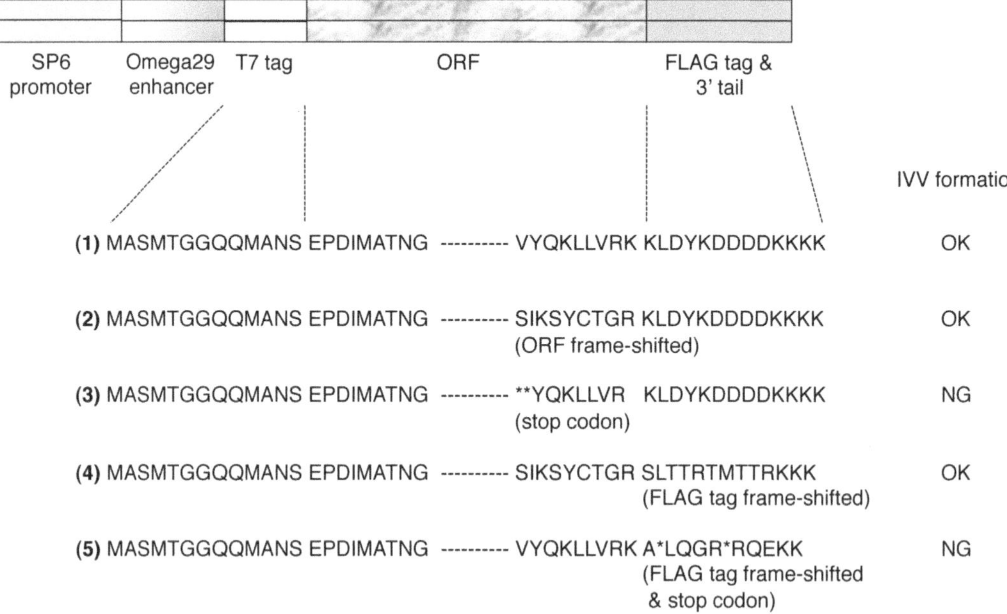

Fig. 5. Example of sequences of selected clones. **(1)** This clone has correct 5′UTR, T7 tag, FLAG tag, and 3′ tail sequences, and the ORF sequence is the same as that registered in databases. This clone can form IVV and be subjected to binding assay. **(2)** This clone has correct 5′UTR and T7 tag, but ORF, FLAG tag, and 3′ tail sequences are frame-shifted. This clone can form IVV, but is eliminated. **(3)** This clone has correct 5′UTR, T7 tag, FLAG tag, and 3′ tail sequences, but the ORF sequence contains a stop codon. This clone cannot form IVV and is eliminated. **(4)** This clone has correct 5′UTR and T7 tag and the ORF sequence is the same as that registered in databases. However, the FLAG tag and 3′ tail sequences are frame-shifted. This clone can form IVV, but cannot be purified with FLAG affinity beads. Alternatively, this clone can be subjected to binding assay after correction of errors by PCR. **(5)** This clone has correct 5′UTR and T7 tag and the ORF sequence is the same as that registered in databases. However, the FLAG tag and 3′ tail sequences are frame-shifted and contain plural stop codons. This clone cannot form IVV, and is eliminated.

3.6. Cloning and Sequencing

Selected clones can be recovered and analyzed by cloning and sequencing.

1. Ligate purified RT-PCR products into cloning vectors using TA cloning kits.
2. Transform the resulting ligation products into competent cells.
3. Select resulting clones randomly (see Note 9).
4. Analyze the selected clones by sequencing.
5. Sequences are subjected to nucleotide–nucleotide BLAST (BLASTN) search to identify the protein represented.
6. Eliminate clones corresponded to either untranslated regions in ORF sequences or incorrect reading frames (Fig. 5). Remaining clones are then clustered and subjected to binding assay.

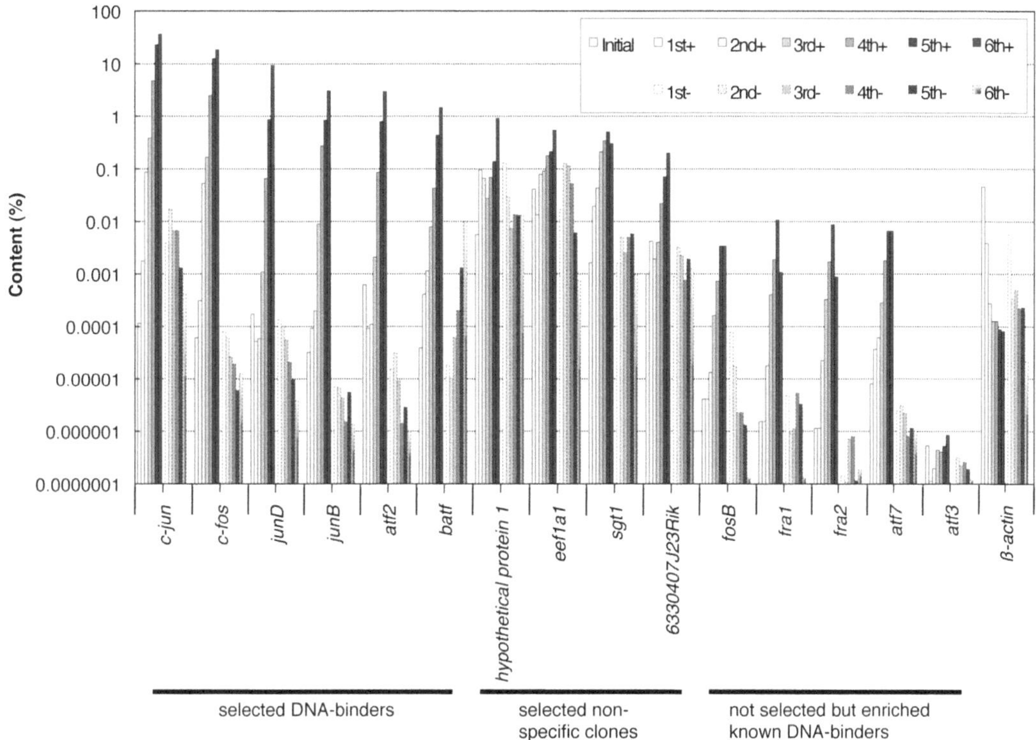

Fig. 6. Example of quantitative real-time PCR analysis of the selected clones in the initial library (*leftmost column*) and libraries after the first to sixth rounds of selection of the IVV template cDNA library in the presence (+) (the following six columns from the *left*, respectively, for each gene) and absence (–) (the remaining six columns from the *left*, respectively, for each gene) of the bait DNA (14). Contents of DNAs of selected clones and β-actin as a negative control in 5 ng (about 10^{10} molecules) of each library were analyzed with specific primers using a LightCycler. Note that not only selected DNA-binders (*c-jun*, *c-fos*, *junD*, *junB*, *atf2*, and *b-atf*), but also other known DNA-binders (*atf4*, *fosB*, *fra1*, *fra2*, and *atf7*) are enriched in each bait (+) library as compared with the initial library. None of the candidates was enriched in any round of the bait (–) libraries. In this experiment, 412 clones were cloned and analyzed, so that only genes whose content is more than 0.25 % would be cloned theoretically. Indeed, the contents of all of the selected DNA-binders quantified by real-time PCR were at least 0.25 %.

3.7. (Option) Quantitative Real-Time PCR Analysis

Real-time PCR can be performed using gene-specific primers to monitor the process of enrichment of selected clones in each round of selection, as well as elimination of false positives (Fig. 6) (see Note 10) (14, 24).

3.8. Preparation of the Selected Clones (Candidate DNA-Binders) for In Vitro Binding Assay

If the IVV selection system works well, many clones will be selected. In vitro preparation of selected clones (proteins) is recommended for either in vitro pull-down assay or EMSA.

1. PCR-amplify using the selected clone (plasmid) as a template, 5′FW and 3′RV36 primers, and a high-fidelity PCR kit.
2. Purify the resulting PCR product with the QIAquick PCR purification kit.

3. Transcribe using the PCR product as a template with a RiboMAX large-scale RNA production system SP6 for several hours.
4. Purify the resulting RNA with a RNeasy RNA purification kit.
5. Calculate the molecular weight of RNA. It is recommended that total yield of the mRNA is more than 5 pmol.
6. Mix 5 pmol of the resulting mRNA, 20 U of a Super RaseIn RNase inhibitor, and a 25 μL aliquot of a cell-free protein synthesis mixture. The final volume of the reaction mixture is 50 μL.
7. Incubate at 24°C for 2 h.
8. Add the resulting protein to 20 μL of anti-FLAG M2 antibody-immobilized agarose beads equilibrated with 75 μL of FLAG binding buffer.
9. Mix on a rotator at 4°C for 1 h.
10. Wash the beads with 400 μL of FLAG binding buffer four times.
11. Add 50 μL of FLAG elution buffer to the beads and mix at 4°C for 1 h.
12. Collect the resulting supernatant containing purified proteins (candidate DNA-binders) and use the proteins for binding assays.

4. Notes

1. Take care when handling cDNA libraries or IVV templates, because trace contamination with DNAs or IVV templates may produce misleading results. In particular, cross-contamination of DNAs or IVV templates encoding proteins with relative DNA-binding activities higher than those of true DNA-binders in the cDNA library is likely to cause serious problems. As the selection rounds proceed, contaminating DNA-binders are enriched in the library pool, as well as true DNA-binders, and compete with each other. Excess rounds of selection may result in enrichment of contaminating DNA-binders and exclusion of the true DNA-binders in the library pool.
2. Magnetic beads can be used instead of agarose-based beads. In some cases, however, nonspecific binding of RNAs (IVV templates) or IVVs to certain magnetic beads may occur, leading to unsuccessful selection. In that case, not only specific DNA-binders (signals), but also nonspecific binders (noise) that may prevent enrichment of specific DNA-binders are enriched, because they compete with each other in the library pool.

3. An in vitro translation system with high translation efficiency should be used. Transdirect insect cell-free protein synthesis mixture (Shimadzu Biotech) and cell-free wheat germ extract (Promega) are now commercially available as kits for IVV selection of DNA–protein complexes. Other in vitro translation systems, such as an *E. coli* cell-free translation system (Puresystem; Wako), may also be suitable. In any case, it is important to check the translation efficiency using an appropriate positive control before starting IVV selection.
4. The ratio of the concentration of mRNA template and 3′ random primer can be varied. However, in some cases, a lower concentration of 3′ random primer (e.g., less than 0.2 pmol per 1 μg of mRNA) can result in unsuccessful selection. Generally, the efficiency of formation of IVV decreases depending on the chain length of the IVV template (in other words, the chain length of cDNA). We have found (unpublished data) that the efficiency of formation of IVV in which the mRNA moiety is longer than 2 kb is greatly decreased, suggesting that cDNA having a chain length of longer than 2 kbp may be unfavorable for IVV selection.
5. Accumulation of artificial by-products (typical amplicon length less than 200 bp) can cause unsuccessful selection. Such by-products are PCR-amplified, transcribed, ligated to the Fluoro-PEG Puro spacer, in vitro-translated, and RT-PCR-amplified more efficiently than IVV template cDNA library clones encoding true DNA-binders, and often disturb enrichment of the true DNA-binders in the library pool. If accumulation of artificial by-products is not observed, this step can be omitted. Size-fractionation using agarose (DNA) or urea-polyacrylamide (RNA) gels is also available.
6. Ethanol precipitation is also available instead of using a RNeasy purification kit.
7. The average molecular weight of RNA or IVV template can be roughly estimated. For example, if the average chain length of RNA is 500 bases, the average molecular weight is about 165,000 Da. In the case of IVV template, add the molecular weight of Fluoro-PEG Puro spacer (about 2,000) to the average molecular weight of the original RNA. For example, if the average chain length of IVV template is 500 bases, the average molecular weight is about 167,000 Da. See also http://www.changbioscience.com/genetics/mw.html.
8. In some cases, a preselection step may not be required before selection. However, we have found that preselection of the IVV library with decoy bait DNA-immobilized beads is very effective to reduce both false-positives and false-negatives.
9. Dozen of sequences should be analyzed.

10. Enrichment of false-negative clones (enriched but not selected, Fig. 6) can be detected by a combination of the IVV system with a DNA microarray (25), as a sensitive and high-throughput alternative to the processes of cloning and sequencing.

Acknowledgments

This work was supported in part by a Grant-in-Aid for Scientific Research, and a Special Coordination Fund from the Ministry of Education, Culture, Sports, Science, and Technology, Japan.

References

1. Trompouki E, Bowman TV, Lawton LN, Fan ZP, Wu DC, DiBiase A et al (2011) Lineage regulators direct BMP and Wnt pathways to cell-specific programs during differentiation and regeneration. Cell 147:577–589
2. Pollack JR, Iyer VR (2002) Characterizing the physical genome. Nat Genet 32:515–521
3. Yu H, Luscombe NM, Quian J, Gerstein M (2003) Genomic analysis of gene expression relationships in transcriptional regulatory networks. Trends Genet 19:422–427
4. Barski A, Frenkel B (2004) ChIP display: novel method for identification of genomic targets of transcription factors. Nucleic Acids Res 32: e104
5. Lengronne A, Katou Y, Mori S, Yokobayashi S, Kelly GP, Itoh T et al (2004) Cohesin relocation from sites of chromosomal loading to places of convergent transcription. Nature 430:573–578
6. Kim J, Bhinge AA, Morgan XC, Iyer VR (2005) Mapping DNA-protein interactions in large genomes by sequence tag analysis of genomic enrichment. Nat Methods 2:47–53
7. Noyes MB, Christensen RG, Wakabayashi A, Stormo GD, Brodsky MH, Wolfe SA (2008) Analysis of homeodomain specificities allows the family-wide prediction of preferred recognition sites. Cell 133:1277–1289
8. Rebar EJ, Pabo CO (1994) Zinc finger phage: affinity selection of fingers with new DNA-binding specifications. Science 263:671–673
9. Nemoto N, Miyamoto-Sato E, Hushimi Y, Yanagawa H (1997) *In vitro* virus: bonding of mRNA bearing puromycin at the 3′-terminal end to the C-terminal end of its encoded protein on the ribosome *in vitro*. FEBS Lett 414:405–408
10. Miyamoto-Sato E, Takashima H, Fuse S, Sue K, Ishizaka M, Tateyama S et al (2003) Highly stable and efficient mRNA templates for mRNA-protein fusions and C-terminally labeled proteins. Nucleic Acids Res 31:e78
11. Horisawa K, Tateyama S, Ishizaka M, Matsumura N, Takashima H, Miyamoto-Sato E et al (2004) *In vitro* selection of Jun-associated proteins using mRNA display. Nucleic Acids Res 32:e169
12. Miyamoto-Sato E, Ishizaka M, Horisawa K, Tateyama S, Takashima H, Fuse S et al (2005) Cell-free co-translation and selection using *in vitro* virus for high-throughput analysis of protein-protein interactions and complexes. Genome Res 15:710–717
13. Miyamoto-Sato E, Ishizaka M, Fujimori S, Hirai N, Masuoka K, Saito R et al (2010) A comprehensive resource of interacting protein regions for refining human transcription factor networks: domain-based interactome. PLoS One 5:e9289
14. Tateyama S, Horisawa K, Takashima H, Miyamoto-Sato E, Doi N, Yanagawa H (2006) Affinity selection of DNA-binding protein complexes using mRNA display. Nucleic Acids Res 34:e27
15. Horisawa K, Imai T, Okano H, Yanagawa H (2009) 3′-Untranslated region of *doublecortin* mRNA is a binding target of the Musashi1 RNA-binding protein. FEBS Lett 583: 2429–2434
16. Tabata N, Sakuma Y, Honda Y, Doi N, Takashima H, Miyamoto-Sato E et al (2009) Rapid antibody selection by mRNA display on a microfluidic chip. Nucleic Acids Res 37:e64
17. Matsumura N, Tsuji T, Sumida T, Kokubo M, Onimaru M, Doi N et al (2010) mRNA display

selection of a high-affinity, Bcl-X_L-specific binding peptide. FASEB J 24:2201–2210

18. Shiheido H, Takashima H, Doi N, Yanagawa H (2011) mRNA display selection of an optimized MDM2-binding peptide that potently inhibits MDM2-p53 interaction. PLoS One 6:e17898
19. Kosugi S, Hasebe M, Matsumura N, Takashima H, Miyamoto-Sato E, Tomita M et al (2009) Six classes of nuclear localization signals specific to different binding grooves of importin α. J Biol Chem 284:478–485
20. Tsuji T, Onimaru M, Doi N, Miyamoto-Sato E, Takashima H, Yanagawa H (2009) *In vitro* selection of GTP-binding proteins by block shuffling of estrogen-receptor fragments. Biochem Biophys Res Commun 390:689–693
21. Tanaka J, Yanagawa H, Doi N (2011) Comparison of the frequency of functional SH3 domains with different limited sets of amino acids using mRNA display. PLoS One 6:e18034
22. Hammond PW, Alpin J, Rise CE, Wright M, Kreider BL (2001) *In vitro* selection and chracterization of Bcl-X_L-binding proteins from a mix of tissue-specific mRNA display libraries. J Biol Chem 276:20898–20906
23. Chinenov Y, Kerppola TK (2001) Close encounters of many kinds: Fos–Jun interactions that mediate transcription regulatory specificity. Oncogene 20:2438–2452
24. Horisawa K, Doi N, Takashima H, Yanagawa H (2005) Application of quantitative real-time PCR for monitoring the process of enrichment of clones on *in vitro* protein selection. J Biochem 137:121–124
25. Horisawa K, Doi N, Yanagawa H (2008) Use of cDNA tiling arrays for identifying protein-interactions selected by *in vitro* display technologies. PLoS One 3:e1646

Chapter 9

Isolation of Intracellular Protein – DNA Complexes Using HaloCHIP, an Antibody-Free Alternative to Chromatin Immunoprecipitation

Danette L. Daniels and Marjeta Urh

Abstract

Mapping of protein binding sites within the genome has been significantly advanced by microarray and sequencing technologies, yet the method traditionally used to isolate protein–DNA complexes, chromatin immunoprecipitation, has remained dependent of the use of antibodies. Furthermore, cross-linking is commonly used to trap protein–DNA complexes and the challenge of using antibodies has come in recognition of the cross-linked epitopes, sometimes limiting the success of the approach. Here we present a method, HaloCHIP, which utilizes a HaloTag protein fusion and corresponding interaction resin, HaloLink, for capture of cross-linked protein–DNA complexes directly from a cellular lysate. This process alleviates the need for using an antibody, yields the DNA fragments bound to a particular protein of interest, and allows for a variety of downstream analyses such as PCR, qPCR, microarrays, and sequencing.

Key words: Transcription factors, DNA binding proteins, HaloTag, HaloLink, HaloCHIP, Chromatin immunoprecipitation, PCR, qPCR, Microarrays, Sequencing

1. Introduction

The study of intracellular protein–DNA interactions has rapidly progressed in recent years from the analysis of a single, known promoter using PCR or qPCR, to analysis of genome-wide binding patterns (1–10). Initiatives are underway to study these patterns for numerous types of cells, different stages of development, as well as large-scale studies of transcription factors. The initial technology developed to study these interactions was called chromatin immunoprecipitation (1). As the name suggests, an antibody is used to enrich the protein bound to the chromatin. As protein–DNA

Minou Bina (ed.), *Gene Regulation: Methods and Protocols*, Methods in Molecular Biology, vol. 977,
DOI 10.1007/978-1-62703-284-1_9, © Springer Science+Business Media, LLC 2013

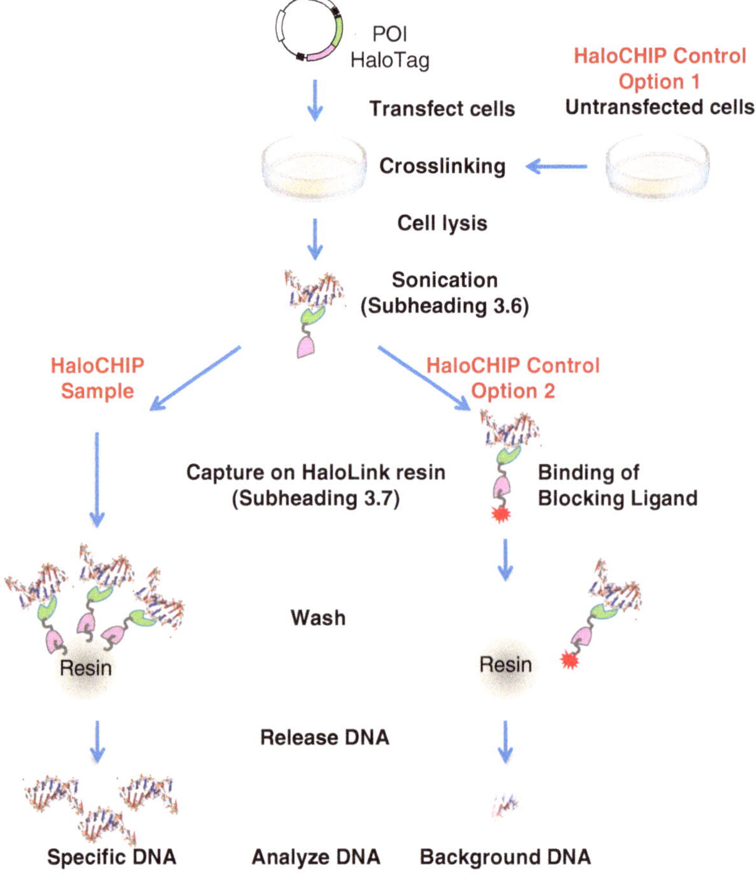

Fig. 1. Schematic of the HaloCHIP Protocol. Schematic of HaloCHIP protocol showing processing of the sample and control; two different options for control are depicted. Control Option 1, nontransfected cells are processed in the same manner as the sample; control option 2, sample is split in half and one half is treated with HaloTag blocking ligand to prevent binding to the HaloLink resin. POI denotes protein of interest (shown as HaloTag fusion); resin designates HaloLink resin which specifically binds HaloTag fusion.

interactions can be either very low affinity or transient in nature, formaldehyde cross-linking, either of cells or tissues, is used to fix these interactions prior to capture (1). Unfortunately formaldehyde cross-linking can potentially block or prevent isotopes from being recognized by the antibody as it cross-links not only protein–DNA complexes but also protein epitopes, domains, and complexes (1). To date, the major challenge of the chromatin immunoprecipitation process is obtaining the optimal antibody for the experiment. To address this challenge, the HaloCHIP process, which does not require the use of antibodies, was developed (11–13) (Fig. 1).

The HaloCHIP method relies upon the use of a protein fusion tag, termed HaloTag (11), which can be genetically fused, either N- or C-terminally, to any protein, including DNA binding or

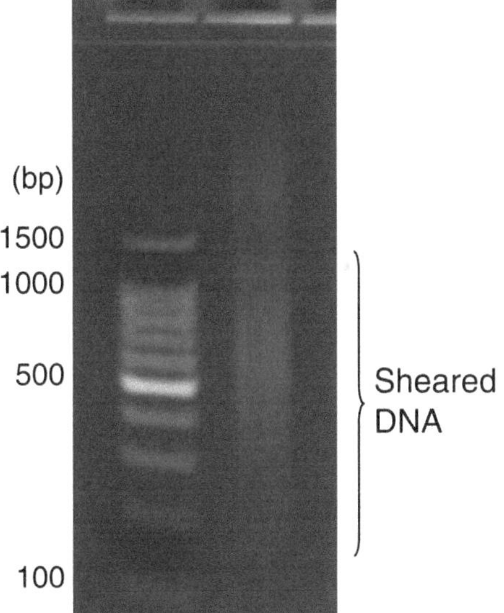

Fig. 2. Sonicated DNA. HeLa cells transfected with p65-HaloTag protein were cross-linked, lysed, and sonicated following the protocol to Subheading 3.3, step 6, then analyzed using the protocol in Subheading 3.6. Lane 1, Promega BenchTop 100 bp DNA Ladder (Cat.# G8291); lane 2, smear of DNA fragments between the range of 200–1,000 bp, which is the target size for sonication.

chromatin associated proteins of interest (Fig. 1). HaloTag has been engineered to form an irreversible, covalent bond with its ligands, which include a resin-based ligand termed HaloLink (11, 14, 15). In the HaloCHIP protocol, a recombinant vector of any DNA binding protein fused to HaloTag is transfected into cells for either stable or transient expression (12, 13). Similar to the chromatin immunoprecipitation method, the cells are treated with formaldehyde to cross-link protein–DNA complexes, lysed, and then sonicated to shear chromatin into smaller fragments (12) (Figs. 1 and 2). The complexes are then covalently captured directly from the lysate onto the HaloLink resin via the interaction with HaloTag. Due to the complete covalent linkage, isolated complexes are washed extensively to remove non-specific interactions. The DNA fragments bound and cross-linked to the starting HaloTag fusion protein are then released through a heat reversal of the formaldehyde-induced cross-links (12) (Fig. 1). These can be further purified and analyzed using existing methods for DNA (Fig. 3). To estimate levels of enriched DNA capture over background, there are two possible controls for the HaloCHIP experiment (12) (Fig. 1). The first is to process untransfected cells throughout the same protocol in parallel (Figs. 1 and 3). The second approach is to split the lysate equally into two tubes prior to capture on the HaloLink resin (Fig. 1). One tube is incubated with a blocking ligand, which covalently binds to the HaloTag protein

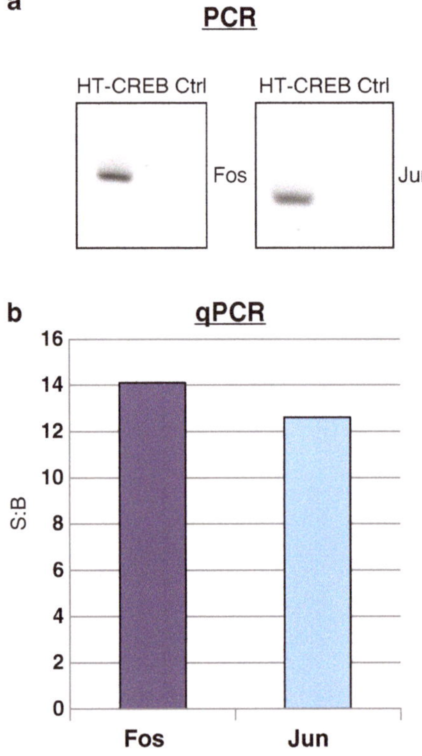

Fig. 3. Amplification of CREB target promoters. HEK293 cells were transiently transfected with HaloTag-CREB (HT-CREB) fusion protein and the HaloCHIP process was performed. Untransfected cells were processed similarly and used as a background control (Ctrl). DNA obtained from both the HT-CREB and Ctrl sample was subject to PCR (**a**) and quantitative PCR (**b**) for known CREB target promoters: Fos and Jun (13). (**a**) 1% Agarose gels stained with ethidium bromide showing relative amplification using standard PCR of Fos and Jun promoters in HT-CREB (lane 1) and Ctrl (lane 2) after HaloCHIP. (**b**) Plots of signal–background (S:B) ratios for Fos and Jun promoters in HT-CREB (S) and Ctrl (B) after quantitative PCR.

in solution and blocks interaction with the HaloLink resin, and can therefore be used as a background control (12) (Fig. 1).

Given the highly specific, irreversible, and rapid binding of HaloTag to its resin, complexes are efficiently isolated, even at low concentrations in mammalian lysates without concern of diffusion (13, 15). This also allows for use of far fewer cells as compared to antibody approaches (12, 13). The covalent capture coupled with the covalent cross-linking of the protein–DNA complexes allows for extensive washing of the resin to remove non-specific proteins and DNA, something which is not possible with an antibody-based capture. HaloCHIP differs from chromatin immunoprecipitation as it is carried out using a fusion protein (12, 13). As with any fusion protein it is important to characterize the protein in advance to ensure it maintains proper physiology in the presence of a protein tag. This is readily possible with the HaloTag technology, as fluorescent

HaloTag ligands can be used to study expression and cellular localization, while the HaloLink resin can be used to isolate complexes (11–15). Examples of these studies have been previously published, including those with p65 (11) and CREB (13) transcription factors (Fig. 3). In the case of CREB, genome-wide binding patterns were compared between the standard chromatin immunoprecipitation process for the endogenous CREB protein to the HaloCHIP process using a Halo-CREB fusion protein (13). These data demonstrated binding patterns were the same for both consensus and non-consensus sequences, showing the utility of HaloCHIP for genome wide studies (13).

2. Materials

2.1. DNA Vectors Encoding HaloTag Protein Fusion to the Chromatin Interacting Protein of Interest

1. If preparing the HaloTag fusion protein clone by PCR amplification, the following mammalian HaloTag fusion vectors are recommended; for an N-terminal HaloTag fusion: pFN21A HaloTag CMV Flexi Vector (Promega #G2821) and for a C-terminal HaloTag construct pFC14A HaloTag CMV Flexi Vector (Promega #G9651) (see Notes 1–3).
2. HaloTag fusion protein vectors are also commercially available from two sources: Kazusa DNA Research Institute: http://www.promega.com/products/pm/halotag-technology/kazusa-collection/ or Genecopoeia: http://www.genecopoeia.com/product/halo/.

2.2. Tissue Culture Reagents, Cross-linker, and Quenchers

1. Eukaryotic cells for transfection or appropriate stable transformed cell line expressing HaloTag fusion protein construct (see Notes 1–3). For each HaloCHIP reaction, $4–8 \times 10^5$ adherent cells (pre-transfection) are used and typically plated in a standard six-well plate. Cross-linking is performed approximately 24 h post-transfection.
2. Cell medium required for growing cells.
3. Appropriate transfection reagent.
4. PBS, tissue culture certified
5. Formaldehyde, 37% stock, A.C.S. reagent Formaldehyde is easily oxidized, reducing its effectiveness in cross-linking. Replace the formaldehyde with fresh every 3 months. Formaldehyde is toxic by inhalation or ingestion and is readily absorbed through the skin. Take suitable precautions when using and disposing of formaldehyde solutions.
6. 1.25 M Glycine, pH 7.0. This is a 10× solution and should be made with sterile water. Store at 4°C for up to 1 month.

2.3. HaloCHIP Materials

Note: All of the materials listed below in detail, with the exception of HaloLink equilibration buffer (#1), are contained in the HaloCHIP System (Promega #G9410).

1. HaloLink equilibration buffer: 1× TE Buffer (10 mM Tris–HCl pH 8.0, 1 mM EDTA) and 0.1% IGEPAL CA-630, also known as Nonidet P-40. Prepare equilibration buffer directly before use by diluting a 10% stock solution of IGEPAL CA-630 made in 1× TE buffer. Store for a maximum time of 24 h at 22°C.
2. HaloLink Resin (Promega #G1914). Store at 4°C.
3. Mammalian Lysis buffer: 50 mM Tris–HCl pH 7.5, 150 mM NaCl, 1% Triton X-100, and 0.1% sodium deoxycholate. Store at 4°C.
4. High Salt Wash Buffer: 50 mM Tris–HCl pH 7.5, 700 mM NaCl, 1% Triton® X-100, 0.1% sodium deoxycholate, and 5 mM EDTA. Store at 4°C.
5. Reversal Buffer: 10 mM Tris–HCl pH 8.0, 1 mM EDTA, and 300 mM NaCl. Store at 22°C.
6. 5 mM HaloCHIP TMR blocking ligand (Promega #G8251). If performing HaloTag binding efficiency analysis, prepare 50–100 μl of 50 μM stock by diluting an aliquot 1:100 in DMSO. Store in amber tubes at −20°C.
7. Nuclease-free water. Store at 22°C.
8. PCR product purification system.

2.4. Hardware

1. Cell scraper or rubber policeman.
2. Low output sonicator equipped with a microtip, e.g., Misonix Sonicator 3000 with Microtip Probe 418.
3. Microcentrifuge.
4. Heat block, e.g., ThermoKool from Barnstead/Thermolyne, capable of stably maintaining temperature at 65°C.
5. Glass dounce homogenizer (0.5–2 ml) or 25- or 27-gauge needle fit for a 1 ml syringe.
6. Microcentrifuge tube rotator, e.g., Barnstead International model # 400110, Scientific Equipment Products Cat.# 60448 or equivalent device.
7. SDS-PAGE gels.
8. 1–2% agarose gels.
9. Fluorimager.

3. Method

This protocol has been optimized for mammalian cells (see Note 4). The following protocol is designed for a single HaloCHIP experiment including an experimental sample and a control sample, using $4–8 \times 10^5$ adherent cells per sample (see Note 5). If using an untransfected cells as a control, follow protocol exactly as recommended for the HaloCHIP experimental sample, yet using untransfected cells instead of transfected. If using the blocking ligand control, follow protocol as recommended to through the Subheading 3.4, step 1. At that point, there will be directions for how to prepare the blocking ligand control and proceed.

3.1. Resin Equilibration

1. Mix the HaloLink Resin by inverting the tube until a uniform suspension is obtained (see Note 6). HaloLink resin is 20% slurry in ethanol. All listed volumes are that of the slurry. Dispense 75 μl of HaloLink Resin into a single 1.5 ml microcentrifuge tube for each HaloCHIP reaction sample (e.g., if performing an experimental HaloCHIP sample along with a control, prepare two tubes of HaloLink Resin) (Fig. 1).
2. Centrifuge at $800 \times g$ for 2 min. Carefully remove and discard the supernatant, leaving the HaloLink Resin at the bottom of the tube.
3. Add 400 μl of HaloLink Resin Equilibration Buffer to each tube of resin. Mix thoroughly by inverting the tube several times.
4. Centrifuge for 2 min at $800 \times g$. Carefully remove and discard the supernatant, leaving the HaloLink Resin at the bottom of the tube.
5. Repeat step 3 two additional times for a total of three washes. After the third wash, do not remove the Equilibration Buffer until you are ready to use the HaloLink Resin. This will prevent the HaloLink Resin from drying out. Store the prepared HaloLink Resin tubes at 4°C for up to 24 h.

3.2. Cross-linking

1. Prewarm an amount of the medium equivalent to the volume of the culture to 22°C.
2. Add formaldehyde, 37% stock, to a final concentration of 0.75–1% and mix well (see Note 7). Tissue-culture-certified PBS may be used in the place of medium.
3. Aspirate the old medium from the cells, and gently add an equivalent amount of formaldehyde-containing medium or formaldehyde-containing PBS. Swirl the plate gently.
4. Incubate for 10 min at 22°C.

5. Slowly, and with constant swirling of the plate containing the cells, add 1.25 M glycine (pH 7.0) to a final concentration of 125 mM to quench the cross-linking, e.g., for a 2 ml culture, add 220 µl of 1.25 M glycine (see Note 8). A uniform yellow color will appear.
6. Incubate for 5 min at 22°C.
7. Aspirate the medium, add 2 ml of ice-cold PBS and gently swirl the plate.
8. Aspirate the PBS, add 2 ml of ice-cold PBS and gently swirl the plate.
9. Aspirate the PBS, add 0.75–2 ml of ice-cold PBS to each well, and collect cells by scraping using a cell scraper or rubber policeman.
10. Pool the cells recovered from each well into a single, prechilled 1.5 ml microcentrifuge tube (1–2 wells) or a 15 ml conical tube, and immediately place the tube on ice.
11. Centrifuge at 2,000 × *g* for 5 min at 4°C.
12. Carefully remove the supernatant.
13. Freeze cell pellets at −70°C for 10 min. At this point cells can be stored for up to 3 months at −70°C.
14. If used directly, then thaw at 22°C and proceed with sonication.

3.3. Sonication

This protocol is written for a Misonix Sonicator 3000 using a MicroTip Probe 418 and can be use as a guideline for use with other sonicators. If you using a different sonicator, optimize sonication to obtain chromatin fragments on average 500 bp in length (Fig. 2). Exercise caution as excessive sonication and heating of the sample during sonication will denature the HaloTag protein, resulting in decreased binding to the HaloLink Resin (see Note 9).

1. Resuspend the thawed cell pellet in 650 µl of cold Mammalian Lysis Buffer. The use of protease inhibitors is not recommended, particularly as most protease inhibitor cocktails contain a HaloTag binding inhibitory compound, AEBSF. If it desired to add protease inhibitors, see Note 10 for recommendations.
2. Vortex briefly and incubate on ice for 15 min.
3. Lyse cells using mechanical disruption (e.g., use a 2 ml glass dounce homogenizer, 25–30 strokes on ice, or pass the cells 4–6 times through a 25- or 27-gauge needle).
4. If optimizing sonication with a different sonicator than the Misonix Sonicator 3000, please use the following steps 4–6 as a guide for sonication conditions and proceed to Subheading 3.6 to assay for sonication fragment size (Fig. 2) (see Note 11). Using the Misonix Sonicator 3000 with the Microtip Probe

418, place the 650 µl of lysate prepared in an ice bath. Submerge the Microtip into the lysate near the bottom of the tube, but not touching the tube itself.

5. Set the output to 2.5 and perform 6 cycles of alternating 10 s on and 10 s off, followed by 2 min on ice.
6. Repeat an additional round of 6 cycles of alternating 10 s on and 10 s off. The total sonication time will be 120 s.

3.4. Capture and Release of DNA

1. Centrifuge the sonicated samples at 14,000 × *g* for 5 min at 4°C. If using the blocking ligand control, pool two samples after centrifugation and then divide into two equal samples of 600 µl each. Label one sample as the experimental sample and the other as the blocked ligand control sample. To the control sample only, add 0.6 µl of the HaloCHIP Blocking Ligand directly to the lysate for a final concentration of 5.0 µM and incubate both samples for 30 min at 22°C with constant mixing before proceeding to the next step. Due to the fluorescent dye attached to the HaloCHIP Blocking Ligand, the solution will be bright pink (see Note 12).
2. Remove and save 10 µl of supernatant from the HaloCHIP sample to analyze the efficiency of HaloTag fusion protein binding and label as starting material (Subheading 3.7). Store sample at 4°C.
3. Carefully remove the HaloLink Resin Equilibration Buffer from the HaloLink Resin prepared in Subheading 3.1, step 5.
4. Add 600 µl of the HaloCHIP sample or the control sample to separate tubes of the HaloLink Resin.
5. Incubate the samples with constant mixing using a tube rotator for 2–3 h at 22°C (see Note 13). Ensure constant mixing since settling of the HaloLink Resin will reduce binding efficiency.
6. Centrifuge the samples at 800 × *g* for 2 min at 22°C.
7. Remove and save 10 µl of the supernatant for analysis of the efficiency of binding, label as unbound fraction (Subheading 3.7), and discard remaining supernatant (see Note 14). Store sample at 4°C.
8. Add 1 ml of Mammalian Lysis Buffer to the resin and mix thoroughly by gently pipetting (see Note 15).
9. Centrifuge at 800 × *g* for 2 min. Carefully remove and discard the supernatant.
10. Add 1 ml of Nuclease-Free Water and mix thoroughly by gently pipetting.
11. Centrifuge at 800 × *g* for 2 min. Carefully remove and discard the supernatant.

12. Repeat steps 10 and 11.
13. Add 1 ml of High Salt Wash Buffer and mix thoroughly by gently pipetting. Incubate at 22°C for 5 min, mixing using a tube rotator.
14. Centrifuge at 800 × *g* for 2 min. Carefully remove and discard the supernatant (see Note 16).
15. Add 1 ml of Nuclease-Free Water and mix thoroughly by gently pipetting.
16. Centrifuge at 800 × *g* for 2 min. Carefully remove and discard the supernatant.
17. Repeat steps 15 and 16.
18. Add 1 ml of Nuclease-Free Water and mix thoroughly by gently pipetting.
19. Incubate for 5 min at 22°C mixing using a tube rotator.
20. Centrifuge for 2 min at 800 × *g*, and carefully remove and discard the supernatant.
21. Add 300 μl of Reversal Buffer to each tube with the HaloLink Resin and gently mix by pipetting.
22. Incubate at 65°C for 6–18 h to reverse the cross-links (see Note 17).
23. Centrifuge the HaloLink Resin at 800 × *g* for 2 min.
24. Transfer the supernatant containing the released DNA to a new tube.
25. Further purify the eluted DNA using a PCR product purification kit in order to reduce the salt and concentrate the DNA prior to use for further analysis (see Note 18).

3.5. Analysis of Recovered DNA

The recovered DNA is compatible with all amplification methods and can be analyzed in many ways including standard PCR, quantitative PCR, microarrays, or high-throughput sequencing. The control reaction, either generated by using the blocking ligand or starting from untransfected cells, should be subjected to similar analysis and will show levels of background DNA. If standard PCR analysis is employed, it is important that the PCR be performed within the linear amplification phase. The appropriate number of cycles must be determined experimentally, but is typically around 30. If the number of cycles is too high, the background signal from the control will be artificially high, resulting in a reduced signal-to-noise ratio. They typical yield from a single HaloCHIP reaction is approximately 1–10 ng of DNA. To obtain higher amounts of DNA, increase the number of HaloCHIP reactions performed, following the protocol as recommended, and pool purified DNA at the end of the process. This DNA can then be lyophilized to the desired concentration.

3.6. Assay for Optimizing Sonicated DNA Fragment Size

1. Following sonication, incubate the lysate at 95°C for 15 min to reverse the cross-links.
2. Centrifuge the sample at 14,000 × *g* for 5 min at 4°C
3. Purify the DNA in the supernatant using a standard method for purifying PCR products.
4. Load 20 μl of the purified DNA onto a 1–2% agarose gel, and determine the fragment size. The goal is to obtain a smear of DNA around 500 bps.
5. If the majority or "smear" of the fragment sizes are not within this range, increase or decrease the output on the sonicator and the number of pulses or cycles during sonication.
6. After you have optimized the sonication conditions, determine the efficiency of HaloTag binding (Subheading 3.7) to ensure that the sonication conditions are optimal for HaloLink Resin capture.

3.7. Analysis of HaloTag Fusion Protein Binding Efficiency to HaloLink Resin

This protocol requires the use of a fluorimager for detection of fluorescent bands within an SDS gel. If such equipment is not available, follow steps 1 and 5, and then perform a standard Western blot using an anti-HaloTag antibody (Promega #G9281).

1. Add 20 μl of nuclease free water to each 10 μl sample collected at the above points in the protocol: Subheading 3.4, step 2 (starting material); step 5 (unbound material).
2. Add 1 μl of 50 μM HaloCHIP Blocking Ligand, and mix thoroughly.
3. Incubate for 15 min at room temperature in the dark.
4. Add 10 μl of 4× SDS-PAGE gel loading buffer, and heat to 70–95°C for 5 min.
5. Analyze 10–15 μl of each reaction by SDS-PAGE.
6. Run the gel until the dye front is near the bottom of the gel.
7. Quantitate the amount of input material and unbound material using a fluorescent detection scanner (excitation: 555 nm; emission: 585 nm). The HaloTag fusion protein will appear as a fluorescent band on a gel when a fluorescence scanner is used to view the gel. Quantitate the amount of input material and the amount of unbound material by determining the fluorescent signal of the HaloTag fusion protein in the two samples. The amount of unbound material should be less than 40% of the input material. If the unbound material exceeds 40% of the input material, then the HaloTag fusion protein was not bound efficiently. This could be due to inactivation of HaloTag fusion protein.

4. Notes

1. It is highly recommended to test for expression of the HaloTag fusion protein prior to proceeding with the HaloCHIP experiment. This can be done using on the HaloTag fluorescent ligands, see http://www.promega.com/resources/protocols/technical-manuals/0/halotag-technology-focus-on-imaging-protocol/ or performing a Western blot against the HaloTag protein (Promega #G9281) or the DNA binding protein of interest. If expression cannot be detected, optimize transfection conditions and or sequence the vector.
2. The placement of tags either on the N-terminus or C-terminus could potentially interfere with physiological function. If possible, perform functional characterization of the fusion protein prior to the HaloCHIP experiment which could include cellular localization http://www.promega.com/resources/protocols/technical-manuals/0/halotag-technology-focus-on-imaging-protocol/ or analysis of protein–protein interactions http://www.promega.com/resources/protocols/technical-manuals/101/halotag-mammalian-pull-down-and-labeling-systems-protocol/.
3. If different expression levels of the HaloTag fusion protein are desired or are necessary for proper physiology, a CMV promoter deletion series is recommended; HaloTag Flexi Vectors—CMV Deletion Series (Promega #G3780).
4. If using other cell types, the cross-linking (Subheading 3.2), lysis, and sonication (Subheading 3.3) steps will need to be optimized. Do not use SDS, IGEPAL CA-630, or Tween-20 detergents in the lysis buffer, as these detergents inhibit capture with the HaloLink Resin.
5. Typical yield from a single HaloCHIP reaction is approximately 1–10 ng of DNA. To obtain higher amounts of DNA, increase the number of HaloCHIP reactions performed, following the protocol as recommended, and pool purified DNA at the end of the process. This DNA can then be lyophilized to the desired concentration.
6. HaloLink resin will settle quite quickly in ethanol, therefore it is important to mix gently right before pipetting to obtain a uniform solution.
7. Excessive cross-linking can reduce the ability of the HaloTag protein to bind to its resin. If this is observed (Subheading 3.7), ensure cross-linking does not extend beyond the 10 min, reagents are at room temperature, and the formaldehyde is well mixed in the media. Cross-linking can be carried out at

lower concentrations of formaldehyde (0.5–0.75%) if there is concern about over-cross-linking.

8. Efficient quenching is very important to prevent over-cross-linking (Note 7).
9. Oversonication and/or overheating the sample during sonication will reduce the ability of the HaloTag protein to bind its resin. If this is observed (Subheading 3.7), decrease sonication time and output settings and keep all samples on ice during sonication to avoid denaturation of the HaloTag fusion protein.
10. Addition of protease inhibitors is not required for this protocol. If adding a protease inhibitor cocktail is desired, we recommend Promega G6521. Avoid cocktails of unknown composition or those known to contain AEBSF (4-(2-aminoethyl benzenesulfonyl fluoride hydrochloride)). AEBSF will significantly reduce binding efficiency or prevent binding altogether.
11. If the fragment size of the chromatin is too high, it is recommended to increase the number of cycles of sonication first. If that is unsuccessful, then increase the output. Please see Note 9 for concerns of oversonication or overheating.
12. A control for the HaloCHIP experiment should always be run to determine the level of background or non-specific DNA within the experiment. Either the blocking ligand control or untransfected control can be used as the proper control.
13. Proteins which degrade rapidly or may be temperature sensitive can be bound to the HaloLink resin at 4°C for a time of 4–6 h.
14. It is highly recommended to test for binding efficiency of the cross-linked HaloTag fusion protein and this can be done without going through the entire HaloCHIP process, i.e., stopping at Subheading 3.4, step 7 and then proceeding directly to Subheading 3.7. If the cross-linked protein is not captured on the HaloLink resin, then see Notes 2, 7–9, and 13 to improve binding to the HaloLink Resin.
15. It is important to mix by gently pipetting the resin in all of the washes to break up the resin particles, allowing for non-specific interactions to be washed away.
16. If high background DNA binding is observed in the control reaction after the HaloCHIP process, a LiCl wash using LiCl Buffer: 100 mM Tris–HCl (pH 8.0) 500 mM LiCl, 1% IGEPAL® CA-360, and 1% sodium deoxycholate could be added at this point.
17. Do no reverse cross-links for longer than overnight, 18 h, as DNA fragments will start to degrade. Also, use tubes which

securely remained closed during heating, otherwise they will open and samples will evaporate.

18. Avoid using DNA purification minicolumns that do not effectively purify smaller fragments (200–500 bp).

References

1. Solomon MJ, Larsen PL, Varshavsky A (1988) Mapping protein-DNA interactions in vivo with formaldehyde: evidence that histone H4 is retained on a highly transcribed gene. Cell 53:937–947
2. Ren B, Robert F, Wyrick JJ et al (2000) Genome-wide location and function of DNA binding proteins. Science 290:2306–2309
3. Pugh BF, Gilmour DS (2001) Genome-wide analysis of protein-DNA interactions in living cells. Genome Biol 2:1013
4. Weinmann AS, Farnham PJ (2002) Identification of unknown target genes of human transcription factors using chromatin immunoprecipitation. Methods 26:37–47
5. Horak CE, Mahajan MC, Luscombe NM et al (2002) GATA-1 binding sites mapped in the beta-globin locus by using mammalian ChIP-chip analysis. Proc Natl Acad Sci USA 99:2924–2929
6. Kurdistani SK, Grunstein M (2003) In vivo protein-protein and protein-DNA crosslinking for genomewide binding microarray. Methods 31:90–95
7. Buck MJ, Lieb JD (2004) ChIP-chip: considerations for the design, analysis, and application of genome-wide chromatin immunoprecipitation experiments. Genomics 83:349–360
8. Kirmizis A, Farnham PJ (2004) Genomic approaches that aid in the identification of transcription factor target genes. Exp Biol Med (Maywood) 229:705–721
9. Euskirchen GM, Rozowsky JS, Wei CL et al (2007) Mapping of transcription factor binding regions in mammalian cells by ChIP: comparison of array- and sequencing-based technologies. Genome Res 17:898–909
10. Johnson DS, Mortazavi A, Myers RM et al (2007) Genome-wide mapping of in vivo protein-DNA interactions. Science 316:1497–1502
11. Los GV, Encell LP, McDougall MG et al (2008) HaloTag: a novel protein labeling technology for cell imaging and protein analysis. ACS Chem Biol 3:373–382
12. Urh M, Hartzell D, Mendez J et al (2008) Methods for detection of protein-protein and protein-DNA interactions using HaloTag. Methods Mol Biol 421:191–209
13. Hartzell DD, Trinklein ND, Mendez J et al (2009) A functional analysis of the CREB signaling pathway using HaloCHIP-chip and high throughput reporter assays. BMC Genomics 10:497–512
14. Ohana RF, Encell LP, Zhao K et al (2009) HaloTag7: a genetically engineered tag that enhances bacterial expression of soluble proteins and improves protein purification. Protein Expr Purif 68:110–120
15. Ohana RF, Hurst R, Vidugiriene J et al (2011) HaloTag-based purification of functional human kinases from mammalian cells. Protein Expr Purif 76:154–164

Chapter 10

A Modified Yeast One-Hybrid System for Genome-Wide Identification of Transcription Factor Binding Sites

Kazuyuki Yanai

Abstract

The yeast one-hybrid system is a powerful genetic method to identify DNA–protein interactions, but there is a major limitation inherent to the system. Namely, frequency of false positives generated by yeast endogenous transcription factors has been thought to be higher than that of true positives by orders of magnitude. However, our modification efficiently can eliminate the false positives. When compared to the other methods for the analysis of DNA–protein interactions on a genome-wide scale, a modified yeast one-hybrid system offers several advantages including low initial and running cost, large-scale output, and easy handling.

Key words: Yeast one-hybrid assay, Transcription factor, DNA–protein interaction, Binding site, DNA element, Target gene, Genome wide, Chromatin, Chromatin immunoprecipitation, ChIP-on-chip

1. Introduction

With the sequencing of the human genome complete, the next great challenge has become the conversion of genomic data into functional information. However, identification of DNA-binding sites for transcription factors on a genomic scale remains a considerable challenge (1). Methods such as electrophoretic mobility shift assays (EMSA), SELEX enrichment, DNase I footprinting assays, ChIP, and reporter assays can be used to identify transcription factor binding sites, but these processes tend to be laborious and not well matched for screening large numbers of DNA elements. Computational methods for the identification of *cis*-regulatory sequences have been successfully applied to simple organisms such as yeast and worm, but such methods have been plagued by high false positive rates in mammals primarily because of the very large quantity of intergenic sequence (2).

Minou Bina (ed.), *Gene Regulation: Methods and Protocols*, Methods in Molecular Biology, vol. 977, DOI 10.1007/978-1-62703-284-1_10, © Springer Science+Business Media, LLC 2013

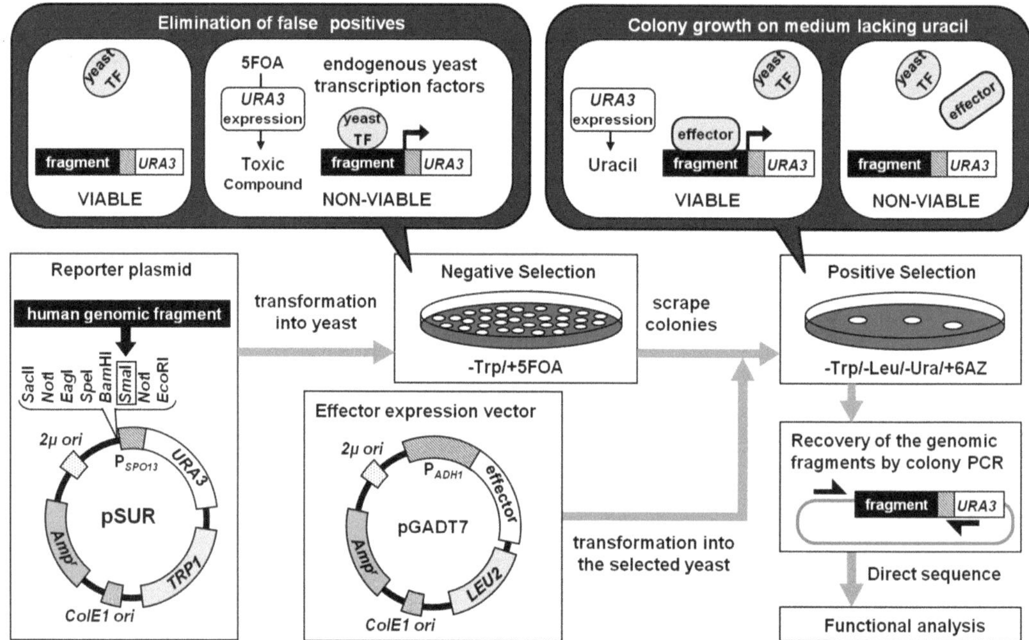

Fig. 1. Overview of the modified yeast one-hybrid system. pSUR contains a bifunctional reporter *URA3* under the tightly controlled SP013 promoter. Yeast cells are transformed with human genomic library and are selected on 5FOA negative selection plates. URA3 expression colonies that contain a genomic fragment recognized by yeast endogenous transcription factor should not be able to grow under these conditions. After the elimination of false-positive colonies, the selected colonies are scraped from plates and transformed with effector plasmids. The transformants are selected on plates lacking uracil. Thus we can obtain effector-dependent URA3 expression colonies that contain genomic fragments recognized by a DNA-binding protein of interest.

Recent techniques including "ChIP-on-Chip" and "ChIP-Seq" provide a new approach to analyze transcription factor-binding sites as a genome-wide scale (3). In chromatin immunoprecipitation (ChIP) analysis, formalin fixation causes the formation of cross-links of not only protein–DNA but also protein–protein. Therefore, the results of ChIP analysis actually include indirect transcription factor-DNA interactions (4). In addition, these technologies have obstacles including high initial cost, high running cost, reproducibility, and quantification. The yeast one-hybrid system is a powerful genetic method to identify DNA–protein interactions, but there is a major limitation inherent to the system. In the genome-wide scale, yeast endogenous transcription factors can bind to a large number of human genomic fragments and activate transcription of the reporter gene in the absence of a transcription factor of interest. Although the frequency of false positives has been thought to be higher than that of true positives by orders of magnitude (5), our modification can efficiently eliminate the genomic fragments recognized by yeast endogenous transcription factors (6) (see Fig. 1).

2. Materials

2.1. Yeast Media and Stock solutions

1. YPD medium: Add about 900 ml water to a glass beaker. Weigh 20 g Difco peptone, 10 g Yeast extract, and 20 g Agar (for plates only). Transfer the powder to the beaker and mix them. Make up to 950 ml with water, adjust the pH to 6.5, and then autoclave. Allow medium to cool to ~55°C and add 50 ml of a sterile 40% dextrose (see Note 1). Pour plates and allow medium to harden at room temperature. Store plates inverted, in a plastic sleeve at 4°C.
2. SD medium: Add about 800 ml water to a glass beaker. Weigh 6.7 g Yeast nitrogen base without amino acids (Difco Cat No. 0919-15-3) and 20 g Agar (for plates only). Transfer the powder to the beaker and mix them. Add 100 ml of the appropriate sterile 10× Dropout Solution. Make up to 950 ml with water, adjust the pH to 5.8, and then autoclave. Allow medium to cool to ~55°C and add 50 ml of a sterile 40% dextrose (see Note 1). Pour plates and allow medium to harden at room temperature. Store plates inverted, in a plastic sleeve at 4°C. For 0.1% 5-Fluoroorotic Acid (5FOA)- or 25 μg/ml 6-azauracil (6AZ)-containing medium, add the appropriate amount of the powder and swirl to mix well (see Note 2).
3. 40% Dextrose (glucose): Autoclave at 121°C for 15 min (see Note 1) and store at 4°C.
4. 10× Dropout Solution: A combination of a Minimal SD Base and a DO Supplement can produce minimal medium lacking one or more specific nutrients. Combine the nutrients list (see Table 1) at the concentrations indicated to prepare a 10× Dropout Solution (see Note 3). A 10× Dropout Solution contains all but one or more of these nutrients. 10× dropout supplements are autoclaved and stored at 4°C.

2.2. E. coli Media

1. SOC: Add about 90 ml water to a glass beaker. Weigh 2 g Bactotryptone and 0.5 g Yeast extract. Transfer the powder to the beaker, add 1 ml of 1 M NaCl and 0.25 ml of 1 M KCl, and mix them. Adjust the pH to 7.0 with NaOH, make up to 96 ml with water, and then autoclave. Allow medium to cool to room temperature and add filter-sterilized 1 ml of 1 M $MgCl_2 \cdot 6H_2O$, 1 ml of 1 M $MgSO_4 \cdot 7H_2O$, and 2 ml of 1 M glucose.
2. LB Media: Add about 900 ml water to a glass beaker. Weigh 10 g Bactotryptone, 5 g Yeast extract, 5 g NaCl, and 18 g Agar (for plates only). Transfer the powder to the beaker and mix them. Adjust the pH to 7.0 with NaOH, make up to 1 L with water, and then autoclave. Allow medium to cool to ~50°C, and add 1 ml of 50 mg/ml ampicillin solution. Store plates inverted, in a plastic sleeve at 4°C.
3. 50 mg/ml Ampicillin (in H_2O). Store at −20°C.

Table 1
10× Dropout Supplements

Nutrient	10× Concentration (mg/L)	Sigma Cat. No.
L-Adenine hemisulfate salt	200	A-9126
L-Arginine HCl	200	A-5131
L-Histidine HCl monohydrate	200	H-8125
L-Isoleucine	300	I-2752
L-Leucine	1,000	L-8000
L-Lysine HCl	300	L-5626
L-Methionine	200	M-9625
L-Phenylalanine	500	P-2126
L-Threonine	2,000	T-8625
L-Tryptophan	200	T-0254
L-Tyrosine	300	T-3754
L-Uracil	200	U-0750
L-Valine	1,500	V-0500

2.3. Solutions for Transformation of Yeast

1. Carrier DNA (10 mg/ml): just prior to use, denature the carrier DNA (Clontech Cat No. 630440) by placing it in a boiling water bath for 20 min and immediately cooling it on ice.
2. 100% DMSO (Dimethyl sulfoxide; Sigma Cat No. D-8779).
3. 10× TE buffer: 0.1 M Tris–HCl, 10 mM EDTA, pH 7.5. Autoclave.
4. 10× lithium acetate (LiAc): 1 M lithium acetate (Sigma Cat No. L-6883). Adjust to pH 7.5 with dilute acetic acid and autoclave.
5. 50% PEG 3350 (Polyethylene glycol, Sigma Cat No. P-3640) prepare with sterile deionized H_2O; if necessary, warm solution to 50°C to help the PEG go into solution.
6. PEG/LiAc solution: Prepare fresh just prior to use. Mix 1 ml of 10× TE, 1 ml of 10× LiAc, and 8 ml of 50% PEG.

2.4. Other Reagents and Kits

1. Yeast: W303-1A (MATa trp1 his3 leu2 ura3 ade2 can1)
2. Effector plasmid: pGADT7 (Clontech Cat No. 630442).
3. Reporter plasmid: pSUR (GenBank, accession number: AB425277) (RIKEN BRC DNA Bank, request ID: RDB 8320).

4. Primers for the recovery of genomic fragments:

   5′-TCGCGTTGCATTTTTGTTCTACAAAATGAAGCAC-3′

   5′-ACTTCCTTTTTGTCGGCGGCTATTTCTCAATAT AC-3′

5. Primers for direct sequences:

   5′-TGTTCTACAAAATGAAGCACAG-3′

   5′-CGGCTATTTCTCAATATACTCC-3′

6. The DNase Shotgun® Cleavage Kit (Novagen Cat No. 69281).
7. Plasmid Giga Kit (QIAGEN Cat No. 12191).

3. Methods

3.1. Constructing Effector and Reporter Plasmids

1. Digest the effector and reporter vector (see Fig. 1) with the appropriate restriction enzyme(s), treat with phosphatase, and purify.
2. Prepare the cDNA fragment encoding the DNA-binding protein of interest and the known binding sequence for the protein.
3. Ligate the appropriate vector and the insert and transform the mixtures into *E. coli*.
4. Identify insert-containing plasmids by restriction analysis. Sequences of the in-frame fusion protein and the synthesized binding sites should be confirmed by sequence analysis.

3.2. Testing DNA–Protein Interactions

1. Prepare SD/–Trp/–Leu plates and SD/–Trp/–Leu/-Ura/+6AZ plates in advance (see Note 4).
2. Inoculate several yeast colonies (W303-1A), 2–3 mm in diameter, into 10 ml of YPD (Colonies should not be >1 weeks old).
3. Vortex vigorously to disperse any clumps and incubate at 30°C for 16–24 h with shaking at 230 rpm.
4. Transfer the overnight culture into 30 ml of prewarmed YPD.
5. Incubate at 30°C for 3 h with rotation at 230 rpm.
6. Centrifuge the cells at 1,000×*g* for 5 min at room temperature.
7. Discard the supernatant and vortex to resuspend the cell pellet in 10 ml of H_2O.
8. Centrifuge the cells at 1,000×*g* for 5 min at room temperature.
9. Decant the supernatant and resuspend the cell pellet in freshly prepared 1 ml of 1× TE/LiAc.

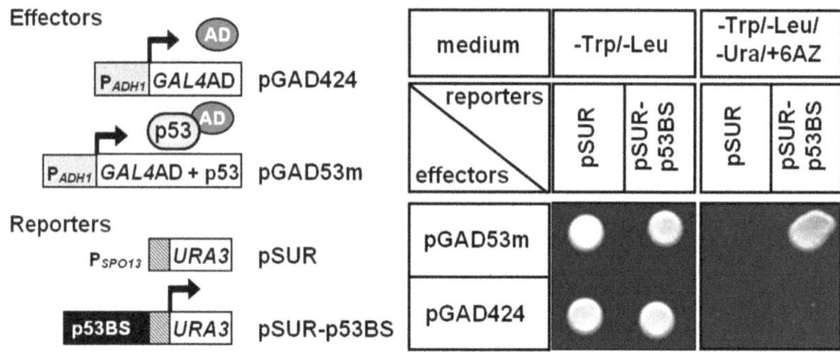

Fig. 2. Confirmation of DNA–protein interaction. Yeast cells transformed with indicated plasmids were grown on the indicated synthetic complete media. The plates were photographed after 3 days growth at 30°C.

10. Add 100–500 ng of effector plasmid, 100–500 ng of reporter plasmid, 100 μg of carrier DNA, and 100 μl of yeast competent cells to each tube and mix well.
11. Add 600 μl of PEG/LiAc solution to each tube and vortex to mix well.
12. Incubate at 30°C for 30 min with shaking (200 rpm).
13. Add 10 μl of DMSO and mix gently by inversion.
14. Heat shock in a 42°C water bath for 15 min.
15. Centrifuge the cells at 1,500×*g* for 5 min at room temperature.
16. Remove the supernatant and resuspend cells in 100 μl of 1× TE buffer.
17. Plate 100 μl of the cells on 100-mm SD/–Trp/–Leu plates.
18. Incubate plates, colony side down, at 30°C for 3days.
19. Pick up large colonies, 2–3 mm in diameter, from each plate and suspended in 200 μl of H_2O.
20. Vortex vigorously to disperse any clumps and adjust the cell density to an OD_{600} of 1.0.
21. Spot 5 μl of yeast cells on the appropriate selection plate (Fig. 2).
22. Incubate plates, colony side down, at 30°C until colonies appear (see Notes 5–8).

3.3. Preparation of a Plasmid Library

1. Digest pSUR with *Sma*I, treat with phosphatase, and purify.
2. To generate random DNA fragments, genomic DNA is digested with DNase Shotgun® Cleavage Kit (see Note 9).
3. Isolate fragments around 300 bp by agarose gel electrophoresis and treat with T4 DNA polymerase in the presence of dNTPs.
4. Ligate the vector and the genomic fragment and transform the mixtures into *E. coli* by electroporation.

5. Spread the transformants on 300 LB/amp plates at a high density (40,000–60,000 colonies/150-mm plate) (see Note 10).
6. Incubate plates at 37°C overnight.
7. Add 10 ml cold LB/amp broth to each plate and scrape the colonies (see Note 11).
8. Incubate at 37°C for 1 h with shaking (200 rpm).
9. Isolate highly purified plasmid by using QIAGEN Plasmid Giga Kit.

3.4. Elimination of False Positives

1. Prepare 150-mm SD/–Trp/+0.1% 5FOA plates in advance.
2. Inoculate several colonies, 2–3 mm in diameter, into 10 ml of YPD (Colonies should not be >1 weeks old).
3. Vortex vigorously to disperse any clumps and incubate at 30°C for 24–48 h with shaking at 230 rpm.
4. Transfer the overnight culture into 200 ml of prewarmed YPD.
5. Incubate at 30°C for 6–12 h with rotation at 230 rpm.
6. Transfer enough overnight culture to produce an $OD_{600} = 0.2$ into 800 ml of prewarmed YPD.
7. Incubate at 30°C for 3 h with rotation at 230 rpm.
8. Centrifuge the cells at 1,000×*g* for 5 min at room temperature.
9. Discard the supernatant and vortex to resuspend the cell pellet in 100 ml of H_2O.
10. Centrifuge the cells at 1,000×*g* for 5 min at room temperature.
11. Decant the supernatant and resuspend the cell pellet in freshly prepared 4 ml of 1× TE/LiAc.
12. Incubate at 30°C for 30 min with shaking at 230 rpm.
13. Add 600 μg of the plasmid library and 4 mg of carrier DNA into 4 ml of yeast competent cells and mix well.
14. Add 12 ml of PEG/LiAc solution and vortex to mix well.
15. Incubate at 30°C for 30 min with shaking (230 rpm).
16. Add 2.8 ml of DMSO and mix gently by inversion.
17. Heat shock in a 42°C water bath for 15 min.
18. Add 50 ml of YPD and vortex to mix well.
19. Centrifuge the cells at 1,500×*g* for 5 min at room temperature.
20. Remove the supernatant and resuspend cells in 40 ml of YPD and vortex to mix well.
21. Repeat steps 19 and 20 two times for a total of three trials.
22. Incubate at 30°C for 1 h with shaking (150 rpm).
23. Centrifuge the cells at 1,500×*g* for 5 min at room temperature.
24. Remove the supernatant and resuspend cells in 40 ml of SD/–Trp and vortex to mix well.

25. Repeat steps 23 and 24.
26. Incubate at 30°C for 4 h with shaking (150 rpm).
27. Centrifuge the cells at 1,500 × *g* for 5 min at room temperature.
28. Remove the supernatant and resuspend cells in 150 ml of SD/–Trp.
29. Plate 500 μl of the transformants on 300 SD/–Trp/+0.1% 5FOA plates by using sterile glass rod, bent Pasteur pipette, or 9-mm glass beads.
30. Incubate plates, colony side down, at 30°C for 7 days.
31. Add 10 ml cold SD/–Trp to each plate, scrape the colonies, and pool them.
32. Centrifuge the cells at 1,500 × *g* for 5 min at room temperature and remove the supernatant.
33. Add 50 ml of YPD and 50 ml of glycerol, and vortex to mix well.
34. Divide cells into 1 ml aliquots and store at –80°C.

3.5. Plasmid Library Titering Protocol

1. Thaw an aliquot of the library and place on ice.
2. Make serial dilutions of 10^{-2}, 10^{-3}, 10^{-4}, … 10^{-10} of the library by adding 10 μl of the stock to 990 μl of SD/–Trp. Vortex the tube gently for 30 s. Repeat using 10 μl of the previous dilution and adding to 990 μl of SD/–Trp until a library dilution of 10^{-10} has been reached.
3. Spread 100 μl of the 10^{-6} to 10^{-12} library dilutions onto separate SD/–Trp plates.
4. Incubate plates, colony side down, at 30°C for overnight.
5. Count the number of colonies and determine the number of colony forming units per ml (see Note 12).

3.6. Library Screening

1. Prepare SD/–Trp/-Leu/+6AZ plates in advance (see Note 4).
2. Thaw the library including $>1 \times 10^7$ colony and transfer into 10 ml of YPD.
3. Incubate at 30°C for 2 h with shaking at 230 rpm.
4. Centrifuge the cells at 1,000 × *g* for 5 min at room temperature.
5. Discard the supernatant and resuspend the cell pellet in 400 ml of SD/-Trp.
6. Incubate at 30°C for 10–15 h with rotation at 230 rpm.
7. Transfer enough overnight culture to produce an $OD_{600} = 0.2$ into 800 ml of prewarmed YPD.
8. Incubate at 30°C for 3 h with rotation at 230 rpm.
9. Centrifuge the cells at 1,000 × *g* for 5 min at room temperature.

10. Discard the supernatant and vortex to resuspend the cell pellet in 100 ml of H_2O.
11. Centrifuge the cells at 1,000×*g* for 5 min at room temperature.
12. Decant the supernatant and resuspend the cell pellet in freshly prepared 4 ml of 1x TE/LiAc.
13. Incubate at 30°C for 30 min with shaking at 230 rpm.
14. Add 600 μg of the plasmid library and 4 mg of carrier DNA into 4 ml of yeast competent cells and mix well.
15. Add 12 ml of PEG/LiAc solution and vortex to mix well.
16. Incubate at 30°C for 30 min with shaking (230 rpm).
17. Add 2.8 ml of DMSO and mix gently by inversion.
18. Heat shock in a 42°C water bath for 15 min.
19. Add 50 ml of YPD and vortex to mix well.
20. Centrifuge the cells at 1,500×*g* for 5 min at room temperature.
21. Remove the supernatant and resuspend cells in 40 ml of YPD and vortex to mix well.
22. Repeat steps 19 and 20 two times for a total of three trial.
23. Incubate at 30°C for 1 h with shaking (150 rpm).
24. Centrifuge the cells at 1,500×*g* for 5 min at room temperature.
25. Remove the supernatant, resuspend cells in 40 ml of SD/–Trp/-Leu, and vortex to mix well.
26. Repeat steps 23 and 24.
27. Incubate at 30°C for 4 h with shaking (150 rpm).
28. Centrifuge the cells at 1,500×*g* for 5 min at room temperature.
29. Remove the supernatant and resuspend cells in 25 ml of SD/–Trp/-Leu/-Ura.
30. Spread 100 μl of a 1:1,000, 1:100, and 1:10 dilution on SD/–Trp/–Leu plates (100-mm) for transformation efficiency controls (see Note 13).
31. Plate 500 μl of the transformants on 50 SD/–Trp/-Leu/-Ura/+6AZ 150-mm plates by using sterile glass rod, bent Pasteur pipette, or 9-mm glass beads.
32. Incubate plates, colony side down, at 30°C for 14–21 days (see Notes 14 and 15).

3.7. Isolation of Genomic Fragments for Further Analysis of Putative Positives

1. Pick positive colonies (>2 mm in diameter) from the original plates, streak out colonies on fresh SD/-Trp/-Leu and SD/–Trp/–Leu/–URA/+6AZ plates, and grow for 2–4 days at 30°C until colonies are at least 1 mm in diameter.

2. Confirm the growth on SD/–Trp/–Leu/–URA/+6AZ plates and transfer several colonies into 100 μl of H_2O (Colonies should not be >5 days old).
3. Perform triple freeze-thaw protocol and centrifuge the lysate at 15,000 × *g* for 5 min at 4°C.
4. Recover the genomic fragments from the supernatants by PCR with primers corresponding to the vector sequences.
5. The PCR fragments will be directly sequenced and used for further experiments (see Note 16).

4. Notes

1. If you add the sugar solution before autoclaving, autoclave at 121°C for 15 min; autoclaving at a higher temperature, for a longer period of time, may cause the sugar solution to darken and will decrease the performance of the medium.
2. 6AZ, a competitive inhibitor of the yeast *URA3* protein, is used to inhibit low levels of *ura3* expressed in a leaky manner in some reporter strains. 5-FOA is toxic to yeast cells that can synthesize the URA3 protein. 5FOA and 6AZ is heat-labile and will be destroyed if added to hot medium.
3. Serine, aspartic acid, and glutamic acid are not included in the nutrient list, because they make the medium too acidic and yeast can synthesize these amino acids endogenously.
4. For protein interactions requiring a ligand, it may be necessary to add it to the media.
5. A low level of leaky URA3 expression may permit slow growth on SD/–Trp/-Leu/-Ura medium in small colonies. Confirm that you have added 6AZ to the selection medium to a final concentration between 12.5 and 50 mM. Perform a 6AZ titration on the target-reporter strain if you have not already done so.
6. If high-level expression of the DNA-binding protein of interest is toxic to the cell, transformants will not grow. Sometimes truncation of the protein will alleviate the toxicity and still allow the interaction to occur. Alternatively, effector vectors which provide a low level of expression may be used (pGAD424, pACT2, etc).
7. If your control reporters grows without the expression of the effector on SD/–Trp/-Leu/-Ura medium even in the presence of >12.5 mM 6AZ, the inserted target element may be interacting with yeast endogenous transcriptional activators. It may be necessary to redesign the target element and construct a new reporter plasmid.

8. Failure to detect the DNA–protein interactions that normally interact in vivo will result in false negatives. In this case, you can use the tandem repeats of DNA fragments to construct positive control reporters.
9. DNase I in the presence of Mn^{2+} causes random double-stranded cleavage of the DNA molecule (7). Fragments of almost any size can be generated by adjusting the amount of enzyme and/or time of reaction.
10. To cover the entire genome, the number of colonies should be $>1\times10^7$. If your library transformation efficiency is very low, check the purity of the DNA and, if necessary, repurify it.
11. To confirm the quality of the library, select 20 colonies at random and analyze their inserts by PCR with vector primers. The average size of amplified fragments should be around 500 bp.
12. The number of colonies should be $>1\times10^7$/ml. Titers are usually stable at −80°C for at least 1 year.
13. The total number of transformants will be between 1×10^6 and 1×10^7.
14. After 2–3 days, some Ura+ colonies will be visible on the library screening (SD/−Trp/−Leu/−URA/+6AZ) plates, but plates should be incubated for 14–21 days to allow slower growing colonies (i.e., weak positives) to appear. Ignore the small colonies that may never grow to >2 mm in diameter. True URA+ colonies are robust and can grow to >2 mm in diameter.
15. The total number of the transformants surviving this selection will be between 100 and 1,000. About 50–95% of the transformants will be true positives.
16. Try the following suggestions for further analysis. (1) Retransform the effector plasmid and the candidate reporter plasmid recovered from the positive colonies, and check the URA3 expression on the selection plates. (2) Perform Electro Mobility Shift Assays using the recovered genomic fragments or synthesized DNA based on the candidate sequences. (3) Perform ChIP analysis using primer sets that cover the candidate gene. (4) Analyze the transcriptional regulation of candidate genes by real-time PCR or microarray analysis.

Acknowledgment

This work was supported by the Research Grant for the faculty of science special grant for promoting scientific research at Toho University.

References

1. Chorley BN, Wang X, Campbell MR, Pittman GS, Noureddine MA, Bell DA (2008) Discovery and verification of functional single nucleotide polymorphisms in regulatory genomic regions: current and developing technologies. Mutat Res 659:147–157
2. Chang LW, Nagarajan R, Magee JA, Milbrandt J, Stormo GD (2006) A systematic model to predict transcriptional regulatory mechanisms based on overrepresentation of transcription factor binding profiles. Genome Res 16:405–413
3. Farnham PJ (2009) Insights from genomic profiling of transcription factors. Nat Rev Genet 10:605–616
4. Wells J, Yan PS, Cechvala M, Huang T, Farnham PJ (2003) Identification of novel pRb binding sites using CpG microarrays suggests that E2F recruits pRb to specific genomic sites during S phase. Oncogene 22: 1445–1460
5. Deplancke B, Dupuy D, Vidal M, Walhout AJ (2004) A gateway-compatible yeast one-hybrid system. Genome Res 14:2093–2101
6. Taniguchi-Yanai K, Koike Y, Hasegawa T, Furuta Y, Serizawa M, Ohshima N, Kato N, Yanai K (2010) Identification and characterization of glucocorticoid receptor-binding sites in the human genome. J Recept Signal Transduct Res 30:88–105
7. Anderson S (1981) Shotgun DNA sequencing using cloned DNase I-generated fragments. Nucleic Acids Res 9:3015–3027

Chapter 11

Identifying Specific Protein–DNA Interactions Using SILAC-Based Quantitative Proteomics

Cornelia G. Spruijt, H. Irem Baymaz, and Michiel Vermeulen

Abstract

A comprehensive identification of protein–DNA interactions that drive processes such as transcription and replication, both in prokaryotic and eukaryotic organisms, remains a major technical challenge. In this chapter, we present a SILAC-based DNA affinity purification method that can be used to identify specific interactions between proteins and functional DNA elements in an unbiased manner.

Key words: Quantitative proteomics, SILAC, Protein–DNA interactions, Mass spectrometry

1. Introduction

The human genome consists of three billion basepairs, but only a small percentage encodes for genes. Apart from well-characterized regulatory sequences such as promoters and enhancers, the rest of the genome used to be considered "junk DNA." However, during the last decade it has become clear that a much larger percentage of the human genome is transcribed in the form of long and short noncoding RNAs. In addition, intergenic DNA sequences contain far more regulatory regions than previously thought (1). Proteins and noncoding RNAs interact with these DNA sequences in a spatio-temporal manner to regulate transcription and replication. A comprehensive characterization of DNA–protein interactions is therefore essential to increase our understanding of the aforementioned processes in the nucleus. To identify sequence specific protein–DNA interactions, researchers have traditionally made use of methods such as the electromobility shift assay (EMSA)

Minou Bina (ed.), *Gene Regulation: Methods and Protocols*, Methods in Molecular Biology, vol. 977,
DOI 10.1007/978-1-62703-284-1_11, © Springer Science+Business Media, LLC 2013

and footprinting. These assays are used to characterize a putative interaction between a candidate protein and a DNA sequence of interest. However, an unbiased identification of interactors for a specific DNA sequence requires other methods. In this regard, mass spectrometry-based proteomics has recently emerged as a powerful tool. Modern instrumentation and software enable the identification of hundreds of proteins in a sample in a few hours (2, 3). Similar amounts of proteins can be identified in DNA affinity purifications from crude nuclear extracts. However, the majority of these proteins are highly abundant background proteins that bind nonspecifically to the beads or DNA and only a small fraction represents sequence-specific interactors. This implies a need for a quantitative filter that can be used to discriminate specific interactors from nonspecific background proteins. In recent years numerous methods have been developed that add a quantitative dimension to mass spectrometry measurements. In most of these methods, proteins or peptides of two conditions are labeled with different, "light" or "heavy," stable isotopes on specific amino acids. The two samples are then combined prior to mass spectrometry analysis. Each peptide that is identified in the mass spectrometer will have a "light" and a "heavy" peak and the ratio of these two signals corresponds to the relative abundance of that peptide (and the corresponding protein) in the two functional states. When applying this technology to protein–DNA interaction studies, by incubating two different DNA sequences with "light" and "heavy" nuclear extracts, the measured peptide ratio indicates the relative affinity of a protein for each of the two DNA probes (Fig. 1).

Recently, we and others have established a DNA affinity purification protocol that makes use of an in vivo stable isotope labeling approach called SILAC (Stable Isotope Labeling by Amino acids in Cell culture) (4). This generic method can be used to identify proteins binding to DNA sequences of interest, including transcription factor binding sites (5), single nucleotide polymorphisms (6) and methylated CpG islands (5, 7–9) (see Note 1). In this chapter we describe the workflow behind this method in detail.

2. Materials

All buffers are prepared with ultrapure water of 18.2 MΩ cm resistance (MilliQ, Millipore). To prevent the accumulation of polymers in the samples, avoid the use of autoclaved pipette tips during the experiment. Furthermore, solvents and buffers are best kept in high quality glass bottles (Schott).

Tabletop centrifuges with cooling capacity for Eppendorfs and 50 ml tubes are required throughout the protocol.

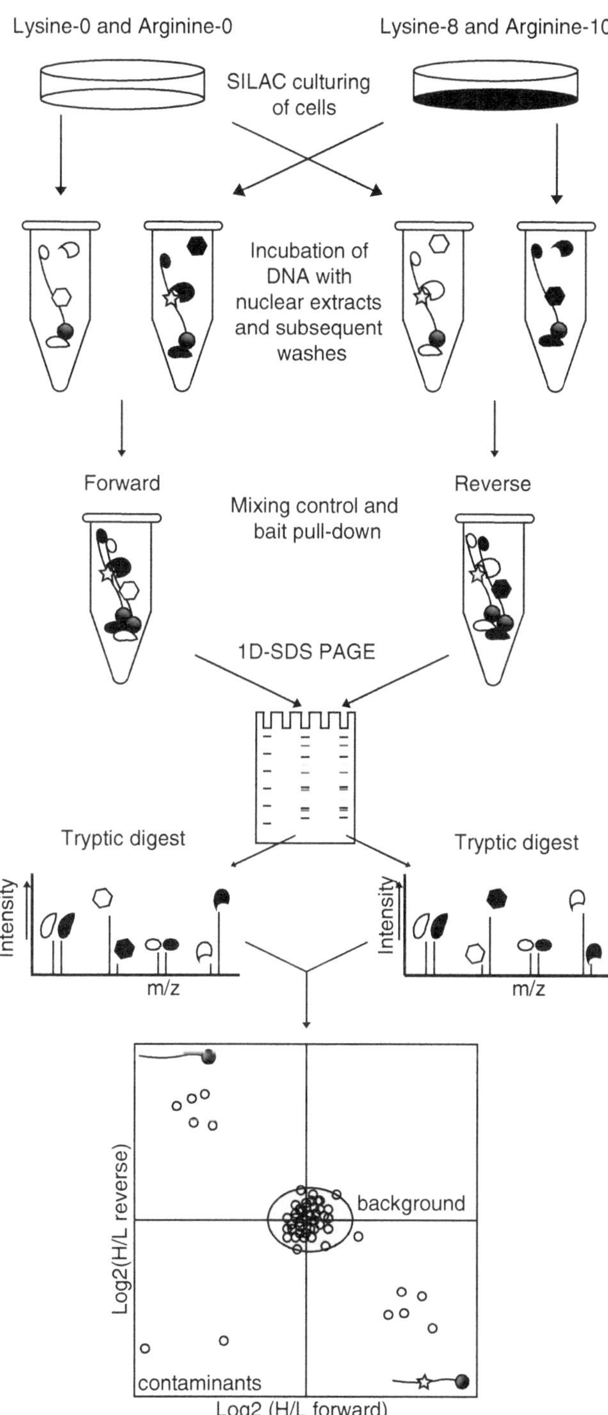

Fig. 1. Schematic representation of the workflow described in this chapter. Bait and control DNA are incubated (separately) with light and heavy nuclear extracts (NE) from cells grown in light or heavy SILAC media. Bait DNA incubated with heavy NE is combined with control DNA incubated with light NE (forward experiment) and bait DNA incubated with light NE is combined with control DNA incubated with heavy NE (reverse experiment). The two experiments are fractionated using 1D SDS-PAGE, followed by in-gel digestion and mass spectrometry. The results can be visualized in a scatterplot. Specific interactors of the bait DNA are located in the lower right quadrant (high forward ratio, low reverse ratio) whereas proteins that are repelled by the bait DNA end up in the upper left quadrant (low forward ratio, high reverse ratio). High-abundant background proteins and nonspecific DNA binders cluster together around the origin of the graph.

2.1. SILAC Culture (See Notes 2–4)

1. SILAC Dulbecco's Modified Eagle Medium without arginine, lysine, and glutamine (PAA, E15-086).
2. Dialyzed serum (Gibco, 26400-044).
3. Glutamine (Lonza, BE17-605E).
4. Penicillin/Streptomycin (Lonza, DE 17-602E).
5. L-Lysine ["light" or "K0" (Sigma, L8662)], dissolved in MilliQ.
6. L-Lysine 4,4,5,5-D4-L-lysine ["medium" or "K4"(Sigma, 616192 or Silantes, 211103912)], dissolved in MilliQ. Only in case of triple labeling (see Note 2).
7. L-Lysine ($^{13}C_6{}^{15}N_2$) ["heavy" or "K8" (Sigma, 608041 or Silantes, 211603902)], dissolved in MilliQ.
8. L-Arginine ["light" or "R0"(Sigma, A6969)], dissolved in MilliQ.
9. L-Arginine $^{13}C_6$-monohydrochloride ["medium" or "R6" (Sigma, 643440 or Silantes, 201203902)], dissolved in MilliQ. Only in case of triple labeling (see Note 2).
10. L-Arginine $^{13}C_6{}^{15}N_4$-monohydrochloride ["heavy" or "R10" (Sigma, 608033 or Silantes, 201603902)], dissolved in MilliQ.
11. Dulbecco's Phosphate buffered saline (DPBS) (Lonza, BE17-512F).
12. Trypsin–EDTA (Lonza, BE17-161E) or, depending on the cell line, Accutase (Sigma-Aldrich, A6964-100).
13. 100× Non essential amino acids (which contains proline, but no lysine or arginine) (Lonza, BE13-114E) (only for specific cell types, see Note 3).
14. 100 mM sodium pyruvate (Lonza, BE13-115E) (only for specific cell types, see Note 3).
15. Leukemia inhibitory factor (LIF), Β-mercaptoethanol and "2i" inhibitors (CHIR99021 and PD0325901) (see Note 3).
16. RPMI without arginine, lysine, and glutamine (PAA, E15-087) (only needed for cells growing in suspension, see Note 3).
17. 50 ml syringes (BD plastikpak, 300865).
18. 0.22 μm filters (Corning, 431219).

2.2. Nuclear Extract Preparation

1. Dulbecco's Phosphate buffered saline (DPBS) (Lonza BE17-512F).
2. Buffer A: 10 mM Hepes KOH pH 7.9, 1.5 mM $MgCl_2$, 10 mM KCl.
3. Buffer C: 420 mM NaCl, 20 mM Hepes KOH pH 7.9, 20% glycerol (v/v), 2 mM $MgCl_2$, 0.2 mM EDTA, 0.1% Igepal CA-630 (v/v) (NP40, Sigma-Aldrich, I8896-100ML). Add fresh

before use: Complete protease inhibitors EDTA-free (Roche, 05056489001, 1 tablet for 50 ml buffer) and 0.5 mM DTT.

4. Glass douncer with type B pestle (tight), available in different sizes: 500 µl (Kimble Kontes, 885300-000), 2 ml (Kimble Kontes, 885303-0002 or 885301-0002) and 7 ml (Wheaton, 357542).

2.3. Bradford Protein Concentration

1. Bio-Rad Protein assay 5× solution (Bio-Rad, 500-0006).
2. Bovine serum albumin (BSA), (1 mg/ml solution in MilliQ) (Sigma-Aldrich, A9647-50G).
3. UV/Vis spectrophotometer.
4. Cuvettes (1 ml).

2.4. DNA Preparation

1. Oligonucleotides (HPLC-purified from any company).
2. TE: 10 mM Tris–HCl pH 8.0, 1 mM EDTA.
3. 2× Annealing buffer: 20 mM Tris–HCl pH 8.0, 100 mM NaCl, 2 mM EDTA.
4. T4 Polynucleotide kinase, (T4 PNK (10,000 U/ml)) (New England Biolabs (NEB), M0201S).
5. T4 DNA ligase (400 U/µl) (NEB, M0202S).
6. 100 mM ATP in MilliQ, pH 7.5 adjusted using NaOH.
7. Phenol–Chloroform (Sigma, P4557).
8. Ice-cold 100% ethanol.
9. Ice-cold 70% ethanol (v/v).
10. 3 M sodium acetate, pH 5.2.
11. Klenow fragment 5′exo- (NEB, M0212S), NEB buffer 2.
12. Biotin-14-dATP (Invitrogen, 19524-016) (make aliquots and store them at −20°C).
13. Sephadex G-50, 50% slurry in 20% ethanol (v/v) (VWR, 17-0043-01).
14. 1 ml syringes (BD plastikpak, 300013) without needle.

2.5. DNA Affinity Purification

1. Magnetic microtube rack.
2. Dynabeads MyOne C1 (Invitrogen, 650.01) (see Note 5).
3. DNA binding buffer: 1 M NaCl, 10 mM Tris–HCl pH 8.0, 1 mM EDTA pH 8.0, 0.05% Igepal CA-630 (NP40, Sigma-Aldrich, I8896-100ML).
4. Poly-dIdC (Sigma-Aldrich, P4929-10UN) or poly-dAdT (Sigma-Aldrich, P0883-10UN) (see Note 6).
5. Protein binding buffer: 150 mM NaCl, 50 mM Tris–HCl pH 8.0, 1 mM DTT, 0.25% Igepal CA-630 (NP40, Sigma-Aldrich, I8896-100ML), and complete protease inhibitors EDTA-free (Roche, 05056489001, 1 tablet for 50 ml).

2.6. In Gel Digestion

1. Gel running system (Invitrogen).
2. NuPage sample buffer (Invitrogen, NP0007).
3. MOPs buffer (Invitrogen, NP0001).
4. NuPAGE Novex 4–12% gradient gels (Invitrogen, NP0321BOX).
5. Colloidal blue stain kit (Invitrogen, LC6025).
6. Methanol (Merck, 1.06009.2500).
7. Acetic acid (Merck, 1.00063.2500).
8. ABC: 50 mM Ammonium bicarbonate (Fluka, 09830).
9. Destain solution: 25 mM ABC/50% ethanol (v/v).
10. Acetonitrile (Biosolve, 01200702).
11. Fixing solution: 50% methanol (v/v), 10% acetic acid (v/v) in MilliQ.
12. Staining solution: 55 ml MilliQ, 20 ml methanol, 20 ml Colloidal Blue Solution A.
13. 1 M 1,4-Dithiothreitol.
14. 0.55 M Iodoacetamide (Sigma, I1149).
15. Sequencing grade modified Trypsin (Promega, V5111).
16. 10% Trifluoric acid (TFA) (v/v) (Sigma, 302031).
17. Vacuum centrifuge.
18. Thermoshaker.

2.7. Peptide Desalting and Purification (Stage Tipping)

1. C18 disks (Empore, 22125-C18).
2. 200 μl pipette tips (Rainin).
3. Hollow needle with a 1.2 mm diameter (BD Microlance 3, 304622). Make the end blunt and use a piece of nano tubing end and plunger.
4. Methanol (Merck, 1.06009.2500).
5. Buffer A: 0.5% acetic acid (v/v) (Merck, 1.00063.2500) in ultrapure water (Biosolve, 232141B1).
6. Buffer B: 0.5% acetic acid (v/v) (Merck, 1.00063.2500), 80% acetonitrile (v/v) (Biosolve, 01200702) in ultrapure water (Biosolve, 232141B1).

2.8. Mass Spectrometry

1. Buffer A: 0.5% acetic acid (v/v) (Merck, 1.00063.2500) in ultrapure water (Biosolve, 232141B1).
2. Buffer B: 0.5% acetic acid (v/v) (Merck, 1.00063.2500), 80% acetonitrile (v/v) (Biosolve, 01200702) in ultrapure water (Biosolve, 232141B1).
3. 96 well thermofast robotic PCR plate (Thermo, 96 AB-1300).
4. Nanoflow HPLC system.

5. Column oven from Sonation (PRSO-V1).
6. Fused silica based emitters (30 cm length, 360 mm OD, 75 μm ID) (New Objective, FS360-75-8-N-5-C30) packed in-house with Reprosil-Pur 120 C18-AQ, 3 μm (Dr. Maisch GMBH, Germany).
7. High performance mass spectrometer such as an LTQ-Orbitrap-Velos or Q-Exactive instrument from Thermo Fisher.

3. Methods

3.1. SILAC Labeling

Cells are SILAC-labeled by culturing them for at least 8 cell doublings in medium containing "light" or "heavy" amino acids. Note that the proliferation rate for some cell types is decreased in SILAC medium compared to normal medium due to the use of dialyzed serum during cell culture. Dialysis is necessary to get rid of non-labeled amino acids in the serum, but this also removes growth factors and other small molecules which may be important for proliferation. For some cell types, such as mouse ES cells, it is important to test whether they grow normally in dialyzed serum-containing medium (see Note 3).

1. Prepare a bottle of "light" and a bottle of "heavy" SILAC medium (see Notes 3 and 4). For each:
   - (a) Take a bottle of 500 ml SILAC Dulbecco's Modified Eagle Medium without arginine, lysine, and glutamine.
   - (b) Transfer 30–40 ml of medium from the bottle into a 50 ml falcon tube and add the appropriate amounts of arginine (light or heavy) and lysine (light or heavy) to this aliquot of DMEM. Add 29.4 μg/ml of arginine and 73 μg/ml of lysine. Filter this medium containing the amino acids using a syringe and a 0.22 μm filter back into the bottle.
   - (c) Add 50 ml dialyzed serum.
   - (d) Add 2 mM glutamine.
   - (e) Add 100 U/ml Penicillin/Streptomycin. Medium can be kept at 4°C for up to 6 weeks.
2. Trypsinize a 10 cm dish of cells grown to ~80–100% confluence in regular medium (not light or heavy).
3. Neutralize the trypsin with regular medium and divide the suspension equally over two tubes.
4. Spin cells for 5 min at 400 × *g*.
5. Resuspend the cell pellet of one tube in 4 ml of light medium and the other cell pellet in 4 ml of heavy medium. Seed 1 ml of this suspension in a 10 cm dish and add 9 ml of light or heavy medium.

6. Grow the cells at 37°C in 5% CO_2 until they reach 80–100% confluency. Split the cells once more in a ratio of 1:8. Make sure to spin down the cells and resuspend them in fresh light or heavy medium after trypsinization since trypsin can be a source of non-labeled amino acids. In some cases, the splitting should be done differently depending on the cell type. Mouse ES cells, for example, are to be split 1:4 only. In this case, cells need to be split more often to ensure the minimal amount of 8 cell doublings required for efficient labeling.
7. Depending on the growth rate of the cells, labeling usually takes between 1 and 2 weeks. During the labeling it is recommended to perform an incorporation check on the heavy cells to make sure that the proteins are completely labeled (see Note 7).
8. When incorporation is complete, cells can be expanded to the desired amount. Typically, around 2 mg of nuclear extract is obtained from five 15 cm dishes, but this may vary depending on the cell line that is used.

3.2. Nuclear Extract Preparation

It is critical to be as consistent as possible when preparing nuclear extracts. Small differences in sample handling, especially during the douncing, can cause proteins to be differentially extracted between different samples. This makes it more difficult to discriminate true outliers from background proteins. This nuclear extraction protocol is based on Dignam et al. (10).

1. Wash cells with 10 ml of PBS and trypsinize them with 2 ml of trypsin per 15 cm dish. Neutralize trypsin by adding 10 ml of SILAC medium to the cells. Collect the cells in a 50 ml tube and rinse the plates once more with PBS to collect the remaining cells. Perform all subsequent steps at 4°C.
2. Centrifuge the cells for 5 min at 400 × *g* and aspirate the supernatant.
3. Wash cells with 50 ml of PBS and centrifuge for 5 min at 400 × *g*, and then aspirate the supernatant.
4. Resuspend cells in 8 ml of PBS and transfer the cells to a 15 ml tube. Rinse the 50 ml tube with 5 ml of PBS and transfer this to the 15 ml tube containing the cell suspension.
5. Centrifuge for 5 min at 400 × *g* (see Note 8) and aspirate the supernatant.
6. Determine the volume of the cell pellet and add 5 volumes of cold buffer A. Resuspend the cells and incubate for 10 min on ice.
7. Centrifuge the cells for 5 min at 400 × *g* and remove supernatant.
8. Determine the volume of the cell pellet (cell volume should increase due to osmotic uptake of buffer A by the cells, see Note 9) and add 2 volumes of buffer A containing complete protease inhibitors and

0.15% Igepal CA-630 (NP40) (v/v). Resuspend cells and transfer the suspension to a dounce homogenizer (see Note 10).

9. Apply 30–40 strokes up and down with a type B pestle (tight) (see Note 11).
10. Transfer the suspension back to a 15 ml tube and centrifuge for 15 min at 3,200 × *g*. The supernatant is the cytoplasmic extract. Collect or discard the supernatant. When keeping the supernatant, add glycerol (10% final concentration) and NaCl (150 mM final concentration).
11. Wash the pellet once with 10 volumes of PBS. Pipette up and down only once, gently.
12. Centrifuge for 5 min at 3,200 × *g* and discard the supernatant.
13. The pellet consists of crude nuclei. Determine the volume and add 2 volumes of buffer C.
14. Resuspend and transfer crude nuclei to an Eppendorf tube. Homogenize the pellet by pipetting up and down (10×). For some cell lines the pellet may be difficult to resuspend.
15. Incubate the suspension for 1 h at 4°Con a rotating wheel. Due to the lysis of the nuclei and the release of chromatin the suspension will become viscous and white clouds of chromatin should appear.
16. Centrifuge the suspension for 45 min at 20,800 × *g* in a tabletop centrifuge at 4°C.
17. Transfer the supernatant to a new tube. This is the nuclear extract (NE) that will be used for DNA pull-downs and it contains soluble nuclear proteins. The pellet contains the insoluble chromatin fraction and consists of DNA and proteins tightly bound to chromatin.
18. Aliquot (approximately 150 μl per Eppendorf tube) and snap-freeze the extracts in liquid N_2. The nuclear pellet can be snap-frozen too. Store at −80°C.

3.3. Protein Concentration Determination

1. Prepare a 1 mg/ml stock solution of BSA in MilliQ.
2. Dilute 2 μl of nuclear extract with 18 μl of MilliQ.
3. Transfer 4 and 10 μl of the diluted nuclear extracts to separate Eppendorf tubes.
4. For the standard curve, pipette 0, 1, 2, 5, 7, and 10 μl of the BSA solution in Eppendorf tubes.
5. Prepare a 1× Bio-Rad protein assay solution by diluting the reagent 5 times with MilliQ.
6. Add 1 ml of 1× Bio-Rad protein assay solution to the Eppendorf tubes containing the standard curve and the nuclear extract samples.

7. Transfer the samples to cuvettes and measure absorbance at 595 nm at the spectrophotometer.
8. Fit a linear curve through the absorbance values of the BSA standard and extract protein concentrations of the nuclear extracts by matching the absorbance of the samples to this curve.

3.4. DNA Preparation

1. Design complementary pairs of oligonucleotides of about 30 bases that contain your sequence of interest (see Note 1). Include two thymidines on the 5′end of one oligonucleotide and two adenosines on the 5′end of the reverse complementary oligonucleotide. For each bait a control pair of oligonucleotides should be designed. For example, a bait containing a methylated CpG dinucleotide should be combined with a control bait that is not methylated.
2. Dissolve the oligonucleotides to a concentration of 0.3 mM in TE buffer by shaking at room temperature (RT) for 1 h. Store DNA at −20°C until use.
3. Combine 12.5 μl of the forward and reverse oligonucleotides and add 25 μl of 2× Annealing buffer in an Eppendorf tube.
4. Incubate the sample at 95°C for 5 min in a water bath or heat block.
5. Spin down the sample and put it back to 95°C.
6. Switch off the heating and let the sample cool down slowly to RT. The oligonucleotides will be annealed at this point.
7. Phosphorylate the annealed oligonucleotides by adding 10 μl of 10× ligase buffer, 5 μl of T4 Polynucleotide kinase (10,000 U/ml) and 35 μl of MilliQ. Incubate for 2 h at 37°C (see Note 12).
8. Ligate the oligonucleotides by adding 10 μl of 100 mM ATP pH 7.5 and 2 μl of T4 DNA ligase (400 U/μl) (see Note 12).
9. Incubate for 4 h at RT and subsequently overnight at 4°C. The ligation efficiency can be investigated by loading 2 μl on a 1.5% agarose gel. The ligation products should form a ladder, as shown in Fig. 2a.
10. Perform a Phenol–Chloroform extraction:
    (a) Adjust the volume of the sample to 200 μl with MilliQ.
    (b) Add 200 μl of Phenol–Chloroform, vortex for 1 min, centrifuge for 2 min at 18,400 × *g* and transfer the upper phase to a new tube.
    (c) Precipitate DNA by adding 500 μl of 100% ice-cold ethanol and 20 μl of 3 M NaAc pH 5.2. Incubate for at least 30 min at −20°C.
    (d) Centrifuge for 10 min at 18,400 × *g* at 4°C.

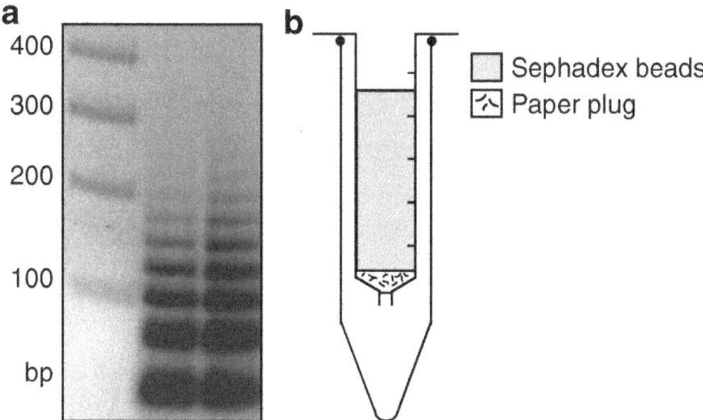

Fig. 2. DNA preparation. (**a**) Shown are one bait and one control DNA sample with similar ligation efficiencies. (**b**) A Sephadex G-50 column prepared by inserting a syringe without a plunger into a 15 ml tube. The syringe contains a paper plug at the bottom and is packed with Sephadex G-50 resin.

(e) Aspirate the supernatant carefully and wash the DNA pellet with 500 μl of ice-cold 70% ethanol (v/v).

(f) Centrifuge for 5 min at 18,400×*g* and aspirate the supernatant.

(g) Air-dry the pellet.

(h) Dissolve the DNA in 35 μl of MilliQ.

11. Add 5 μl of 10× NEB buffer 2, 2 μl of Klenow exo- (50 U/μl) and 8 μl of Biotin-14-dATP (0.4 mM) to the DNA and incubate for 3 h at RT.
12. Prepare Sephadex G-50 columns (see Fig. 2b and Note 13). Add 100 μl of TE to the DNA strands and load them onto the column. Centrifuge at 490×*g* for 1 min at 4°C. This step is performed to separate the DNA strands from the free biotin-ATP.
13. Measure the DNA concentration of the eluent.

3.5. DNA Affinity Purification

The protocol below describes a so-called forward and reverse experiment. In the forward experiment, the control DNA is incubated with light extract, while the bait of interest is incubated with heavy extract. In the reverse experiment a label-swap is performed in which the control DNA is incubated with heavy extract and the bait DNA is incubated with light extract. This setup constitutes a biological replicate.

1. Take four Eppendorf tubes and pipette 75 μl of Dynabeads MyOne C1 in each of them (see Note 6).
2. Add 0.5 ml of DNA binding buffer and place the tubes in a magnetic Eppendorf holder.
3. Aspirate the supernatant once the solution has cleared.
4. Take the tubes out of the holder, add 0.5 ml of DNA binding buffer, and invert the tubes until the beads are completely resuspended.

5. Centrifuge briefly and place the tubes back into the magnetic holder. Aspirate the supernatant.
6. Take two times 10 μg of bait DNA and two times 10 μg of control DNA (obtained in Subheading 3.4) and adjust the salt concentration to 1 M NaCl. Add the DNA in a total volume of 350 μl of DNA binding buffer to each of the tubes. Two tubes should contain bait DNA and the other two should contain control DNA.
7. Incubate for 1 h at RT on a rotation wheel.
8. Briefly centrifuge and place the samples in the magnetic rack. Check the coupling of the DNA to the beads by assessing the depletion of the DNA from the solution (see Note 14).
9. Wash the beads two times with 0.5 ml of DNA binding buffer as described in steps 4 and 5.
10. Wash the beads two times with 0.5 ml of Protein binding buffer.
11. Add 400 μg of nuclear extract to the beads in a total volume of 600 μl protein binding buffer, including 10 μg of poly-dIdC or poly-dAdT (see Note 7). **NB**: Add "light" NE to one tube with control DNA and to one tube containing bait DNA. Add "heavy" NE to the other two tubes.
12. Incubate for 90 min at 4°C on a rotation wheel.
13. Wash the beads three times with 0.5 ml protein binding buffer. After the last wash, remove supernatant completely, also from the lid.
14. Resuspend the beads of the control DNA pull-down in 30 μl 2× NuPAGE loading buffer containing 20 mM DTT. Add the control DNA pull-down suspension to the beads containing the bait DNA, as such that the heavy bait pull-down is mixed with the light control pull-down (forward experiment) and vice versa (reverse experiment).
15. Incubate the samples for 5 min at 95°C.

3.6. In Gel Digestion

The DNA pull-down procedure described in Subheading 3.5 results in two samples to be processed for mass spectrometry using in-gel trypsin digestion (11) (one forward and one reverse experiment). Wear gloves at all times and work as cleanly as possible. Keratin contamination of the samples can compromise the identification of proteins in the experiment.

1. Load the samples on a precast 4–12% gradient gel. Keep a blank lane between all the samples, including the lane between the molecular weight marker and the first sample.
2. Run the gel at 200 V.
3. Fix the gel for 10 min in fixing solution in a clean plastic box (50% methanol (v/v), 10% acetic acid (v/v) in MilliQ) on a shaker.

4. Incubate the gel for 5 min in 55 ml MilliQ, 20 ml methanol, and 20 ml Colloidal Blue Solution A.
5. Add 5 ml of Colloidal Blue solution B and incubate for 1 h.
6. Destain the gel in MilliQ for at least 2 h and refresh MilliQ a couple of times. It is also possible to store the gel at 4°C in MilliQ for up to a week.
7. Clean a glass plate with MilliQ and absolute ethanol and air-dry the plate.
8. Put the gel on the glass plate and cut out one lane at a time. Divide each lane into 6–10 slices depending on the protein amount in the sample. Make sure to cut the lanes of the forward and the reverse experiment in a similar pattern and try to isolate very abundant proteins in a single gel slice (see Fig. 3).
9. Cut each gel slice up into smaller pieces of about 1 mm^3 and transfer them to an Eppendorf tube.
10. Incubate the gel pieces two times for 1 h in 1 ml of destain solution (50% ethanol, 25 mM ABC) in a thermoshaker at RT at 1,200 rpm. Aspirate all the liquid after each incubation step.
11. Dehydrate the gel pieces in 1 ml of acetonitrile for 5–10 min in a thermoshaker. The gel pieces will shrink and become white opaque.
12. Swell the gel pieces in 1 ml 50 mM ABC for 5–10 min.
13. Dehydrate the gel pieces two times with 1 ml of acetonitrile for 5–10 min in a thermoshaker. Aspirate the supernatant after each incubation.
14. Vacuum centrifuge the gel pieces (with lids of Eppendorf tubes open) for 5–10 min until all the liquid has evaporated. At this point it is possible to store the samples at 4°C up to a week.
15. Add 200 μl of reducing buffer (10 mM DTT in 50 mM ABC) and incubate for 45 min at 55°C without shaking.
16. Carefully remove all the liquid.
17. Add 300 μl of 55 mM iodoacetamide in 50 mM ABC and incubate for 30 min at RT in the dark (iodoacetamide is light-sensitive).
18. Carefully remove all the liquid.
19. Wash the gel pieces for 15 min in 1 ml of 50 mM ABC.
20. Wash the gel pieces twice with 1 ml of acetonitrile in a thermoshaker and remove the supernatant after each incubation.
21. Vacuum centrifuge the gel pieces.
22. Add 30 μl of sequence grade trypsin at 10 ng/μl in 50 mM ABC.
23. Incubate for 10 min at RT until the gel pieces have absorbed the trypsin solution.

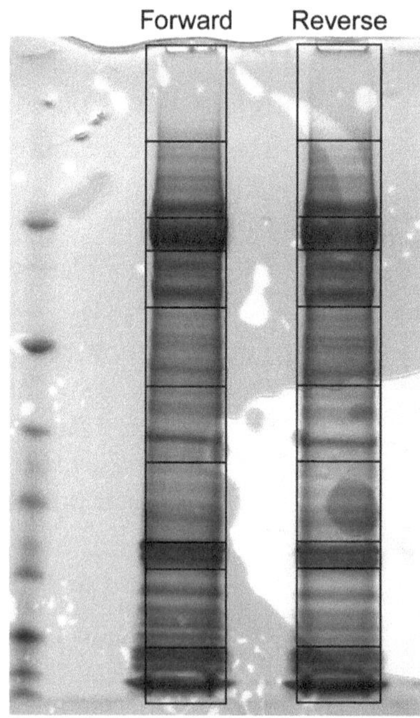

Fig. 3. 1D SDS-PAGE fractionation of proteins obtained in a forward and a reverse DNA pull-down. The lines around the proteins indicate how these lanes should be sliced into 6–10 pieces (10 slices in this case), isolating abundant proteins in a single slice.

24. Add 50 mM ABC until the gel pieces are completely covered and incubate overnight at 37°C.
25. Add 100 μl of 30% acetonitrile (v/v) and 3% TFA (v/v) in MilliQ to the gel pieces and shake for 10–15 min in a thermoshaker.
26. Transfer all the liquid to new tubes and repeat steps 25 and 26.
27. Add 100 μl of 100% acetonitrile to the gel pieces and shake for 10–15 min.
28. Centrifuge briefly and transfer the liquid to the collection tubes. Repeat steps 27 and 28.
29. Vacuum centrifuge the collected supernatants (45–90 min, depending on the number of samples) until ~100 μl of liquid is left (the acetonitrile in the sample should be completely evaporated).

3.7. Peptide Desalting and Purification

Following vacuum centrifugation of the tryptic peptides, it is common practice to desalt and purify the peptides using self-made or commercial C18 columns (see Note 15) called "stop and go extraction" or "stage tips" (Fig. 4) (12). During this procedure residual salt and small contaminants are removed and the peptides are captured on a 200 μl tip containing a small plug of C18 material. Peptides bound to stage tips can be stored for months at 4°C.

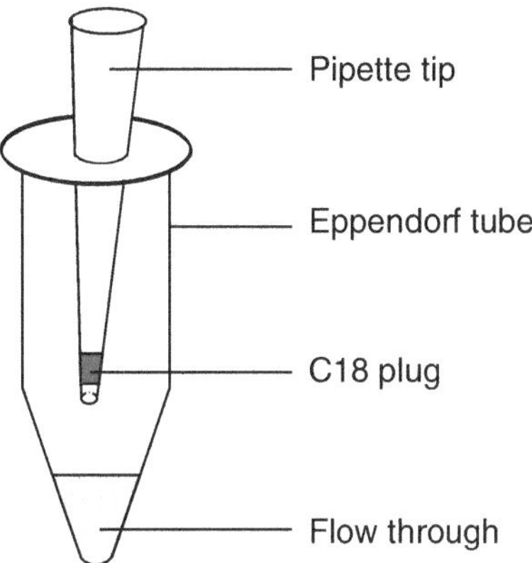

Fig. 4. Stage tips. Schematic representation of a stage tip (200 μl pipette tip with C18 plug) inserted into a 2 ml Eppendorf to collect the flow through.

1. Prepare stage tips by stamping out small disks from a double layer of C18 Empore filter using a blunt ended syringe needle. Eject the C18 disks from the needle into a 200 μl pipette tip and fix the material at the narrow end of the tip. Do not apply too much force since this will hinder buffer flow through the column. Prepare one stage tip per gel slice.
2. Punch a hole into the lid of 2 ml Eppendorf tubes and place the stage tips into the holes. Activate the stage tips by applying 50 μl of methanol and centrifuge at 1,500 × *g* for about 2–5 min. Make sure that all the liquid has passed through the column.
3. Wash the stage tips by applying 50 μl of buffer B and centrifugation at 1,500 × *g*.
4. Wash the stage tips twice with 50 μl of buffer A and centrifuge at 1,500 × *g*.
5. Load the samples on the stage tips and centrifuge at 380 × *g* until all the liquid has passed through the column. This takes about 10–20 min.
6. Wash the stage tips with 50 μl of buffer A and centrifuge at 1,500 × *g*.
7. Store stage tips at 4°C or proceed with elution and mass spectrometry as described below.

3.8. Mass Spectrometry

Modern mass spectrometers are very sensitive, fast and are able to sequence thousands of peptides in a short period of time. However, their operation and raw data analysis require extensive expertise and training. Therefore, the protocol below only provides a rough guideline for the liquid chromatography coupled to tandem mass spectrometry (LC-MS/MS) following in gel digestion.

1. Wash the stage tips once with 30 μl of buffer A and centrifuge at 1,500 × *g*.
2. To elute the peptides, load 30 μl of buffer B onto the stage tips and elute into a new Eppendorf tube by centrifugation at 380 × *g*.
3. Dry down the samples in a vacuum centrifuge to about 4 μl. Do not dry down the sample completely since this will result in a loss of peptides.
4. Add 4 μl of buffer A to the samples and transfer it to a 96-well plate that is compatible with the nano-HPLC.
5. Program the autosampler of the nano-HPLC to inject 4 μl onto the nano-HPLC column.
6. The peptides are eluted from the nano-HPLC column using a ~120 min, 5–30% acetonitrile (v/v) gradient followed by a sharp increase to 60% acetonitrile (v/v) in 10 min. Setting up these gradients requires extensive expertise, supervision by an experienced mass spectrometrist is highly recommended.
7. When using an LTQ-Orbitrap-Velos mass spectrometer, the following basic data acquisition settings are recommended: Acquire precursor MS spectra at an m/z range of 300–1,750 at a resolution of 60,000 and a target value of one million ions per full scan. MS/MS spectra can be acquired in HCD or CID mode. For protein identification experiments we usually obtain the highest number of protein identifications in CID mode. When using CID, select the 15 most intense precurser ions of every full scan for fragmentation at a minimal ion count target value of 500. Fragment and record peptides in the dual pressure linear ion trap using a normalized collision energy of 35% and acquire these spectra in centroid mode. Enable dynamic exclusion (repeat duration 30 s, list size 500, exclusion duration 30 s, early expiration enabled (count 2, S/N threshold 2)).

3.9. Raw Data Processing and Data Analysis

We make use of the MaxQuant software to process and analyze the raw data generated by the LTQ-Orbitrap-Velos mass spectrometer (13, 14). This software is freely available and can be downloaded at www.maxquant.org. Installation instructions and guidelines regarding the basic recommended settings during data processing can also be obtained through this Web site. Also available at www.maxquant.org are a suit of downstream data analysis tools embedded in the Maxquant software called Perseus. In addition, there is an online MaxQuant google group where practical questions regarding usage of the software are posted and answered.

When analyzing the raw data from the forward and reverse DNA pull-downs, it is important to specify the forward and reverse mass spectrometry runs in the "experimentalDesignTemplate.txt" file that is generated by MaxQuant. In the "experiment" column in the experimental design file simply name all the forward runs "forward" and all the reverse runs "reverse." Alternative names may also be used, but avoid numbers. Maxquant will now report separate protein ratios for the forward and the reverse pull-down. The Proteingroups.txt output table that MaxQuant generates contains all the basic information regarding identified proteins and their ratio in the forward and the reverse pull-down. This table should be filtered for contaminants and reverse hits. Furthermore, we recommend a minimal ratio count of 3 for each protein, both in the forward and the reverse pull-down. The ratios are then log2 transformed and eventually the ratios of all the proteins in the forward and reverse pull-down are plotted against each other in a two dimensional graph (see Fig. 1). In this graph the *x*-axis and *y*-axis represent the H/L ratio in the forward and the reverse experiment, respectively. Background proteins will cluster together at the origin of the graph, showing roughly a 1:1 ratio in both experiments. Proteins that specifically bind to the bait of interest cluster in the bottom right quadrant, whereas proteins that are repelled appear in the upper left quadrant. Proteins that are significant outliers from the background population can be deduced using boxplot statistics or by making use of the "significance B" value that can be calculated using the Perseus software.

4. Notes

1. The described protocol is optimized to identify proteins binding to a transcription factor binding site, a single nucleotide polymorphism (SNP) or to an epigenetic DNA modification such as cytosine methylation. Since the method makes use of synthetic oligonucleotides, the bait length is restricted to about 60 basepairs. In principle, the method can be adapted to identify specific interactions for any given DNA sequence such as enhancers or locus control regions. However, these sequences are generally longer and a single point mutation and/or a modification may not be sufficient to abolish all interactions with these elements. Therefore, designing the control DNA sequences is not straightforward. As a general guideline, the control sequence should be of the same length as the bait and should have roughly the same nucleotide composition. Palindromic sequences should be avoided. Note that sequences longer than 60 basepairs have to be cloned into a plasmid for amplification and digestion prior to biotinylation.
2. To study interactions with two DNA sequences (bait and control DNA), culturing of cells in light (K0R0) and heavy (K8R10) medium is required. Interactions with three different

DNA sequences can also be studied in a single experiment using a so-called triple pull-down. This requires growing cells in light (K0R0), medium (K4R6), and heavy (K8R10) medium. Nuclear extracts from these cells are incubated with the three different stretches of DNA and combined prior to mass spectrometry analysis. Each peptide identified in the mass spectrometer will now appear as a triplet and the ratio between the three peptide peaks will indicate the relative affinity of a protein for each of the three DNA sequences.

3. This SILAC medium can be used for most commonly used cell lines such as HeLa, MCF7 and HEK293T cells. However, certain cell lines, like mouse embryonic stem cells (ESC), are less easy to grow in medium containing dialyzed serum. This may depend on the batch of dialyzed serum and the conditions in which the cells are normally grown. The use of 2i compounds may overcome the necessity to culture in serum. A bottle of mouse ESC SILAC medium consists of the following components: 500 ml SILAC Dulbecco's Modified Eagle Medium without arginine, lysine, and glutamine, (90 ml dialyzed serum), 3.3 mM Glutamine, 100 U/ml Penicillin/Streptomycin, 1× non-essential amino acids (which contains proline, but no lysine or arginine), 1 mM sodium pyruvate, 73 μg/ml l-Lysine (light or heavy), and 29.4 μg/ml arginine (light or heavy), LIF (1,000 U/ml), 4.2 μl B-mercaptoethanol, and 2i inhibitors (CHIR99021 and PD0325901, 3 and 1 μM respectively). For cells growing in suspension, SILAC medium can be made on the basis of RPMI medium.

4. In some cell types, arginine to proline conversion may cause problems during SILAC labeling. Arginine10, when converted, results in a proline that is 6 Da heavier compared to normal proline. In addition to the "normal" heavy peak containing a labeled arginine or lysine on the C terminus, a third peak can now be observed for proline containing peptides. This peak contains a heavy proline and a heavy arginine or lysine. During quantification, this results in an underestimation of the peptide ratio. The extent of conversion can be minimized by titrating the amount of arginine and proline in the medium. Adding too much proline, however, may reduce arginine labeling, since conversion can take place in the other direction as well. Another solution is to make use of lysine-only labeling combined with Lys-C digestion instead of trypsin.

5. Different types of streptavidin-conjugated beads are available, for example Dynabeads M280; each of them may require optimization of the protocol (amount of DNA and nuclear extract, etc).

6. Different types of competitor DNA may influence the proteins that will be identified. When using CG-rich strands for the pull-down, it is recommended to use poly-dAdT as competitor.

7. A label check can be done by running a small amount of cell lysate on a gel and performing an in gel trypsin digestion and mass spectrometry analysis (on one gel slice) as described in Subheading 3.6. The incorporation of the heavy amino acids can be deduced from the observed peptide ratios in the mass spectrometer (for example a peptide ratio of 49 indicates 98% incorporation). When the incorporation of the heavy amino acids is 90%, the maximum observable ratio in a pull-down is 9, at 80% incorporation 4, etc. It is therefore recommended to strive for at least 95% incorporation of the heavy amino acids. Note that labeling never reaches 100% due to impurities in the "heavy" amino acids and small amounts of non-labeled amino acids in the culture medium.
8. After this step it is possible to leave the cells on ice for 30–60 min. Centrifuge the cells again before continuing with the protocol. When continuing without a break, the first next step that can be prolonged is the incubation of the nuclei in buffer C (step 16 of Subheading 3.2).
9. The increase of cell volume after incubation with buffer A is cell type-specific. HeLa cells increase their volume about two-fold, while HEK293T cells hardly swell.
10. Precool the dounce homogenizers on ice and rinse them with buffer A before usage. Clean and rinse the douncer with cold buffer A between different samples. Depending on the cell volume that is obtained after harvesting the cells, different douncer sizes should be used. For swollen cell pellet volumes up to 50 μl, use a 100 μl douncer. For volumes between 50 and 600 μl, use a 2 ml douncer and for volumes between 600 and 2.5 ml, use a 7 ml douncer. Douncers for even larger volumes are also available.
11. Keep the douncer on ice while douncing and do so in a slow steady rhythm. Wait for 45 s after every 10 strokes. Friction during the douncing results in an increase in temperature which may affect protein stability.
12. Both the T4 Polynucleotide kinase and the T4 DNA ligase exhibit 100% activity in 1× T4 DNA Ligase Buffer which contains 1 mM ATP. However, because the kinase uses most of the ATP in the buffer (step 7 in Subheading 3.4), when adding the T4 DNA ligase (step 8 in Subheading 3.4) make sure to add fresh ATP to the solution.
13. Commercial Sephadex G-50 columns are available: Illustra ProbeQuant G-50 micro columns (GE Healthcare, 28-9034-08) or Illustra NAP-10 columns (GE Healthcare, 17-0854-02). Use these columns according to the manufacturer's protocol. Alternatively, prepare your own: Take a 1 ml syringe without a needle, put a paper plug (tissue) in the bottom of the syringe

and put it into a 15 ml tube. Fill the syringe with Sephadex G-50 slurry, centrifuge for 1 min at 490×*g* and add more slurry. Repeat this step a couple of times until the column is filled with ~1 ml of beads. Wash the column twice with 0.5 ml of TE buffer and centrifuge for 1 min at 490×*g*. Put the column into a new 15 ml tube, add 100 μl of TE buffer to the DNA and load it onto the column. Centrifuge for 1 min at 490×*g* and measure the DNA concentration of the eluent (see Fig. 4).

14. Load 0.5 μg of input DNA and 17.5 μl of the supernatant on an agarose gel. Adjust the NaCl concentration of the input DNA to 1 M. High salt concentrations affect DNA migration and equalizing the salt concentration makes it easier to see the extent of DNA depletion from the solution.
15. Companies offering commercial stage tips include Proxeon and Millipore.

Acknowledgments

We thank members of the Vermeulen lab for critical reading of the manuscript. The Vermeulen lab is supported by grants from the Netherlands organization for scientific Research (NWO-VIDI), the Dutch Cancer Society (KWF), and the Cancer Genomics Centre (CGC-NGI).

References

1. Alexander RP, Fang G, Rozowsky J, Snyder M, Gerstein MB (2010) Annotating non-coding regions of the genome. Nat Rev Genet 11:559–571
2. Cox J, Mann M (2011) Quantitative, high-resolution proteomics for data-driven systems biology. Annu Rev Biochem 80:273–299
3. Beck M, Claassen M, Aebersold R (2011) Comprehensive proteomics. Curr Opin Biotechnol 22:3–8
4. Ong SE, Blagoev B, Kratchmarova I, Kristensen DB, Steen H, Pandey A, Mann M (2002) Stable isotope labeling by amino acids in cell culture, SILAC, as a simple and accurate approach to expression proteomics. Mol Cell Proteomics 1:376–386
5. Mittler G, Butter F, Mann M (2009) A SILAC-based DNA protein interaction screen that identifies candidate binding proteins to functional DNA elements. Genome Res 19: 284–293
6. Butter F, Kappei D, Buchholz F, Vermeulen M, Mann M (2010) A domesticated transposon mediates the effects of a single-nucleotide polymorphism responsible for enhanced muscle growth. EMBO Rep 11:305–311
7. Bartels SJ, Spruijt CG, Brinkman AB, Jansen PW, Vermeulen M, Stunnenberg HG (2011) A SILAC-based screen for Methyl-CpG binding proteins identifies RBP-J as a DNA methylation and sequence-specific binding protein. PLoS One 6:e25884
8. Spruijt CG, Bartels SJ, Brinkman AB, Tjeertes JV, Poser I, Stunnenberg HG, Vermeulen M (2010) CDK2AP1/DOC-1 is a bona fide subunit of the Mi-2/NuRD complex. Mol Biosyst 6:1700–1706
9. Bartke T, Vermeulen M, Xhemalce B, Robson SC, Mann M, Kouzarides T (2010) Nucleosome-interacting proteins regulated by DNA and histone methylation. Cell 143:470–484
10. Dignam JD, Lebovitz RM, Roeder RG (1983) Accurate transcription initiation by RNA polymerase II in a soluble extract from isolated mammalian nuclei. Nucleic Acids Res 11: 1475–1489

11. Shevchenko A, Tomas H, Havlis J, Olsen JV, Mann M (2006) In-gel digestion for mass spectrometric characterization of proteins and proteomes. Nat Protoc 1: 2856–2860
12. Rappsilber J, Mann M, Ishihama Y (2007) Protocol for micro-purification, enrichment, pre-fractionation and storage of peptides for proteomics using StageTips. Nat Protoc 2:1896–1906
13. Cox J, Mann M (2008) MaxQuant enables high peptide identification rates, individualized p.p.b.-range mass accuracies and proteome-wide protein quantification. Nat Biotechnol 26:1367–1372
14. Cox J, Matic I, Hilger M, Nagaraj N, Selbach M, Olsen JV, Mann M (2009) A practical guide to the MaxQuant computational platform for SILAC-based quantitative proteomics. Nat Protoc 4:698–705

Chapter 12

Electrophoretic Mobility-Shift and Super-Shift Assays for Studies and Characterization of Protein–DNA Complexes

Elsie I. Parés-Matos

Abstract

Gene expression is in part regulated by transcription factors that bind specific sequence motifs in genomic DNA. Transcription factors cooperate with the basal machinery to upregulate or downregulate transcription. Experimental data have revealed the importance of interactions among members of distinct families of transcription factors to form complexes that regulate gene expression. Thus, a full characterization of protein–DNA complexes is essential to understanding of gene regulation in a more complex cellular environment. Electrophoretic mobility shift assay (EMSA) is a powerful technique to resolve nucleic acid–protein complexes formed with transcription factors in nuclear extracts. Herein is described how EMSA and super-shift assays were used to characterize several complexes produced from binding of transcription factors to a regulatory DNA sequence upstream from the promoter region of the human NF-IL6 gene.

Key words: Electrophoretic mobility shift assay, Super shift assay, DNA–protein interaction, Nuclear extract, Promoter of NF-IL6 gene

1. Introduction

Numerous proteins control transcription of genes in human chromosomes. These proteins can be broadly divided into three groups: proteins that compact the DNA, transcription factors that bind to specific DNA sequences known as control elements, and proteins that function through protein–protein interactions. All these three groups act in concert to regulate various aspects of gene expression (1–3). The DNA binding domain of transcription factors interacts with specific sequences in regulatory modules, in control region of genes, to activate or repress transcription (4, 5). Electrophoretic mobility-shift assay (EMSA) is a rapid and sensitive method to characterize protein–DNA interactions (6–12). In this method, a labeled probe (usually DNA or RNA) is incubated with nuclear extracts. Subsequently, nucleic acid–protein complexes are resolved

Minou Bina (ed.), *Gene Regulation: Methods and Protocols*, Methods in Molecular Biology, vol. 977,
DOI 10.1007/978-1-62703-284-1_12, © Springer Science+Business Media, LLC 2013

in a non-denatured polyacrylamide gel (13, 14). These complexes migrate slower than protein-free DNA. Competitor DNA fragments (e.g., cold probe or mutated fragments) are often used to distinguish specific from nonspecific bands, and to identify sequences responsible for DNA–protein interactions. Protein components in the complexes can be identified by adding antibody to a specific protein, to EMSA reactions. Antibody either blocks complex formation or form an antibody–protein–DNA complex, resulting in further reduction in the mobility (a super-shift) in electrophoresis experiments (11, 15–17).

2. Materials

All solutions should be prepared with ultrapure water (by distilling deionized water to attain a sensitivity of 18 MΩ cm at 25°C). The solutions should be stored at 4°C (unless indicated otherwise). All biotechnology grade chemicals were obtained from Sigma Chemical Company, except where noted.

2.1. Nuclear Extracts

1. Buffer A: 10 mM HEPES, 10 mM KCl, 0.1 mM EDTA, 0.1 mM EGTA, 1 mM DTT (the pH of the solution should be about 7.9–8.0).
2. 10% (v/v) Nonidet NP-40 Stock made by dilution of Nonidet NP-40 with water.
3. Leupeptin: dissolved in water at 10 mM. Aliquots are stored at −20°C.
4. Aprotinin: dissolved in water at 1 μg/mL. Aliquots are stored at −20°C.
5. Pepstatin: dissolved in methanol at 1 mg/mL.
6. Phenylmethylsulfonyl fluoride (PMSF): dissolved in absolute ethanol at 100 mM.
7. Buffer C (pH 7.9): 20 mM HEPES, 0.4 M NaCl, 1 mM EDTA, 1 mM EGTA, 1 mM DTT (17).
8. Pierce BCA Protein Assay Kit (Pierce Biotechnology).

2.2. For Isolation of DNA Fragments

1. 30 % (w/v) Acrylamide–Bis Stock: dissolved in water at a 29:1 Acrylamide–Bis ratio.
2. 1× TBE: 89 mM Tris Base, 89 mM boric acid, 2 mM EDTA (the pH of the solution should be about 8.0).
3. Ethidium bromide: dissolved in water at 5 mg/mL.
4. Ammonium acetate: dissolved in water at 4 M.
5. Razor blade.
6. Plastic wrap.

7. Siliconized glass wool.
8. Pipette tip, 1.0-mL.
9. Ethanol (200 proof).
10. TE buffer: 10 mM Tris–HCl, pH 7.4–8.0, 1 mM EDTA.

2.3. For ^{32}P-Labeling DNA Fragments

1. T4 polynucleotide kinase (New England Biolabs). Stored at –20°C.
2. γ-^{32}P ATP: 6,000 Ci/mmol, 10 mCi/mL (GE Healthcare Biosciences).
3. EDTA: dissolved in water at 20 mM.
4. Elutip-d purification minicolumns (Schleicher and Schuell, Midwest Scientific).
5. Low salt buffer: 20 mM Tris–HCl, pH 7.4, 1.0 mM EDTA, 20 mM NaCl.
6. High salt buffer: 20 mM Tris–HCl, pH 7.4, 1.0 mM EDTA, 1 M NaCl.
7. tRNA: dissolved in water at 10 μg/μL. Aliquots are stored at –20°C.
8. Ethanol (200 proof).
9. TE buffer: 10 mM Tris–HCl, pH 7.4–8.0, 1 mM EDTA.
10. Scintillation fluid.

2.4. Non-denaturing Polyacrylamide Gel

1. 30% (w/v) Acrylamide–Bis Stock: dissolve Acrylamide–Bis (at a 29:1 ratio) in water.
2. 5× TGE: 125 mM Tris-Base, 615 mM glycine, 5 mM EDTA (adjust the pH to 8.3).
3. 1× TGE: made by dilution of 5× TGE in water and supplemented with 2.5% (v/v) glycerol.
4. Ammonium persulfate (APS): dissolved in water at 10% (w/v).
5. *N*,*N*,*N*,*N*′-tetramethylethylenediamine (TEMED).

2.5. For Protein Binding Reactions

1. 5× Binding buffer: 15 mM HEPES, pH 7.9, 80 mM NaCl, 1 mM EDTA, 1 mM DTT, 0.3 mg/mL BSA, 10% (v/v) glycerol.
2. Poly dI-dC (Amersham Pharmacia Biotech): dissolved in water at 2 μg/μL. Aliquots were stored at –20°C.

2.6. For Electrophoretic Mobility Shift and Supershift Assays

1. Protein specific antibody (Santa Cruz Biotechnology).
2. Whatman filter paper.
3. Plastic wrap.
4. Gel dryer and vacuum pump.
5. Kodak X-OMAT AR film.
6. Kodak Biomax MS intensifying screen.
7. Autoradiogram exposure cassette.

3. Methods

3.1. Preparation of Nuclear Extracts

1. Pellet cells by centrifugation, at 1,500×*g* for 5 min. Discard the supernatant.
2. Suspend the pellet in 500 μL of ice-cold buffer A. Subsequently add 2 μg/mL leupeptin, 0.03 TiU/mL aprotinin, 1 μg/mL pepstatin, and 1 mM PMSF (see Note 1).
3. Allow cells to swell on ice for 20 min.
4. Add 50 μL of Nonidet NP-40 solution and vortex vigorously for 10 s.
5. Centrifuge the homogenate in a microfuge at top speed for 30 min at 4°C.
6. Discard the supernatant.
7. Suspend the nuclear pellet in 200 μL of ice-cold buffer C and subsequently add 2 μg/mL leupeptin, 0.03 TiU/mL aprotinin, 1 μg/mL pepstatin, and 3 mM PMSF (see Note 1).
8. Mince the nuclear pellet slightly with a notched pipette tip to ensure complete elution of nuclear proteins from the nuclei (see Note 2).
9. Rock the tube gently on a shaking platform for 15 min at 4°C.
10. Centrifuge the suspension in a microfuge, at top speed for 5 min at 4°C.
11. Collect the supernatant.
12. Determine protein concentration by the BCA protein assay (see Note 3).
13. Aliquot the supernatant (ca. 40 μL) and store at −70°C.

3.2. Purification of Double-Stranded DNA Fragments

1. Digest approximately 250 mg of a recombinant plasmid either with one or two different restriction enzymes, according to manufacturer's instructions.
2. Heat inactivate for 20 min at 65°C.
3. Add 0.1 volume of glycerin to the digestion mixture.
4. Separate the DNA fragments through a non-denaturing polyacrylamide gel containing 1× TBE (see Note 4).
5. Soak the gel in 2 mg/L ethidium bromide solution for 30 min.
6. Wash briefly and transfer the gel to a piece of plastic wrap.
7. Detect ethidium-bromide-stained DNA fragments using a UV transilluminator.
8. Mark on the plastic wrap the position of the band of interest with a Sharpie pen.

9. Cut the desired band with a new razor blade and transfer the gel piece to a 2.0-mL Eppendorf tube.
10. Soak the gel piece in the ammonium acetate solution and incubate for 24–48 h at 37°C with vigorous shaking.
11. Load the ammonium acetate solution (that now contains the DNA fragment) onto a pipette tip containing a small amount of siliconized glass wool. Collect the filtrate into a clean 2.0-mL Eppendorf tube (see Note 5).
12. Add 2.5 volumes of ethanol and incubate for 1 h at −20°C.
13. Centrifuge in a microfuge at top speed for 15 min at 4°C.
14. Speed-vacuum to dry the pellet for 25 min. Subsequently, dissolve the pellet in 35 μL of TE buffer.
15. Determine the concentration of the DNA fragment solution by UV analysis and by agarose gel electrophoresis.

3.3. ^{32}P-Labeling of Purified Double-Stranded DNA Fragment

1. Prepare a 20 μL labeling mixture containing 50–100 ng of a purified double-stranded DNA fragment, 2 μL of 10× T4 polynucleotide kinase buffer, 50 μCi of γ-^{32}P ATP, and 10 units of T4 polynucleotide kinase in a 0.5-mL Eppendorf tube (see Note 6).
2. Incubate the mixture for 2 h at 37°C (see Note 7).
3. Quench the reaction with 3 μL of EDTA solution and incubate further for 5 min.
4. Purify the ^{32}P-labeled DNA fragment by the Elutip-d Purification Minicolumn procedure (see Note 8).
5. Precipitate the ^{32}P labeled DNA fragment with 1 μL of tRNA solution and 2 volumes of ethanol for 16 h at −20°C.
6. Microfuge the pellet at top speed for 20 min at 4°C.
7. Wash the pellet twice with 70 % ethanol and speed-vacuum dry for 20 min at room temperature.
8. Dissolved the ^{32}P labeled DNA fragment in 100 μL of TE buffer.
9. Determine the efficiency of the ^{32}P-labeling by reading the counts per minute (cpm) in a Liquid Scintillation counter: add 1 μL of the ^{32}P-labeled DNA fragment to 3.0 mL of the scintillation fluid.

3.4. Preparation of 5 % Non-denaturing Polyacrylamide Gel

3.4.1. Plugging Gel

1. Mix 2.4 mL of 30 % (w/v) Acrylamide-Bis Stock, 0.6 mL of 5× TGE, 20 μL of APS, and 10 μL of TEMED in a 10-mL Erlenmeyer flask.
2. Pour immediately the polymerizing mixture into a siliconed glass plate assembly, avoiding formation of air bubbles (see Note 9).
3. Allow the gel to polymerize for 30 min in a vertical position.

3.4.2. Main Gel

1. Mix 5.0 mL of 30 % (w/v) Acrylamide-Bis Stock, 6.0 mL of 5× TGE, 19.0 mL of water, and 750 μL of glycerol in a 125-mL suction Erlenmeyer flask. Degas the mixture for 45 min (see Note 10).
2. Add 200 μL of APS followed by 25 μL of TEMED and swirl the mixture gently.
3. Pour immediately the polymerizing mixture on top of the plugging gel and insert the well-forming comb (see Note 11).
4. Allow the gel to polymerize in a vertical position (see Note 12).

3.5. Electrophoretic Mobility Shift and Super Shift Assays

1. Prepare a 20 μL mixture containing 4 μL of 5× binding buffer, 10 μg of nuclear extract, 0.5 μL of poly dI-dC, and about 30,000 cpm of ^{32}P-labeled DNA fragment in a 0.5-mL Eppendorf tube (see Note 13).
2. Incubate the binding mixture on ice for 30 min.
3. For super-shift assays, add 1–2 μL of an antibody to the pre-incubated mixture and incubate the solution for an additional 30 min, on ice.
4. Load 18 μL aliquot of the binding mixture onto each well of the pre-equilibrated gel (see Note 14).
5. Electrophorese at a constant voltage of 300 V for 3.5 h at 4°C.
6. Vacuum-dry the gel for 30 min at 75°C (see Note 15).
7. Make a record of resolved bands by exposing the dried gel to Kodak X-OMAT AR film (see Note 16); Fig. 1 shows an example of results of an electrophoretic mobility-shift and super-shift assays.

4. Notes

1. Protein inhibitors should be added to cell suspensions in the same order as described. Keep samples in ice at all times.
2. Nuclear proteins can be easily degraded by temperature fluctuations. For better results, perform steps 8–10 in a cold room.
3. A typical range of protein concentration is 2–6 μg/μL.
4. Percentage of polyacrylamide gel depends on fragment size (18). All gels were electrophoresed at 250 V (or 20 mA), using DNA loading dye as a marker and 1× TBE as the running buffer. Electrophoresis should be continued until the bromophenol blue traveled a distance of three-quarters from the starting position.
5. Siliconization involves placing a thin layer of dimethyldichlorosilane onto glass surfaces to make them extremely hydrophobic.

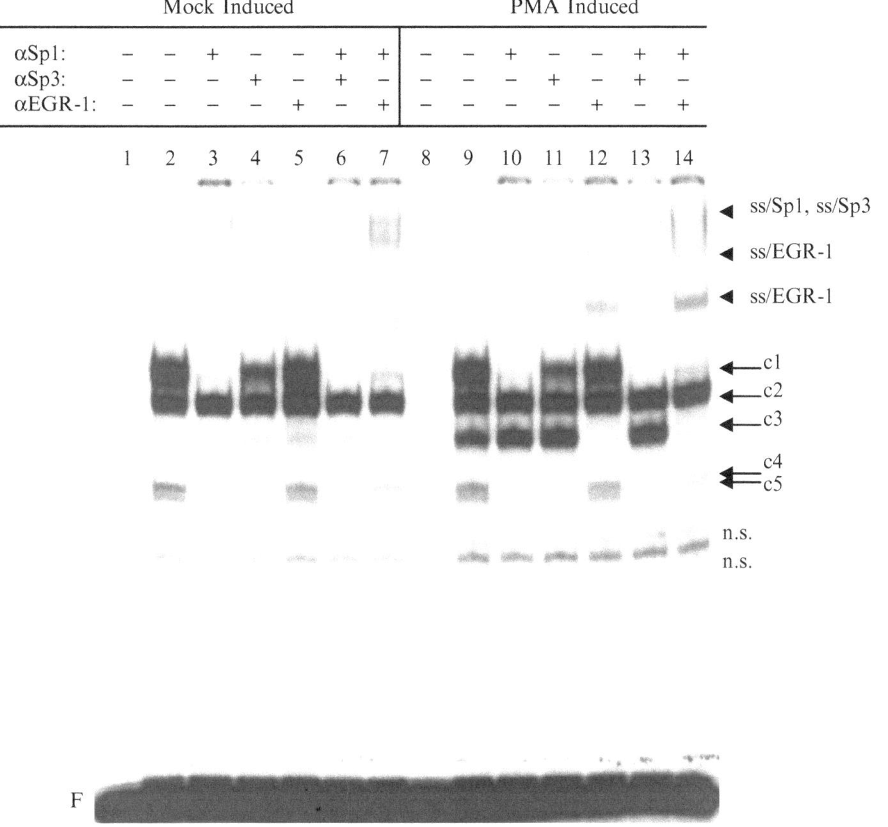

Fig. 1. Electrophoretic mobility and super shift assays. Nuclear extracts prepared from U937 cells, mock induced (lanes 2–7) and PMA induced (lanes 9–14), were incubated with a probe containing 5′GAAAGTGCCTGTTCCCTCCCCGCCCCCGC CTGGCTCCTGGC3′ and its complement. This sequence was derived from an upstream regulatory region (−3626/−3588) in the human NF-IL6 gene (19). In EMSAs, free and protein-bound probe were separated in a 5 % non-denaturing polyacrylamide gel. After electrophoresis, the gel was dried and autoradiographed. Lanes 1 and 8 have no nuclear extract. Sp1 antibody (αSp1), Sp3 antibody (αSp3) and EGR-1 antibody (αEGR-1) were added to the binding mixture either alone (lanes 3–5 or lanes 10–12, respectively) or by combination of two (lanes 6–7 or lanes 13–14) after 30 min of incubation. Three major DNA–protein complexes (c1, c2, c3) and two minor DNA–protein complexes (c4, c5) were seen and are indicated by *solid arrows*. Three *arrowheads* indicate the position of supershifted complexes; *n.s.* nonspecific band, *F* free probe.

The glass wool is first soaked in dimethyldichlorosilane for 30 min. Then, the siliconization reagent is removed with a Pasteur pipette and the glass wool is allowed to air-dry. This filtration step is necessary to remove any remaining small gel particles out of the DNA solution.

6. Work with caution, γ-^{32}P ATP has a high emission energy. Wear gloves, and use eye and body protection while handling the material. Follow the safety and administrative rules provided by the NRC, OSHA, or any regulatory agency from your institution.
7. A temperature-controlled water bath is strongly recommended.

8. Unincorporated nucleotide bases and α-^{32}P dATP, as well as the enzyme, were washed out from the Elutip-d minicolumn with the low salt buffer. The ^{32}P-labeled DNA fragment was eluted with the high-salt buffer.
9. Ensure that gel plates, spacers and comb are thoroughly cleaned. Remove any fingerprints or other residues with a 70 % ethanol solution and wipe out any remaining with a large Kimwipe.
10. This recipe includes the amount of reagents required for one gel with dimensions $17 \times 0.1 \times 18$ cm (nominal volume 30.6 mL).
11. To avoid air bubbles, tilt the gel-plate assembly 45° from the vertical during pouring. Air bubbles disrupt electrophoretic migration.
12. Since polymerizing mixture contains glycerol, it may take 15–30 min longer to polymerize than a conventional gel does.
13. Amount of water added to binding mixture depends on actual concentration of nuclear extracts and the efficiency of ^{32}P labeling of the DNA fragment. As a control, a reaction mixture without nuclear extract was also prepared. Addition of poly dI-dC minimizes nonspecific binding (20). The binding mixture must be kept in ice at all times.
14. A gel pre-run is strongly recommended. The casted gel is mounted with clamps on a vertical electrophoresis apparatus. Wells are gently washed with the running buffer (1× TGE) and, then, loaded with 5 μL of 1× gel loading buffer. The gel is pre-equilibrated at a constant voltage of 300 V for 15 min, in a cold room at 4°C.
15. At the completion of electrophoresis, glass plates are carefully pulled apart while avoiding stretching the gel. A piece of whatman filter paper is laid on top of the gel, detached from the glass plate, and then, transferred to a vacuum-dryer with the gel facing up. A plastic wrap is used to cover and protect the gel from contaminants or to avoid any damage that may come from the plastic cover of the gel dryer. The gel typically dries within 30 min, but sometimes requires longer depending on gel thickness. Once the gel is dried, attach the periphery of plastic wrap to the back of the filter paper with a clear tape. At this stage, it is recommended to keep the gel away from unclean surfaces; otherwise, you will find dark spots in developed X-ray films.
16. Radioactivity levels of about 5–10 counts/s usually require an overnight (16 h) exposure at −70°C using a Kodak Biomass MS intensifying screen. Extended exposure times may also be needed depending on the amount of the targeted nuclear protein in nuclear extracts and stability of oligonucleotides after the ^{32}P-labeling step.

Acknowledgment

This work was supported by Patricia Roberts Harris Fellowship, NIH Trainee Research and fellowships from the Department of Chemistry, Purdue University, IN (USA).

References

1. Struhl K (1999) Fundamentally different logic of gene regulation in eukaryotes and prokaryotes. Cell 98:1–4
2. Emerson BM (2002) Specificity of gene regulation. Cell 109:267–270
3. Fowler T, Sen R, Roy AL (2011) Regulation of primary response genes. Mol Cell 44:348–360
4. Villard J (2004) Transcription regulation and human diseases. Swiss Med Wkly 134:571–579
5. Stower H (2012) Gene regulation: resolving transcription factor binding. Nat Rev Genet 13. doi:10.1038/nrg3153
6. Gagnon KT, Maxwell ES (2011) Electrophoretic mobility shift assay for characterizing RNA-protein interaction. Methods Mol Biol 703:275–291
7. Lin JJ, Grosskurth SE, Harlan SM, Gustafson-Wagner EA, Wang Q (2007) Characterization of cis-regulatory elements and transcription factor binding: gel mobility shift assay. Methods Mol Biol 366:183–201
8. Gaudreault M, Gingras M-E, Lessard M, Leclerc S, Guérin SL (2009) Electrophoretic mobility shift assays for the analysis of DNA-protein interactions. Methods Mol Biol 543:15–35
9. Hellman LM, Fried MG (2007) Electrophoretic mobility shift assay (EMSA) for detecting protein-nucleic acid interactions. Nat Protoc 2:1849–1861
10. Hestekin CN, Barron AE (2006) The potential of electrophoretic mobility shift assays for clinical mutation detection. Electrophoresis 27:3805–3815
11. Buratowski S, Chodoshy LA (2001) Mobility shift DNA-binding assay using gel electrophoresis. Curr Protoc Mol Biol Chapter 12: Unit 12.2
12. Varshavsky A (1987) Electrophoretic assay for DNA-binding proteins. Methods Enzymol 151:551–565
13. Sidorova N, Hung S, Rau DC (2010) Stabilizing labile DNA-protein complexes in polyacrylamide gels. Electrophoresis 31:648–653
14. Holden NS, Tacon CE (2011) Principles and problems of the electrophoretic mobility shift assay. J Pharmacol Toxicol Methods 63:7–14
15. Parés-Matos EI, Milligan JS, Bina M (2006) Exploring transcription factor binding properties of several non-coding DNA sequence elements in the human NF-IL6 gene. J Mol Biol 357:732–747
16. Chan RJ, You M, Feng GS (2004) Identification of trans-acting factors by electrophoretic mobility shift assay. Methods Mol Biol 249:7–20
17. Doyle K, Zhang Y, Baer R, Bina M (1994) Distinguishable patterns of protein-DNA interactions involving complexes of basic helix-loop-helix proteins. J Biol Chem 269: 12099–120105
18. Sambrook J, Fritsch EF, Maniatis T (1989) Polyacrylamide gel electrophoresis. In: Molecular cloning: a laboratory manual, 2nd edn. Cold Spring Harbor Laboratory, New York, pp 6. 36–6.48
19. Yang Y, Pares-Matos EI, Tesmer VM, Dai C, Ashworth S, Huai J, Bina M (2002) Organization of the promoter region of the human NF-IL6 gene. Biochim Biophys Acta 1577:102–108
20. Schreiber E, Matthias P, Muller MM, Schaffner W (1989) Rapid detection of octamer binding proteins with 'mini-extracts', prepared from a small number of cells. Nucleic Acids Res 17:6419

Chapter 13

Combination of Native and Denaturing PAGE for the Detection of Protein Binding Regions in Long Fragments of Genomic DNA

Kristel Kaer and Mart Speek

Abstract

In traditional electrophoresis mobility shift assay (EMSA) a single ^{32}P-labeled double-stranded DNA oligonucleotide or a restriction fragment bound to a protein is separated from the unbound DNA by polyacrylamide gel electrophoresis (PAGE) under nondenaturing conditions. An extension of this method uses a population of DNA restriction fragments derived from long genomic regions for the identification of fragments containing protein binding regions. Although the method allows simultaneous analysis of large fragments, it is relatively laborious and can be used to detect only fragments containing high affinity protein binding sites. Here we describe an alternative and straightforward strategy which is based on a combination of native and denaturing PAGE. With this strategy restriction fragments, derived from genomic DNA (<10 kb), containing high as well as low affinity protein binding regions may be easily identified.

Key words: EMSA, Protein binding site, DNA–protein interaction

1. Introduction

Electrophoretic mobility shift assay (EMSA), developed by Fried and Crothers (1), and Garner and Revzin (2), is a popular method for detection of protein–DNA interactions (3). It is highly sensitive and may be used to obtain qualitative as well as quantitative information about protein binding to various DNA molecules (4–6). In traditional EMSA, a DNA oligonucleotide or a restriction fragment, generally within the size range of 20–400 bp (7), is radiolabeled and incubated with purified protein or a mixture of proteins (nuclear or whole cell extract). Nucleoprotein complexes are separated from naked DNA by polyacrylamide gel electrophoresis (PAGE) under native conditions. Because of the "caging" effect of gel matrix (8, 9), DNA–protein interactions are stabilized; the corresponding shifted complexes are resolved as discrete bands.

Minou Bina (ed.), *Gene Regulation: Methods and Protocols*, Methods in Molecular Biology, vol. 977,
DOI 10.1007/978-1-62703-284-1_13, © Springer Science+Business Media, LLC 2013

Although in some cases, complexes may dissociate and thus do not produce detectable shifted bands.

Previously, two similar high-throughput methods were developed for identification of protein binding regions using a large population of fragments derived from DNAs (plasmids, bacteriophages, bacterial chromosome, and human genome fragment) ranging in size from 3 to 4,700 kb (10, 11). These methods are relatively laborious and require large number of procedures (including initial screening by two-dimensional PAGE, linker addition, PCR amplification, cloning, and sequencing), which could complicate the fragment identification.

Here we describe an alternative and straightforward strategy which is based on the combination of EMSA selection step with subsequent denaturing PAGE for identification of protein binding regions in long (up to10 kb) fragments of genomic DNA (12). Our strategy consists of the following steps: digestion of genomic DNA with a four-cutter restriction enzyme (*Alu*I, *Bsu*RI, *Tru1*I, etc.), separation of low (L) and high (H) molecular weight fractions of resultant DNA fragments, ^{32}P-labeling with Klenow fragment, traditional EMSA, gel elution and identification of the shifted bands (or zones) by denaturing PAGE. Identification of DNA fragments containing protein binding sites is carried out by running gel-eluted fragments alongside with the full "spectrum" of initial restriction fragments of known size. With this strategy, unique protein bound fragments, giving rise to shifted bands, can be eluted and identified. Moreover, DNA fragments that dissociate from the complexes during electrophoresis can also be identified.

2. Materials

2.1. Buffers and Solutions

All buffers were made from sterilized stock solutions using deionized water. CL, 50× BAAL, WCE, 10× K, dNTPs were stored at −20°C. All other solutions, unless indicated otherwise, are kept at room temperature or according to manufacturer instructions. 50× BAAL, 10× TBE, 5× SBa, and 10× K are used at final concentrations of 1×.

1. Cell washing buffer, Phosphate Buffered Saline (PBS): 10 mM Na_2HPO_4, 2 mM KH_2PO_4, pH 7.4, 137 mM NaCl, and 2.7 mM KCl.
2. Cell lysis buffer (CL): 10 mM Tris–HCl, pH 8.0, 10 mM NaCl, 1 mM EDTA-Na_2, 10 mM DTT, 10% glycerol, 0.5% NP-40 supplemented with 1 mM PMSF and 1× BAAL. Protease inhibitors (PMSF and BAAL) and DTT are added just prior to use.

3. Protease inhibitor cocktail (50× BAAL): 0.5 M benzamidine, 0.5 mg/ml antipain, 100 μg/ml aprotinin, and 0.5 mg/ml leupeptin.
4. Whole cell or nuclear protein extraction buffer (WCE): 20 mM Hepes–KOH, pH 7.9, 400 mM KCl, 1 mM EDTA-Na_2, 10 mM DTT, 10% glycerol, 1 mM PMSF, and 1× BAAL.
5. Klenow reaction buffer (10× K): 500 mM Tris–HCl, pH 8.0, 50 mM $MgCl_2$, and 100 mM DTT.
6. dNTPs: 10 mM of each dNTP.
7. (α-^{32}P)dCTP 3,000 Ci/mmol, 10 μCi/μl.
8. TE: 10 mM Tris–HCl, pH 7.5, 1 mM EDTA-Na_2.
9. Electrophoresis buffer 10× TBE: 500 mM Tris–borate, pH 8.3, 10 mM EDTA-Na_2.
10. Sample loading buffer (5× SBa) for agarose gels: 5× TBE, 20% glycerol, and 0.03% bromophenol blue.
11. Sample loading buffer (1× SBd) for denaturing polyacrylamide gels: 90% ultrapure formamide, 10 mM EDTA-Na_2, 0.01% bromophenol blue, and 0.01% xylene cyanol.
12. 3 M sodium acetate, pH 5.5 (NaOAc) and 5 M ammonium acetate (NH_4OAc).
13. Poly (dI-dC) 2 mg/ml (see Note 1).
14. Liquid phenol (see Note 2).
15. 20% (v/v) glycerol.
16. Polyacrylamide gel components: Acrylamide (AA, >99%), *N*,*N*′-methylene bisacrylamide (BAA, >98%), *N*,*N*,*N*′,*N*′-tetramethyl-ethane-1,2-diamine (TEMED), 20% ammonium persulfate (APS) (see Note 3). Stock solutions containing 40% AA (AA–BAA/29:1) for EMSA gels and 5% (AA–BAA/25:1) denaturing gel containing 8 M Urea. Both solutions are kept at 4 °C. For gel preparation see Note 4.
17. Regular and low gelling temperature agarose (see Note 5).
18. Ethidium bromide (EtBr) 10 mg/ml.
19. 100 bp DNA marker.
20. Klenow fragment of DNA Polymerase I 2 U/μl.
21. Carrier tRNA 2 μg/μl.
22. 70 and 95% ethanol.

2.2. Gel Electrophoresis and Detection Systems

1. Agarose gel electrophoresis system. This system consists of a comb, a removable tray and a gel caster for the preparation of agarose gel, and a submarine type electrophoresis tank connected to a power supply (output range 10–300 V) for running gels.
2. Ultraviolet light transilluminator and image capture system for recording DNA in agarose gels stained with ethidium bromide.

3. Polyacrylamide gel electrophoresis system. This system consists of combs, glass plates with dimensions 14×20×0.3 cm for EMSA and 14×40×0.3 cm for high resolution denaturing polyacrylamide gels, vertical electrophoresis apparatus (preferably with adjustable anode and cathode chambers to fit 20 and 40 cm long gels) and a high voltage power supply (output range 0.1–2 kV) for running long and thin (<1 mm) gels.
4. Phosphoimager system. This system is required for recording images of ^{32}P-labeled DNAs in polyacrylamide gels. Alternatively, gels containing radioactive DNA images can be exposed to X-ray films (traditional autoradiography).
5. Nanodrop spectrophotometer for determination of DNA concentration in small volumes.
6. Gel drier for drying short and long polyacrylamide gel.

3. Methods

We have developed a strategy for the detection of protein binding regions in long (<10 kb) fragments of genomic DNA (Fig. 1) (12). With this strategy restriction fragments, containing transcription factor binding sites, derived from promoter region of any gene may be identified by combination of EMSA and denaturing PAGE. The following is a detailed description of the methods used in achieving this goal.

3.1. Preparation of Whole Cell and Nuclear Protein Extracts

1. Wash monolayer cells twice with 10 ml 1× PBS at room temperature.
2. For whole cell extract preparation, add 1 ml 1× PBS per dish (diameter 8 cm) and scrape cells with rubber policeman. For nuclear extract preparation, lyse cells directly on plate by adding 1 ml ice-cold CL buffer.
3. Transfer cell suspension or nuclei to a 1.5 ml tube and centrifuge at 600×*g* for 4 min at 4 °C.
4. Remove supernatant, estimate the volume of pellet and suspend it in 3–5 volumes of ice-cold WCE. Keep on ice.
5. Snap-freeze cells or nuclei in liquid nitrogen and thaw on ice.
6. Mix and repeat freeze–thaw cycle three times.
7. Pellet nuclear or cellular debris by centrifugation at 13,000×*g* for 5 min at 4 °C.
8. Remove supernatant (containing protein extract) and aliquot it into 50–100 μl amounts.
9. Snap-freeze and store at −80 °C.

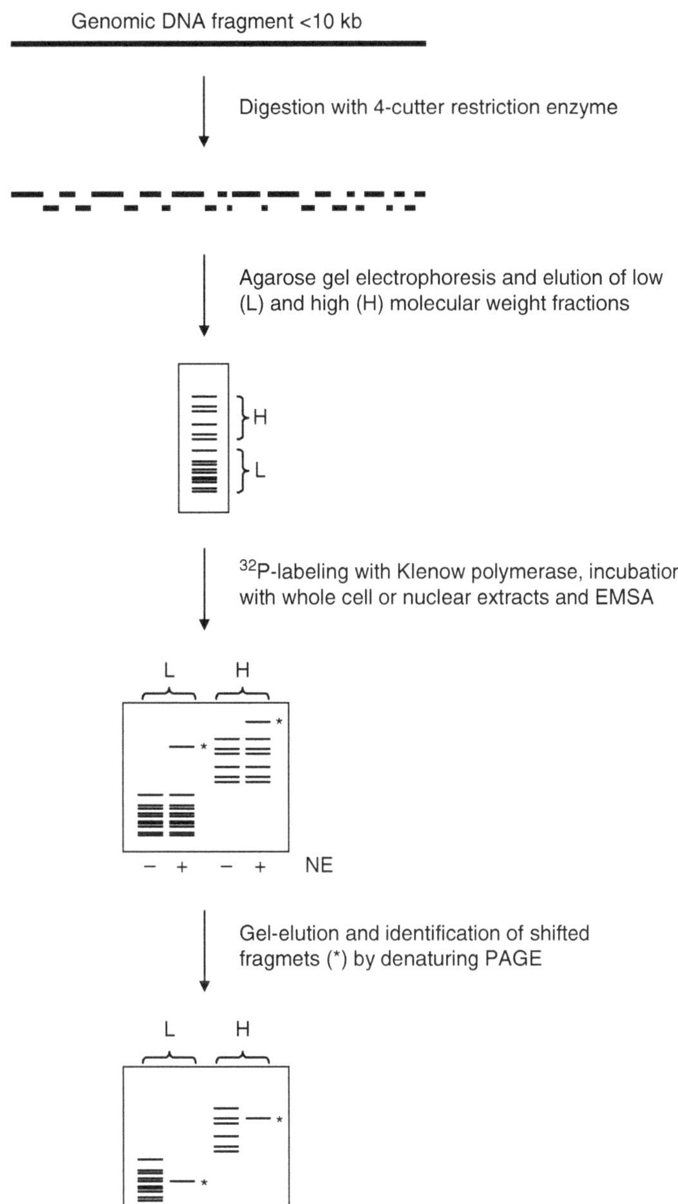

Fig. 1. A strategy for the detection of protein binding regions in genomic DNA fragments. (reproduced from ref. 12 with permission from BioMed Central Ltd).

3.2. Digestion of Genomic DNA

1. First, use computer program (DNAMAN, Gene Inspector or similar) to carry out restriction analysis of the cloned genomic DNA fragment containing putative binding sites of proteins (see Note 6).
2. Digest about 2 μg genomic DNA in 20 μl volume by incubating it with a given restriction enzyme under conditions recommended by manufacturer.

3.3. Separation, Size Selection, and Isolation of Genomic Fragments

1. Prepare 1% low gelling or regular agarose 1× TBE gel containing EtBr at final concentration 0.1 μg/ml (see Note 5).
2. Stop genomic DNA digestion by adding 5 μl 5× Sba.
3. Load each digest onto agarose gel. Also load 100 bp DNA marker for size estimation of DNA fragments.
4. Carry out electrophoresis (10 V/cm).
5. Stop electrophoresis after bromophenol blue has migrated about 5 cm from the start.
6. Using DNA marker as a reference, locate regions containing L and H molecular weight fractions with approximate sizes of fragments 30–300 bp and 300–700 bp by using long ultraviolet light lamp of transilluminator (see Note 7).
7. Use scalpel to cut out gel slices containing DNA fragments of interest.
8. Cut slices into 2–3 smaller pieces to facilitate diffusion and place them into 1.5 ml tubes.
9. Centrifuge tubes to collect gel slices to the bottom of tubes.
10. Estimate the volume of gel slices by tube graduation and then add about 3 volumes of TE.
11. Mix and incubate at 37 °C for 8–12 h.
12. Transfer TE containing DNA fragments into a new tube.
13. Add 1/20 volume of 3 M NaOAc and 3 volumes of 95% ethanol to precipitate DNA.
14. Centrifuge at max speed for 5 min, remove supernatant, and dissolve the DNA pellet in 10–20 μl TE.
15. Measure the DNA concentration (see Note 8).

For representative example of this section see Fig. 2.

3.4. ^{32}P End-Labeling of DNA Fragments

Perform 3′ end-labeling of blunt-ended DNA fragments by 3′-5′ exonuclease and 5′-3′ polymerase activities of the Klenow fragment of DNA polymerase I according to the following protocol (see Note 9).

1. Mix the following components:

   200–400 ng (2–4 μl) of DNA fraction

   1 μl 10× K buffer

   20 μCi of (α-^{32}P)dCTP

   H_2O to make a final volume of 10 μl

   1 U of Klenow fragment
2. Incubate at 37 °C for 15 min.
3. Add 1 μl of 1 mM diluted unlabeled dNTP solution and 1 U of Klenow fragment.

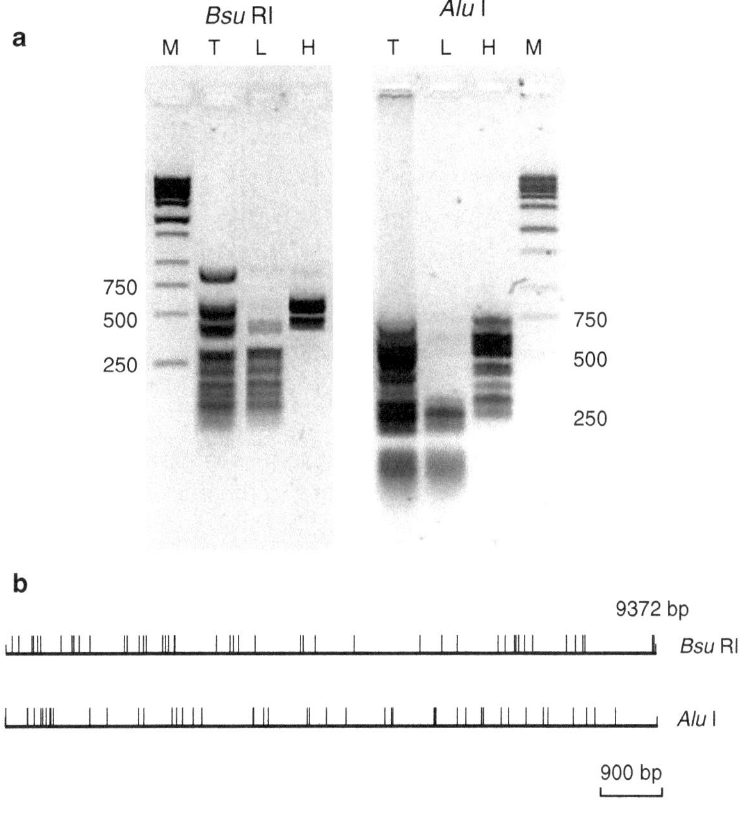

Fig. 2. Digestion of a 9,372 bp genomic DNA fragment of the rat p75NTR gene (cloned in pBS KS vector) with *Bsu*RI and *Alu*I restriction enzymes. (**a**) Agarose gel electrophoresis of total digests (T) and gel-isolated fractions 30–300 bp (L) and 300–700 bp (H). M, 1 kb DNA ladder. (**b**) Restriction enzyme digestion pattern generated by sequence analysis with DNAMAN software (reproduced from ref. 12 with permission from BioMed Central Ltd).

4. Continue the reaction for 30 min.
5. Stop reaction by adding 25 μl TE and 35 μl neutralized phenol.
6. Mix and centrifuge at maximum speed for 2 min.
7. Transfer supernatant (35 μl) to the new tube containing 15 μl of 5 M NH_4OAc.
8. Mix and add 2.5 volumes of 95% ethanol.
9. Mix and centrifuge for 5 min.
10. Carefully remove supernatant (do not disturb the small pellet!) to the new tube marked "radioactive waste."
11. Wash the pellet with 150 μl of 70% ethanol, mix gently.
12. Centrifuge for 2 min.
13. Remove washing solution and dissolve ^{32}P end-labeled DNA in TE at a concentration of 40 ng/μl.

3.5. EMSA for the Detection of Protein-Bound DNA Fragments

1. Prepare 3% polyacrylamide gel (see Note 4).
2. Carry out pre-electrophoresis for 30–60 min (see Note 10).
3. Mix the following components (see Note 11):

   80 ng (2 μl) of ^{32}P-labeled DNA fraction

   5.5 μl of H_2O

   1 μl poly (dI-dC)

   1.5 μl of nuclear or whole cell extract
4. Incubate at room temperature for 20 min.
5. Add 2.5 μl of 20% glycerol, mix, and load onto polyacrylamide gel (see Note 12).
6. Electrophorese with predetermined voltage (for example 100 V) at room temperature for 2 h or until the bromophenol dye reaches to the middle of the gel.
7. Terminate the electrophoresis, open the glass plates, and cover the gel with Saran wrap.
8. Expose the wet gel to phosphoimager screen at room temperature for 1 h (see Note 13).

   A representative example of this electrophoresis is shown in Fig. 3a.
9. Align the phosphoimager image to the wet gel panel (see Note 14).
10. Using scalpel, cut out the marked bands or zones through the wrap.
11. Transfer gel pieces into centrifuge tubes.
12. Add 3–5 volumes of TE and incubate at 37 °C for 1–2 h (see Note 15).
13. Recover labeled DNA fragment(s) as described above (Subheading 3.3, steps 14–16), except add carrier tRNA (2 μg) to facilitate precipitation.
14. Dissolve the DNA in 5–10 μl of 1× Sbd.

3.6. Identification of Shifted DNA Fragments by Denaturing PAGE

1. Prepare a high resolution denaturing 5% polyacrylamide gel (see Note 4).
2. Carry out pre-electophoresis at 1 kV for 30–60 min.
3. Heat treat gel isolated DNA fragments and original ^{32}P-labeled DNA fractions at 95 °C for 2 min and place on ice.
4. Flush the lane pockets with 1× TBE electrophoresis buffer and load DNAs and ^{32}P-labeled 100 bp DNA marker onto polyacrylamide gel in parallel lanes.
5. Start electrophoresis at 2 kV for 5 min (see Note 16).
6. Run the gel until bromophenol blue reaches to the bottom of the gel (see Note 17).
7. Stop the electrophoresis and disassemble the system.

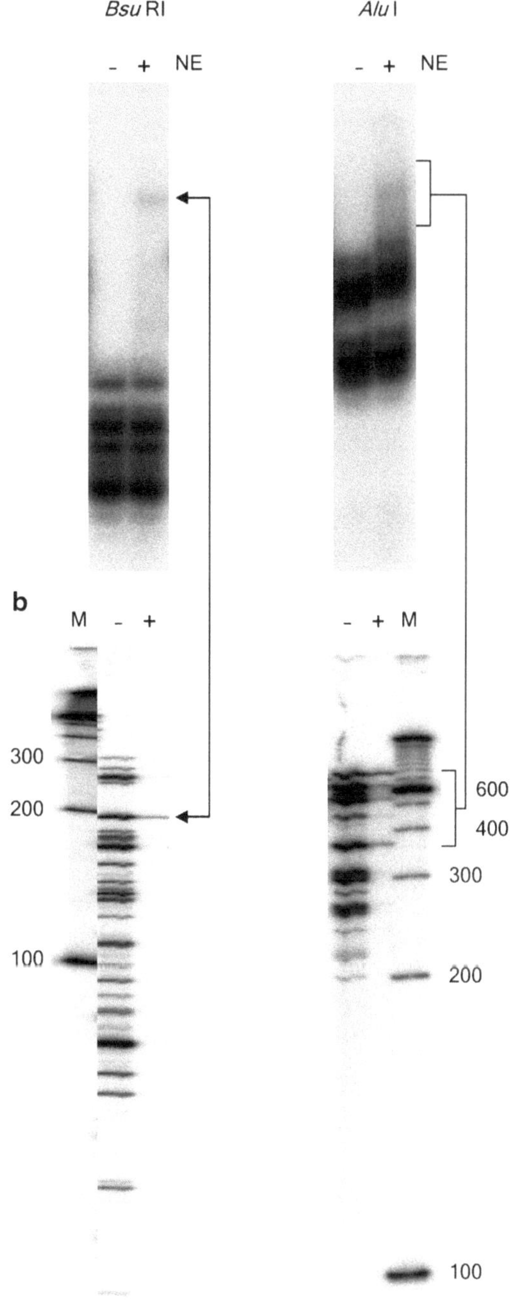

Fig. 3. Identification of the protein binding regions in DNA fragments by a combination of EMSA and denaturing PAGE. (**a**) L and H molecular fractions of corresponding *Bsu*RI and *AluI* fragments were incubated with (lane +) and without (lane –) PC12 nuclear extract (NE) and analyzed by EMSA. Direct phosphoimaging of the wet 1 mm-thick gel is shown. (**b**) Denaturing PAGE of the shifted fragments. Lanes – and + show the fraction of DNA fragments used in EMSA and shifted bands (or smear), respectively. Connecting *lines* indicate the bands or zones analyzed on two different gels. M, ^{32}P-labeled 100 bp DNA ladder (reproduced from ref. 12 with permission from BioMed Central Ltd).

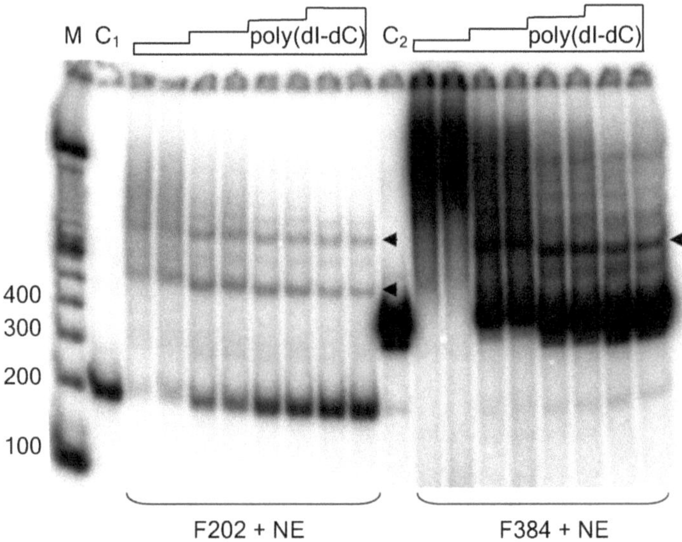

Fig. 4. The effect of the poly (dI-dC) concentration on the mobility shift of restriction fragments. A poly (dI-dC) concentration range from 100 to 800 μg/ml was tested with twofold differences. At each concentration, the incubation of labeled restriction fragment with PC12 nuclear extract (NE) was performed using two different poly (dI-dC) preparations with the average sizes 250 and 500 bp. C1, a 202 bp *Bgl*I–*Nde*I fragment and C2, a 384 bp *Nde*I–*Mph*I fragment derived from −1.8 and −1.6 kb regions of the rat *p75NTR* promoter, respectively. *Arrowheads* point at the major specific DNA–protein complexes. M, ^{32}P-labeled 100 bp DNA ladder (reproduced from ref. 12 with permission from BioMed Central Ltd).

8. Carefully separate the glass plates with scalpel.
9. Place the same size Whatman 3 M filter paper (or similar strong filter paper) on top of the gel.
10. Lift the gel with filter paper and cover the gel with Saran wrap.
11. Dry the gel in gel drier at 80 °C for 30–60 min.
12. Expose the dried gel to phosphorimager screen.
13. Scan the gel image and analyze the data (see Note 18).
    A representative example of this analysis is shown in Fig. 3b.

4. Notes

1. Poly (dI-dC) dissolved in H_2O was heat treated by boiling at 100 °C for 5 or 10 min. This treatment produced fragments with average sizes of polymer about 500 bp and 250 bp, respectively. With these preparations no difference in mobility shift experiments was observed (Fig. 4).
2. Solution of the neutralized phenol was prepared according to the following protocol. First, H_2O, 1/10 of the phenol volume was added. Second, to this solution 10 N NaOH was added in

small aliquots (~200 μl). After each addition solution was mixed and pH was checked by pH paper. This neutralization was repeated until the pH 6–7 was obtained. Third, to this solution, 8-hydroxyquinoline (final concentration of 0.1% w/v) was added and mixed thoroughly. Finally, neutralized phenol was aliquoted in 1 ml amounts and kept at −20°C for long term storage. Before use, phenol solution was melted, 130 μl of the TE was added and mixed thouroughly.

3. When kept at 4 °C, this solution is stable at least 2 months.
4. To get good quality and regular gel with excellent DNA banding performance, polymerization time of the gel must be checked before preparation of the gel. This is particularly important when any new solution is used. We routinely check polymerization time with a small aliquot (1–2 ml) taken from the main gel solution containing the following ingredients: 40 ml 3–5% AA–BAA solution, 4 ml 10× TBE buffer and 40 μl TEMED. To this aliquot, kept in a test tube, add 20–40 μl APS, mix carefully, switch on timer and check gel hardening by inclining the tube after every ~30 s. Polymerization time should be within 1–3 min. Make appropriate calculation or extrapolation of the APS volume for the main gel solution. Keep in mind the linear relationship between the amount of APS and polymerization time. It is important to note that for low-porous (3–5%) gels gel polymerization time must be within 3–5 min. Longer polymerization times often yield irregular, less stable gels containing more unpolymerized AA/BAA monomers.
5. Regular preparative agarose gels (instead of low melting gels) may be used with commercial kits for preparation of DNA fragments. However, in this case it should be kept in mind that conformations of DNA fragments may be changed and this may or may not affect the results of EMSA (12). The diffusion method described with low gelling agarose gives a higher yield and does not appear to disturb the conformations of DNA fragments.
6. Four-cutter restriction enzymes, such as *Alu*I, *Bsu*RI, *Tru1*I, or other, preferably generating 3′ recessed or blunt ends, which can be labeled by fill-in or combination of 3′-5′ exonuclease and 5′-3′ polymerase activities of Klenow fragment (see below). To increase the chance of finding binding sites of proteins, two restriction enzymes with different base specificities (AT vs. GC) in parallel reactions may be used.
7. Other regions, for example 50–150 bp, 150–300 bp, 300–450 bp, etc. may be used. An important aspect here is that shorter fragments should be separated from the longer fragments because shifted bands containing the former may comigrate with the latter. In general, mobility shifts of DNA fragments bound by proteins show strongly reduced mobility and do not fall within the same range of unbound DNA fragments.

8. It is expected that at least 25% yield will be achieved when DNA isolation by the diffusion method is used. Quality of the isolated DNA may be checked by loading an aliquot (~1/5) of the sample onto the separate analytical gel.
9. Also, 3′ end-labeling of the 100 bp marker DNA can be carried out by using the same protocol. It is important to note that some fragments, having with G-C-rich structures at their termini, may be resistant or difficult to label by 3′-5-exonuclease of the Klenow fragment. This could result in labeled fragments with different specific activities. To overcome this limitation, end-labeling with bacteriophage T4 DNA polymerase, with more efficient exonuclease, may be used. Also, end-filling of 3′ recessed ends may be carried out with Klenow fragment and appropriate radioactive dNTPs according to the manufacturer protocol.
10. During pre-electrophoresis check the temperature of the gel. Usually, running gels at 100 V and with 1× TBE buffer will not cause significant heating, which could result in dissociation of DNA–protein complexes. However, it is always a good idea to test the temperature during pre-electrophoresis and make appropriate adjustments, if necessary.
11. It is important to note that nonspecific interactions can frequently occur between long (>200 bp) DNA fragments and proteins used in EMSA. Nonspecific binding may be reduced at higher concentrations of competitor (Fig. 4). The occurrence of nonspecific interactions also depends on the DNA concentration (at higher concentrations more competitor is required). It is recommended that appropriate amount of competitor is determined experimentally for each fraction of fragments or individual fragment and protein extract. Generally, for fragments >200 bp, higher concentrations (up to 1 mg/ml) of competitor must be used.
12. To follow electrophoresis load 1× SBa onto one lane.
13. Weak mobility shift signals may not be detected with wet gels due to radioactivity quenching effect of these gels. Therefore, to increase the sensitivity of detection, duplicate panels may be prepared, one of which may be dried and exposed to phosphoimager screen, and the other (wet gel) used for preparative purpose. Sometimes smear, instead of bands, may be detected suggesting DNA protein complex dissociation during electrophoresis. In this case zones, containing the smear, may be cut out and analyzed. To facilitate the alignment of radioactive images to the wet gel, use radioactive labels to mark the margins/lanes of the gel.
14. Copy the phosphoimager 1:1 image to the transparent sheet, which can be used to locate shifted bands of interest. Mark the shifted bands or zones on Saran wrap with marker pen.

15. Diffusion of the DNA may be monitored by Geiger counter.
16. Starting electrophoresis at higher voltage (~2 kV) will help to concentrate samples and obtain sharper bands. Do not remember to reduce the voltage to 1.0–1.5 kV after this 5 min period.
17. In 5% polyacrylamide gel bromophenol blue corresponds roughly to 25 nucleotides. This run can resolve fragments up to 300 nucleotides. For analysis of longer DNA fragments (up to 700 bp) run xylene cyanol to the bottom of the gel.
18. Note that weak background bands of the DNA fraction are visible in the gel-eluted material (Fig. 3b, compare - and + lanes). This is most likely due to impurities in the gel, which adsorb some of the faster migrating DNA fragments. To prove that EMSA positive DNA fragments contain protein binding sites of interest, combination of EMSA and denaturing PAGE may be repeated with individual fragments.

Acknowledgments

This work was supported by grants from the Estonian Science Foundation (ETF8381) and the Estonian Ministry of Education and Research (SF0140143s08).

References

1. Fried M, Crothers DM (1981) Equilibria and kinetics of lac repressor-operator interactions by polyacrylamide gel electrophoresis. Nucleic Acids Res 9:6505–6525
2. Garner MM, Revzin A (1981) A gel electrophoresis method for quantifying the binding of proteins to specific DNA regions: application to components of the *Escherichia coli* lactose operon regulatory system. Nucleic Acids Res 9:3047–3060
3. Hellman LM, Fried MG (2007) Electrophoretic mobility shift assay (EMSA) for detecting protein-nucleic acid interactions. Nat Protoc 2:1849–1861
4. Fried MG (1989) Measurement of protein-DNA interaction parameters by electrophoresis mobility shift assay. Electrophoresis 10: 366–376
5. Carey J (1991) Gel retardation. Methods Enzymol 208:103–117
6. Rasimas JJ, Pegg AE, Fried MG (2003) DNA-binding mechanism of O6-alkylguanine-DNA alkyltransferase. Effects of protein and DNA alkylation on complex stability. J Biol Chem 278:7973–7980
7. Laniel MA, Bergeron MJ, Poirier GG, Guerin SL (1997) A nuclear factor other than Sp1 binds the GC-rich promoter of the gene encoding rat poly(ADP-ribose) polymerase in vitro. Biochem Cell Biol 75:427–434
8. Cann JR (1989) Phenomenological theory of gel electrophoresis of protein-nucleic acid complexes. J Biol Chem 264:17032–17040
9. Vossen KM, Fried MG (1997) Sequestration stabilizes lac repressor-DNA complexes during gel electrophoresis. Anal Biochem 245:85–92
10. Boffini A, Prentki P (1991) Identification of protein binding sites in genomic DNA by two-dimensional gel electrophoresis. Nucleic Acids Res 19:1369–1374
11. Chernov IP, Akopov SB, Nikolaev LG, Sverdlov ED (2006) Identification and mapping of DNA binding proteins target sequences in long genomic regions by two-dimensional EMSA. Biotechniques 41:91–96
12. Kaer K, Mätlik K, Metsis M, Speek M (2008) Combination of native and denaturing PAGE for the detection of protein binding regions in long fragments of genomic DNA. BMC Genomics 9:272

Chapter 14

Quantitative NanoProteomics Approach for Protein Complex (QNanoPX) Using Gold Nanoparticle-Based DNA Probe

Shu-Hui Chen and Mei-Yin Lin

Abstract

Affinity purification by pulldown methods using target-bound gel beads provides a powerful approach for purifying endogenous protein complexes. Such methods can be improved by using nanoparticle-based probe, coupled with immunoblot analysis or quantitative proteomics method using stable isotope labeling via liquid chromatography-mass spectrometry (LC-MS). Here, we describe sample preparation and a pull-down method using gold nanoparticle-based DNA probe for characterizing the transcriptional complex of estrogen response element (ERE). The described protocol includes the fabrication of gold nanoparticle-based probe, nuclear extract preparation, and affinity purification for the analysis by immunoblotting, as well as the subsequent trypsin digestion and stable isotope dimethyl labeling for the analysis by LC-MS.

Key words: Affinity purification, Pulldown, Protein complex, Transcription, Nanoparticle, Stable isotope labeling, Dimethyl labeling, Quantitative proteomics

1. Introduction

To fine-tune transcription, large multiprotein complexes are assembled in vivo. These complexes may include transcription factors, activators, repressors, chromatin remodeling proteins, and mediators present at response elements in promoter regions of target genes. It is often challenging to identify and quantify protein components of transcription complexes using protein binding sites as bait because conventional gel-bead-methods for protein immunoprecipitation (IP) (1) do not work well for characterization of large complexes. Although the porous nature of the traditional gel beads used for IP experiments provides high binding capacity, the pore size can exclude large complexes and many species can stick on the

Minou Bina (ed.), *Gene Regulation: Methods and Protocols*, Methods in Molecular Biology, vol. 977, DOI 10.1007/978-1-62703-284-1_14,

porous surface nonspecifically. Moreover, lengthy and continuous rotation is required for gel-bead-methods because beads precipitate very quickly. Mono-dispersed nonporous nano-materials such as modified nanoparticles are good substitutes for gel beads (2, 3) because large complexes are not excluded by solid nonporous supports. Moreover, binding capacity of nanoparticles is comparable to porous gel beads due to a high surface area to volume ratio. Additionally, nanoparticles can be efficiently suspended in liquid phase to enhance interfacial interactions. The pulled down proteins can be directly loaded onto gels and further characterized by PAGE/immunoblotting techniques.

Affinity purification can be further improved to differentiate specific and nonspecific binders by coupling probe-based affinity purification with quantitative proteomics protocols, stable isotope labeling, in which proteins derived from control and experimental probe are individually labeled with light and heavy isotopes (4–6). After labeling, the solutions are combined for protein identification via LC-MSMS based shot gun proteomics. Specific and nonspecific binders are distinguished by determining the ratio of ion intensities of isotopic pairs. We designate the nanoparticle functionalized probe, coupled with stable isotope labeling method for resolving protein complexes, as Quantitative NanoProteomics approach for Protein Complex (QNanoPX) (6).

Here, we describe sample preparation protocol for QNanoPX using estrogen response element (ERE) functionalized with AuNPs (gold nanoparticles). AuNPs are relatively inert and nontoxic (7, 8), with comparable size to biomolecules. Therefore, AuNPs provide excellent materials for protein enrichment (9). The surface of AuNPs can be easily modified, with high uniformity, through thiol-Au binding chemistry (10) in order to minimize nonspecific binding. Without digestion and stable isotope labeling, the described method can be applied for protein detection by PAGE/immunoblotting techniques. General methods for PAGE/immunoblotting and LC-MS analysis for protein identification and quantification, such as those described in ref. 6, can be adopted and will not be included in this protocol.

2. Materials

All solutions were prepared by dissolving analytical or higher grade reagents in ultrapure de-ionized (DI) water (with 18 MΩ cm resistance at 25°C). Prepare the solutions at room temperature and follow all waste disposal regulations.

2.1. Double Stranded ERE (dsDNA)

HS-T_{25}ERE solution: 500 μM HS-T_{25}DNA (HS-5′-T_{25}-GGTCAGAGTGACC-3′).

Add about 28 μL water (see Note 1) to the synthesized sequence (MDBio Inc., Taipei, Taiwan). Store at −20°C.
cERE solution: 500 μM cDNA (5′-GGTCACTCTGACC-3′).

Add about 84.4 μL water (see Note 1) to the synthesized sequence (MDBio Inc., Taipei, Taiwan). Store at −20°C.

Binding buffer: 0.01 M Tris–HCl, 0.01 M EDTA, 0.05 M pH 7.5.

2.2. AuNP Solution

To prepare a stock solution, add 1 g of $HAuCl_4$ (Alfa Aesar, Johnson Matthey Co., London, UK), to a 15-mL brown bottle, and bring the volume to 10 mL with water. Store the stock solution at 4°C in a dark place. To prepare the $HAuCl_4$ solution (0.65 mM $HAuCl_4$), pipette 256 μL of the stock solution into a 100-mL flask. Bring the volume to 100 mL with water.

To prepare 38.8 mM sodium citrate, weigh 0.114 g of sodium citrate, transfer it to a 10-mL flask, and bring the volume 10 mL with water (see Note 2).

2.3. AuNP-DNA Probe

PEG solution: 500 μM HS-PEG. Transfer 0.001 g of HS-PEG (MW 750) (Rapp Polymere GmBH, Tübingen, Germany) to a 1.5-mL Eppendorf tube, bring the volume to 1.33 mL with water. Subsequently, pipette 500 μL of the solution to a 1-mL vial and bring the volume to 1 mL with water. Store at −20°C.

Sodium chloride solution: 2 M sodium chloride. Transfer 3.4932 g of sodium chloride to a 50-mL bottle. Bring the volume to 30 mL with water.

2.4. Preparation of Nuclear Extracts

Lysis buffer: 10 mM HEPES, 10 mM KCl, 0.5 mM EDTA, 0.5 mM EGTA, 1 mM DTT (pH 7.9).

Nuclear buffer: 20 mM HEPES, 10 mM KCl, 1 mM EDTA, 1 mM EGTA, 1 mM DTT (pH7.9).

PBS buffer: 2.68 mM KCl, 10.14 mM Na_2HPO_4, 1.76 mM KH_2PO_4, and 136.89 mM NaCl (pH 7.4). Transfer the solution into two 500-mL serum bottles (autoclaved by a sterilizer and dried in an oven before use). Store the bottles at 4°C.

1× Trypsin-EDTA: Transfer 5 mL of 10× 0.5% Trypsin-EDTA (GIBCO; Grand Island, N.Y., U.S.A.) to a 50-mL flask. Bring the volume to 50 mL with PBS buffer.

2.5. Affinity Pulldown

PBS buffer

PBST buffer: 70 mM NaCl, 0.74 mM NaH_2PO_4, 9.26 mM Na_2HPO_4 (pH 7.4). Store the stock solution at room temperature. Transfer 100 mL of the stock solution to a 1-L flask. Add 0.5 mL of Tween-20 (Sigma, St. Louis, MO, USA), bring the volume to 1 L with water.

Sample loading buffer: 10% SDS, 350 mM Tris (pH 6.8), 30% (v/v) Glycerol, 6% 2-mercaptoethanol. Weigh 4.728 g of Tris–HCl; bring the volume to 30 mL with water, adjust the pH to 6.8. Transfer 3.5 mL of the Tris–HCl solution into a 10-mL flask. Add 1 g of SDS, 0.6 mL of 2-mercaptoethanol and 3 mL Glycerol to the bottle. Bring the volume to 10 mL with water.

SDS: 7.5% SDS. Weigh 3.75 g of SDS; bring the volume to 50 mL with water. Store the solution at room temperature.

DTT: 1 M DTT. Weigh 0.0309 g of DTT; bring the volume to 200 μL with water (see Note 3).

Elution buffer: 10 mM DTT and 1% SDS. Transfer 67 μL of SDS and 5 μL of DTT to a 1.5-mL Eppendorf tube and bring the volume to 500 μL with water.

2.6. Trypsin Digestion

Ammonium bicarbonate buffer (NH_4HCO_3): 0.1 M ammonium bicarbonate. Weigh 0.3953 g of ammonium bicarbonate and transfer to a 50-mL bottle. Bring the volume to 50 mL with water. Store at 4°C.

IAM: 0.5 M iodoacetamide in 0.1 M ammonium bicarbonate. Weigh 0.0185 g of IAM into a 1.5-mL brown Enppendorf tube and add 200 μL of ammonium bicarbonate buffer (see Note 3).

TCA: 50% trichloroacetic acid. Weigh 25 g of TCA (Sigma, St. Louis, MO, USA); bring the volume to 50 mL with water.

Trypsin: 1 μg/μL trypsin (Promega, Madison, WI, USA). Add 20 μL water to the tube that contains dried trypsin (20 μg for each tube), containing 50 mM acetic acid in order to reconstruct the solution. Store at −20°C.

2.7. Stable Isotope Dimethyl Labeling

Sodium acetate buffer: 100 mM sodium acetate buffer (pH about 5 to 6). Weigh 0.41 g of sodium acetate and transfer to a 50-mL tube. Add some water. Adjust pH to about 5 to 6 with 1 N HCl. Bring the volume to 50 mL with water. Store at 4°C (see Note 4).

H_2- or D_2-formaldehyde (4% in water): Pipette 109.6 μL of H_2-formaldehyde (36.5%, Riedel-deHaen, St. Louis, MO, USA) to a 1.5-mL Eppendorf tube. Bring the volume to 1 mL with water. Pipette 200 μL of D_2-formaldehyde (*20%*, Isotec, Great neck NY, USA) to a 1.5-mL Eppendorf tube. Bring the volume to 1 mL with water (see Note 3).

Sodium cyanoborohydride: 600 mM sodium cyanoborohydride. Weigh 0.0378 g of sodium cyanoborohydride (Fluka, St. Louis, MO, USA) to a 1.5-mL Eppendorf tube. Bring the volume to 1 mL with water (see Note 5).

Ammonium hydroxide (4% in water) solution: Transfer 142.9 μL of ammonium hydroxide (Mallinckrodt Baker. Inc., Phillipsburg, NJ, USA) to a 1.5-mL Eppendorf tube. Bring the volume to 1 mL with water (see Note 3).

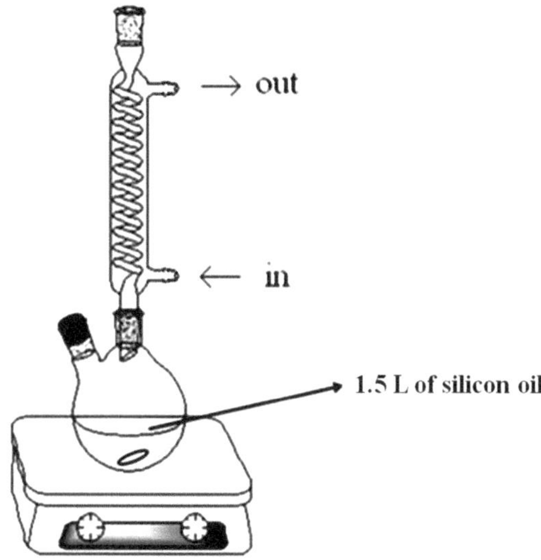

Fig. 1. Configuration of the reflux system for the synthesis of AuNPs.

3. Methods

3.1. Preparation of dsERE Solution

1. Add 5 μL of HS-T_{25}ERE and 5 μL of cERE to 60 μL of binding buffer in a 1.5-mL Eppendorf tube.
2. Heat the mixture at 90°C for 4 min.
3. Cool the solution to 40°C (see Note 6) for 1h. Perform agarose gel electrophoresis to confirm the formation of dsERE (see Note 7).

3.2. Preparation of AuNP Solution

1. Place 100 mL of $HAuCl_4$ solution in a 250-mL round-bottom flask containing a stirring bar (see Note 8).
2. Attach the round-bottom flask to a reflux condenser and place the whole apparatus on a magnetic heater (Fig. 1).
3. Rapidly add 10 mL of sodium citrate solution to the flask. Quickly turn on the heater, stir vigorously; boil and reflux the solution.
4. Boil the solution for an additional 10 min, until the color of the solution is changed from pale yellow to brightly red.
5. Turn off the heater but keep stirring.
6. Cool the solution to room temperature. Store the solution at 4°C.

3.3. Preparation of AuNP-ERE Affinity Probe

1. For the experimental probe, add 70 μL of dsERE solution (prepared via Subheading 3.1) and 1 μL of HS-PEG solution to 1 mL of AuNP solution in a 1.5-mL Eppendorf tube.
2. Incubate the experimental probe solution at room temperature for 2 h or 4°C overnight (see Note 9).
3. Add 50 μL of 2 M sodium chloride stock solution to the experimental probe solution to obtain a final concentration of 0.1 M NaCl (see Note 10).
4. Incubate the solution (experimental probe) overnight at 4°C.
5. For the control probe, add 10 μL of HS-PEG solution to 1 mL of the AuNP solution in a 1.5-mL Eppendorf tube. Incubate the solution overnight at 4°C.

3.4. Nuclear Extract Preparation

1. Grow the cells in a 10-cm culture dish to 100% confluency.
2. Wash the cells with 5 mL of ice-cold PBS.
3. Add 1 mL of 1× trypsin-EDTA to the dish; incubate in a CO_2 (5%) incubator at 37°C, for 5 min. Transfer the dish to the bench top; tap the bottom of the dish to detach cells from the dish.
4. Transfer the suspension into a centrifuge tube.
5. Centrifuge at 600×*g*, at 4°C for 5 min. Decant the supernatant; add 1 mL of the ice-cold PBS to resuspend the pellet. Transfer the cell lysate to a 1.5-mL Eppendorf tube.
6. Wash the collected cells with 1 mL of ice-cold PBS and centrifuge at 600×*g* at 4°C for 5 min. Decant the supernatant and add 1 mL of the ice-cold PBS and centrifuge again. Repeat the washing procedure two to three times.
7. Resuspend the pellet (10× cell volume) in lysis buffer (add 500 μL if the pellet is 50 μL). Place the resuspended pellet on an oscillator for 15 min at 4°C.
8. Centrifuge at 1,000×*g* for 15 min at 4°C. Separate the pellet (nuclear fraction) from the supernatant (non-nuclear fraction).
9. Resuspend the pellet (10× cell volume) in the nuclear buffer. Place the resuspended pellet on an oscillator for 2 h at 4°C.
10. Pass the pellet through a 1-mL 23G needle for 8~10 times.
11. Centrifuge at 16,000×*g* for 10 min at 4°C. The resulting supernatant contains the nuclear extract.

3.5. Affinity Pull Down from Cell Lysates

1. For experimental probe, add the same volume (~100 μL) of PBS to the AuNP-ERE probe solution and briefly vortex the solution for washing.
2. Centrifuge at 20,400×*g*, at 15°C for 20–30 min (see Note 11).
3. Discard the supernatant.

4. Add an equal volume (~100 μL) of PBS to the pellet. Resuspend the probe by sonication.
5. Repeat steps 2–4.
6. Place 100 μL of the probe solution in a 1.5-mL Eppendorf tube. Bring the volume to 1 mL with PBS.
7. Add the nuclear fraction (see Subheading 3.4) extracted from cells containing 100 μg of total protein (see Note 12) to the probe solution. Mix the solution by a brief vortexing.
8. Fix the tube on a homogenizer. Vortex under a speed of 4~6 (see Note 9) at 4°C for 18 h or overnight.
9. Centrifuge at 20,400 × *g*, at 15°C for 30 min. Separate the supernatant and the pellet.
10. Centrifuge the supernatant at 20,400 × *g*, at 15°C for 10 min. Discard the supernatant; combine the pellet with the pellet produced in Step 9.
11. Add 1 mL of PBST solution for wash (see Note 13).
12. Repeat steps 9–10
13. For subsequent analysis by gel electrophoresis/immunoblotting, add 10 μL of the sample loading buffer to the pellet. For the subsequent analysis by trypsin digestion and LC/MSMS, add 10 μL of the elution buffer to the pellet. Resuspend the probe by sonication.
14. Heat the resuspended solution to 95°C for 5–10 min (see Note 14).
15. Cool the solution in an ice bath for 3 min.
16. Centrifuge at 20,400 × *g* under 15°C for 30 40 min. Collect the supernatant which contains the pulled down proteins.
17. The supernatant, resuspended by the sample loading buffer, can be directly loaded onto PAGE for separation followed by detection using immunoblotting (see Note 15). The supernatant resuspended by the elution buffer can be subjected to trypsin digestion (Subheading 3.5) and stable isotope dimethyl labeling (Subheading 3.6) for LC/MSMS analysis.
18. For the control probe, add the same volume (~100 μL) of PBS buffer to the AuNP-PEG control probe solution and briefly vortex the solution for washing. Repeat Steps 2–17.

3.6. TCA Precipitation and Trypsin Digestion

1. Add 60 μL of water to the pellet resuspended in elution buffer (see Note 16).
2. Add 8 μL of IAM solution (final conc. of IAM is 50 mM); vortex the solution in dark, at room temperature, for 30 min.
3. Add 52 μL of 50% TCA solution (final conc. of TCA is 20%) and place the solution in an ice-bath for 15 min. Centrifuge at 20,400 × *g* (at 0°C), for 30 min to collect the pellet (see Note 17).

4. Add 150 μL of 10% TCA. Centrifuge at 20,400×g (at 0°C), for 5 min and collect the pellet.
5. Add 200 μL of water. Centrifuge at 20,400×g (at 0°C), for 5 min and collect the pellet. Repeat this procedure three times.
6. Add 1 μL of trypsin solution (enzyme to protein ratio of 1:100) to an Eppendorf tube. Bring the volume to 50 μL with ammonium bicarbonate buffer.
7. Incubate the solution at 37°C for 18 h.
8. Repeat Steps 1–7 for elution using the control probe.

3.7. Stable Isotope Dimethyl Labeling

1. Dry the tryptic digest under vacuum at 37°C.
2. Add 100 μL of the sodium acetate buffer to resuspend the digest.
3. Add 20 μL of H_2-formaldehyde solution to the digest pulled down by the control probe. Add 20 μL of D_2-formaldehyde solution to the digest pulled down by the experimental probe.
4. Add 20 μL of the freshly prepared sodium cyanoborohydride solution (see Note 5) to each pull down digest; vortex for 30 min.
5. Add 20 μL of ammonium hydroxide solution to each pull down digest to quench the un-reacted formaldehyde; vortex for 5 min.
6. Acidify each solution to reach a pH 2~3 with 0.1% Formic acid.
7. Vortex briefly and wait for 1 h (see Note 18).
8. Combine the D_2- and or H_2-formaldehyde labeled solution. Dry the sample by vacuum, at room temperature.
9. Resuspend the sample with 12 μL of the starting mobile phase used for LC-MS analysis (see Note 19).

4. Notes

1. Required volume of DI water varies with different batches depending on the amount of DNA synthesized. Follow the datasheet specification provided by the manufacturer.
2. Sodium citrate solution must be freshly prepared before use.
3. DTT, IAM, H_2- or D_2-formaldehyde, ammonium hydroxide solutions must be freshly prepared before use.
4. Make sure the pH of sodium acetate is between 5 and 6 before use.

5. Sodium cyanoborohydride powder must be stored in a moisture-proof box. Weigh the reagent as quickly as possible to minimize exposure to air. The solution must be freshly prepared before use.
6. Cooling takes about 1 h.
7. Agarose gel (1.7%) stained with ethidium bromide can be used to detect single and double stranded DNA (ssDNA and dsDNA); ssDNA can be stained by ethidium bromide due to self- hybridization of two complementary half sequences. A gel shift is observed as a consequence of a slower mobility associated with dsDNA, as compared to shorter self-hybridized ssDNA. Poly(dT) or poly(dA) oligonucleotide can be loaded as control since they do not stained due to a lack of hybridization.
8. All apparatuses including flasks, condensers, and stirring bars must be cleaned by aqua regia solution before use. Immerse all glassware in aqua regia bath for at least 3 h. Rinse each apparatus with water and dry it in an oven.
9. Control the vortexer power to prevent splashing to the cap. Some Eppendorf tubes may cause AuNPs coating onto the wall. Tubes purchased from Scientific Specialties Inc., USA seemed to work well.
10. Test a small aliquot of the AuNP-ERE solution before adding NaCl to 0.1 M final concentration. If the color of the aliquot remains wine red, add NaCl to reach 0.1 M. If its color turns to purple, slowly increase the NaCl concentration to 0.02 M, 0.05 M, 0.07 M, to 0.1 M.
11. Time for centrifugation varies with the volume of AuNP solution. Longer centrifugation is applied to larger volumes.
12. The amount of protein contained in total cell-lysate can be determined by Lowry or BCA assay using commercial kits.
13. Use a tip to wash out AuNP-DNA probe coated on the wall.
14. The color turns to purple black after resuspension.
15. 10% SDS-PAGE is commonly used for protein separation; 0.22-μm PVDF membrane is commonly used for protein transfer and detection by Western blotting (6).
16. The solution contains at least 30 μg protein.
17. Always keep the pellet facing opposite to the axle center (Fig. 2), when placing the Eppendorf tube onto a centrifuge. Do not disturb the pellet when decanting the supernatant.
18. Some bubbles may form.
19. The collected MS data can be processed by any search engine such as Mascot for protein identification and any quantification software such as Distiller for calculating the intensity ratios for

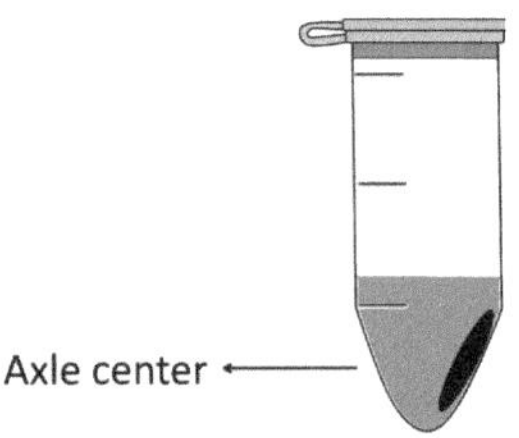

Fig. 2. Orientation of placing Eppendorf tube in centrifuge.

the isotopic pair (*D/H*). Proteins identified with ratio value greater than 1 beyond the confidence level (ex. 95%) are likely to be specific binders see ref. 6 for additional tips.

Acknowledgment

This work was supported by National Science Council in Taiwan, Republic of China, Grant NSC 99-2627-M-006-002.

References

1. Bonifacino JS, Dell'Angelica EC, Springer TA (2001) Immunoprecipitation. Curr Protoc Mol Biol 10:16.1–10.16.29
2. Ugelstad J, Berge A, Ellingsen T, Bjørgum J, Schmid R, Stenstad P, Aune O, Nilsen TN, Funderud S, Nustad K (1987) Biomedical applications of monodisperse magnetic polymer particles. In: El Asser MS, Fitch RM (eds) Future directions in polymer colloids. Polymer colloids in biomedical field, vol 138, NATO ASI Series E. M. Nijhoff Publ, Dodrecht, The Netherlands, pp 355–370
3. Baron R, Willner B, Willner I (2007) Biomolecule-nanoparticle hybrids as functional units for nanobiotechnology. Chem Commun 28(4):323–332
4. Wepf A, Glatter T, Schmidt A, Aebersold R, Gstaiger M (2009) Quantitative interaction proteomics using mass spectrometry. Nat Methods 6:203–205
5. Liao LJ, Park SK, Xu T, Vanderklish P, Yates JR (2008) Quantitative proteomic analysis of primary neurons reveals diverse changes in synaptic protein content in fmr1 knockout mice. PNAS 105:15281–15286
6. Cheng P-C, Chang H-K, Chen S-H (2010) Quantitative Nano-Proteomics for Protein Complexes (QNanoPX) Related to Estrogen Transcriptional Action. Mol Cell Proteomics 9:209–224
7. Khan JA, Pillai B, Das TK, Singh Y, Maiti S (2007) Molecular effects of gold nanoparticles in HeLa cells. Chembiochem 8: 1237–1240
8. Murphy CJ, Gole AM, Stone JW, Sisco PN, Alkilany AM, Goldsmith DC, Baxter AC (2008) Gold nanoparticles in biology: beyond toxicity to cellular imaging. Acc Chem Res 41(12):1721–1730
9. Wang A, Wu CJ, Chen SH (2006) Gold nanoparticle-assisted protein enrichment and electroelution for biological samples containing low protein concentrations—A prelude of gel electrophoresis. J Proteome Res 5: 1488–1492
10. Jadzinsky PD, Calero G, Ackerson CJ, Bushnell DA, Kornberg RD (2007) Structure of a Thiol Monolayer–Protected Gold Nanoparticle at 1.1 Å Resolution. Science 318:430–433

Chapter 15

Chromatin Assembly and In Vitro Transcription Analyses for Evaluation of Individual Protein Activities in Multicomponent Transcriptional Complexes

Takayuki Furumatsu and Hiroshi Asahara

Abstract

Eukaryotic DNA and core histones form the fundamental repeating units of chromatin. Condensed chromatin, which has higher-order structures, prevents transcriptional complexes from accessing their target genes. Epigenetic regulation, including structural changes of chromatin, histone modification, and DNA methylation, strictly controls the pattern of gene expression and silencing. Recent studies have revealed that histone acetylation plays a crucial role in relaxing chromatin structure for initiation of transcription. Crosstalk between DNA-binding transcription factors and histone acetyltransferases (HATs) serves as a key mechanism for regulating gene expression and developmental processes. However, the precise roles of multicomponent transcriptional complexes have not been fully elucidated because of technical difficulties in using in vitro experimental systems. Previously we demonstrated that the DNA-binding transcription factor Sox9, HAT coactivator p300, and other regulatory factors (Smad3/4) cooperatively activate Sox9-dependent transcription on chromatin. Here, we describe an experimental approach to investigate the function of each component on reconstructed chromatin in vitro. Our methods offer a useful system for analyzing the additional effect of a third component in a transcriptional complex on chromatin structure.

Key words: Chromatin assembly, In vitro transcription, Sox9, p300, Smad3, Histone acetylation, Epigenetic regulation

1. Introduction

Regulatory codes in DNA play key roles in normal development and appropriate patterns of gene expression. Tissue-specific transcription factors regulate spatiotemporal patterns of gene expression via association with *cis*-acting elements in promoters,

Minou Bina (ed.), *Gene Regulation: Methods and Protocols*, Methods in Molecular Biology, vol. 977,
DOI 10.1007/978-1-62703-284-1_15, © Springer Science+Business Media, LLC 2013

enhancers, and silencers in control regions of genes (1). Transcriptional coactivators and corepressors also play important roles in transcriptional regulation. Another layer of control processes includes epigenetic modes of regulation (i.e., histone acetylation, chromatin remodeling, and DNA methylation). Epigenetics is defined as regulatory processes that do not involve changes in the underlying genomic DNA sequence. The fundamental unit of eukaryotic chromatin—the nucleosome—consists of 146 bp of genomic DNA wrapped around an octamer of core histones: 2 sets each of H2A, H2B, H3, and H4 (2). Chromatin structure has an important role in transcriptional regulation in eukaryotes. Histone acetylation can alter condensed chromatin (heterochromatin) structures by influencing histone–DNA and histone–histone interactions (3). Epigenetic marks "histone code" could exert synergistic, or antagonistic, effects on chromatin-associated proteins to regulate gene expression during coordinated developmental processes including chondrogenesis (4, 5).

Chondrogenesis derived from mesenchymal condensation is a dynamic process in endochondral bone formation. The sequential differentiation and maturation steps of chondrocytes are regulated by transcription factors and growth factors such as the Sry-type high mobility group box (Sox) and the transforming growth factor (TGF)-β superfamily (6, 7). Sox9, which has a high mobility group DNA-binding domain, has been identified as a master transcription factor in chondrogenesis (8). In our previous studies, we showed that Sox9 activated expression of its target genes via association with the transcription coactivator p300. This coactivator has an intrinsic histone acetyltransferase (HAT) activity (9). In addition, Sox9-dependent transactivation is induced by p300-mediated histone acetylation of chromatin (10).

The TGF-β superfamily includes multifunctional growth factors that participate in many cellular processes including proliferation, differentiation, and apoptosis (11). Following the activation of TGF-β receptors, several pathways, such as those mediated by Smad2 and Smad3, relay key intracellular signals in response to TGF-β treatments (11). We previously demonstrated that TGF-β-regulated Smad3 induces primary chondrogenesis via association with Sox9 and p300 (12, 13). On the other hand, chondrocyte maturation is inhibited by TGF-β in the late chondrogenic stage (14). These conflicting effects of TGF-β during chondrogenesis may depend on chromatin structure, on the epigenetics of each differentiated stage, or both (5). However, the precise mechanism by which TGF-β-regulated Smad3 and Sox9 participate in the epigenetic regulation has not been elucidated because technical difficulties arise when using in vitro chromatin assembly systems to evaluate the functions of each protein in multicomponent transcriptional complexes. We investigated, quantitatively, transcriptional activity of a chromatinized DNA template in the presence or absence of Sox9, p300, and/or the third component Smad3, using a chromatin assembly system, followed by in vitro transcription and S1 nuclease assays (15).

In this paper, we describe our strategy to investigate the function of each transcriptional component on chromatin reconstructed in vitro. Our strategy can analyze additional effects that a third molecule may have on chromatin structure, in a multicomponent transcriptional complex.

2. Materials

2.1. DNA Templates, Buffers, and Reagents

1. Plasmids: Supercoiled plasmids were purified using QIAGEN Plasmid Midi/Mega Kit (Qiagen) or NucleoBond Xtra Midi/Maxi Kit (Macherey-Nagel). Solutions containing the plasmids should be aliquoted and stored at −20°C. For chromatin assembly (see Note 1) we used a plasmid (designated 12×48-pGL3-Promoter) that contained 12 copies of the 48-bp Sox9-binding Col2a1 enhancer element (10). We used pRL-CMV (Promega) as an internal control for in vitro transcription analysis.
2. HEG buffer: 10 mM HEPES (pH 7.6), 50 mM KCl, 10% glycerol, and 0.1 mM EDTA. Store at −20°C.
3. 6× His elution buffer: 50 mM sodium phosphate (pH 8.0), 300 mM NaCl, 10% glycerol, and 500 mM imidazole (see Note 2). Store up to 6 months at 4°C.
4. FLAG elution buffer: 10 mM Tris (pH 7.5), 150 mM NaCl, 0.1% Triton X-100, and 200 μg/mL FLAG peptide (Sigma). Store at 4°C.
5. Buffer D: 20 mM HEPES (pH 7.9), 100 mM KCl, 20% glycerol, 0.2 mM EDTA, 0.5 mM PMSF, and 0.5 mM DTT (add DTT and fresh PMSF to the buffer just before use). Store at 4°C.
6. R buffer: 10 mM HEPES-KOH (pH 7.5), 10 mM KCl, 1.5 mM $MgCl_2$, 0.5 mM EGTA, 10% glycerol, 10 mM β-glycerophosphate, 1 mM DTT, and 0.2 mM PMSF. Store aliquots at −20°C.
7. PVOH/PEG mix: 16.8 mM HEPES-KOH (pH 7.6), 3.3% polyvinyl alcohol (PVOH, molecular weight 30,000–70,000 Da), and 3.3% polyethylene glycol (PEG, molecular weight 15,000–20,000 Da). Store aliquots at −20°C.
8. ATP/phosphocreatine/$MgCl_2$ mix: 10 mM HEPES-KOH (pH 7.6), 21 mM ATP, 210 mM phosphocreatine, and 15 mM $MgCl_2$. Store aliquots at −80°C.
9. Creatine phosphokinase: 10 mM potassium phosphate (pH 7.0), 50 mM NaCl, 50% glycerol, and 50 mg/mL creatine phosphokinase. Store aliquots at −80°C.

10. Acetyl-coenzyme A (AcCoA): 200 μM AcCoA (Sigma) in distilled water. Store aliquots at −20°C.
11. Buffer H: 66 mM HEPES-KOH (pH 8.0), 234 mM KCl, 15 mM $MgCl_2$, and 0.66 mM EDTA. Store at 4°C.
12. rNTP mix: 10 mM HEPES-KOH (pH 7.6) and 5 mM each of ATP, CTP, GTP, and UTP. Store aliquots at −80°C.
13. Transcription-stop solution: 200 mM NaCl, 20 mM EDTA (pH 8.0), 1% SDS, and 0.25 mg/mL glycogen (Sigma). Store at room temperature. Add 5% volume of 2.5 mg/mL Proteinase K (Invitrogen) just before use.
14. Phenol–chloroform–isoamyl alcohol: Prepare a 25:24:1 solution just before use.
15. Chloroform–isoamyl alcohol: Prepare a 24:1 solution just before use.
16. 3M CH_3COONa (NaOAc)
17. 100% ethanol

2.2. Components for Chromatin Assembly

Purified chromatin assembly molecules are evaluated by Western blot using specific antibodies. Concentrations of purified proteins are determined from staining density of their isolated bands in SDS-polyacrylamide gel (see Note 3) using Silver Stain Plus (Bio-Rad).

1. Histones: Core histones are purified from HeLa nuclear extracts (16) and dialyzed in HEG buffer. Purified histones are aliquoted and stored at −80°C.
2. Nucleosome assembly protein-1 (NAP-1): NAP-1, a histone-shuttling protein, recruits histones onto the DNA template. Histidine-tagged NAP-1 (17) is produced in baculovirus-infected Sf9 cells (Invitrogen). Recombinant NAP-1 is purified by Ni-NTA Agarose (Invitrogen) and eluted in 6× His elution buffer at 4°C (10). Store aliquots at −80°C.
3. ATP-utilizing chromatin assembly and remodeling factor (ACF): ACF includes two subunits (ACf1 and ISWI). ACF assembles periodic nucleosome arrays in an ATP-dependent process (18). Simultaneous infection of baculoviral ACf1 and ISWI is performed in Sf9 cells. The ACF complex (FLAG-tagged ISWI and untagged Acf1) is purified by EZview Red ANTI-FLAG M2 Affinity Gel (Sigma) and eluted in FLAG elution buffer at 4°C (10). To elute ACF complex, add glycerol and DTT at a final concentration of 10% and 1 mM, respectively. Store aliquots at −80°C.

2.3. Components for In Vitro Transcription

Purified recombinant proteins are evaluated by Western blot using specific antibodies. Protein concentrations are estimated by SDS-PAGE (see Note 3) and silver staining.

1. p300: the multifunctional coactivator p300 acts as a protein scaffold and a bridging factor for the assembly of the transcriptional apparatus (19). The intrinsic HAT activity of p300 can facilitate transcriptional activity by modulating the chromatin structure (10, 13, 19). FLAG-tagged p300 is synthesized in baculovirus-infected Sf9 cells (10). Recombinant p300 (see Note 4) is precipitated with ANTI-FLAG Affinity Gel and purified by FLAG elution buffer at 4°C (10). Add glycerol (10%) and DTT (1 mM final concentration). Store aliquots at –80°C.
2. Sox9: FLAG-tagged Sox9 (10) is produced in Sf9 cells using BaculoDirect Baculovirus Expression Systems (Invitrogen). Sox9 is purified by ANTI-FLAG Affinity Gel and stored in glycerol- and DTT-supplemented FLAG elution buffer at –80°C (see Note 5).
3. Smad3 and Smad4: TGF-β induces early chondrogenesis through the phosphorylation of Smad3 (12). Phosphorylated Smad3 is transferred into the nucleus by associating with Smad4 (15). Activated Smad3/4 complex associates with Sox9 and p300 and enhances Sox9-dependent transcription on chromatin (15). Simultaneous infection of FLAG-tagged Smad3 and untagged Smad4 is performed (see Note 6). Smad3/4 complex is purified from baculovirus-infected Sf9 cells after 15-min treatments of TGF-β3 (5 ng/mL, R&D Systems). Activated Smad3/4 complex is prepared by ANTI-FLAG Affinity Gel and eluted in FLAG elution buffer at 4°C (10, 15). Glycerol (10%) and DTT (1 mM final concentration) are added. Store aliquots at –80°C (see Note 7).
4. Nuclear extracts: Nuclear extracts are prepared from chondrocytic SW1353 cells and dialyzed in buffer D (16). HeLa cell nuclear extracts are prepared with Buffer D (see Note 8). Protein concentrations are measured by BCA Protein Assay Kit (Pierce). Store aliquots at –80°C.

3. Methods

3.1. Chromatin Assembly

Chromatin is generally assembled from circular plasmids and core histone octamers, using histone chaperone NAP-1 and the ACF (Acf1 and ISWI) assembly factors in the presence of ATP (Fig. 1). Handle purified proteins and mixed solutions carefully.

1. Chromatin assembly reactions are performed essentially as described (10, 17, 18, 20, 21) using 12×48-pGL3-Promoter (10).

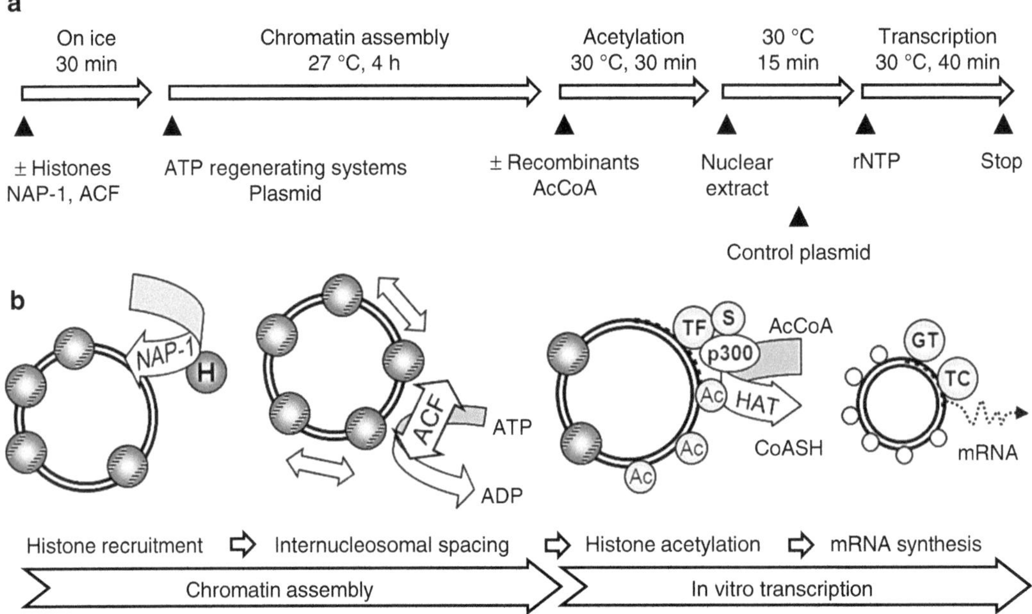

Fig. 1. (**a**) Sequential steps for chromatin assembly and in vitro transcription analyses. Arrow-heads indicate the addition of purified proteins and reagents. (**b**) Schematic illustrations of chromatinized DNA templates during chromatin assembly and in vitro transcription analyses. NAP-1 recruits histones (H) to unchromatinized plasmids (double-lined circle). ACF assembles periodic nucleosome arrays on histone-attached plasmids in an ATP-dependent manner. Transcription factor (TF) interacts with the inserted binding sequences (small dots on circulated plasmids) on the promoter. HAT coactivator p300 induces histone acetylation of chromatin templates through the association with TF. Stimulator (S) modulates the activity of multi-component transcriptional complex (TC) on reconstructed chromatin. After the histone acetylation step, general transcription factors (GTs) included in nuclear extracts progress the TC-dependent gene expression. Ac acetylated histone.

2. Standard reactions (20 μL) containing plasmid (150 ng), histones (100 ng), NAP-1 (500 ng), ACF complex (Acf1/ISWI, 0.65 ng each), ATP (3 mM), and ATP regenerating systems (30 mM phosphocreatine and 20 ng of creatine phosphokinase) are used for chromatin assembly.

3. Incubate closed circular plasmids at 94°C for 3 min, and then keep solutions on ice. Centrifuge.

4. Make a reaction of R buffer, PVOH/PEG mix, NAP-1, histones, and ACF in the indicated order to a single tube (see Note 9). Total amounts of reactions can be scaled up, depending on the number of samples. Mix gently. Another reaction mixture, without histones, is prepared to generate histone-free unchromatinized DNA.

5. Keep the samples on ice for 30 min.

6. To the samples, add ATP regenerating systems (ATP/phosphocreatine/$MgCl_2$ mix and creatine phosphokinase) and circular DNA templates for chromatin assembly (see Note 10). Mix gently.

7. Incubate reaction tubes at 27°C for 4 h (see Note 11).

   Micrococcal nuclease (MNase) digestion assays should be performed to assess assembled chromatin quality and demonstrate regularly spaced bands on 1.5% agarose gels stained with ethidium bromide (15, 20–22). These analyses ensure complete assembly of plasmid DNA into a chromatin template. A nucleosome-repeated pattern (approximately 165-bp repeat) indicates that assembled chromatin is suitable for subsequent in vitro transcription analysis.

3.2. In Vitro Transcription

Purified components and nuclear extracts activate transcription from reconstructed chromatin templates in the presence of HAT coactivator and AcCoA (Fig. 1). Purified proteins and mixed solutions must be handled carefully under RNase-free conditions.

1. Thaw purified transcriptional components, AcCoA, PVOH/PEG mix, control plasmid, nuclear extract, and rNTP mix. Keep all reagents on ice.
2. Combine chromatinized DNA template (or histone-free DNA template), transcriptional components, and AcCoA in 1.5 mL tubes (see Notes 8 and 12).
3. Incubate at 30°C for 30 min to allow histone acetylation of assembled chromatin templates.
4. To samples add Buffer H, PVOH/PEG mix, internal control plasmid (pRL-CMV), and nuclear extract (derived from SW1353 cells) in the indicated order.
5. Incubate at 30°C for 15 min.
6. Add rNTP mix to samples to initiate in vitro transcription.
7. Incubate at 30°C for 40 min to allow mRNA transcription from a chromatin (or histone-free) template and an internal control plasmid.
8. Stop in vitro transcription by the addition of transcription stop solution with Proteinase K.
9. Incubate at 37°C for 15 min to digest the proteins and nuclear extracts.
10. Add an equal volume (approximately 100 μL) of phenol–chloroform–isoamyl alcohol solution to the samples. Invert the mixture several times.
11. Centrifuge for 5 min. Transfer the upper aqueous phase to a new tube.
12. Add an equal volume (approximately 100 μL) of chloroform–isoamyl alcohol solution to the tube. Invert the mixture several times.
13. Centrifuge for 5 min and then transfer the upper aqueous phase to a new tube.

14. Add 10% volume of 3 M NaOAc and 250 μL of 100% ethanol to each sample. Mix well.
15. Keep the samples at −20°C for 10 min.
16. Centrifuge at 4°C for 1 h and carefully aspirate the supernatant (see Note 13).
17. Dry the mRNA-containing pellet at 75°C for 3 min. Store at −20°C.
18. S1 nuclease assay (10, 15) and/or primer extension analysis (20, 21) are required to evaluate the amount of RNA product derived from chromatin templates and internal control plasmids (see Note 14). Primer labeling and annealing steps should be performed carefully with (γ-^{32}P) ATP. The fragments protected from S1 nuclease digestion (or the cDNA products obtained by primer extension) are analyzed on 8% denaturing polyacrylamide gels and visualized by autoradiography.

4. Notes

1. Plasmid condition affects the efficiency of chromatin assembly and in vitro transcription. Repeated freezing and thawing of plasmid is not recommended. In our preliminary experiments, lower copies of Sox9-binding elements on the promoter region reduce the transcription activity derived from chromatinized template. The copy number of a target sequence should be optimized according to the activity of the chosen DNA-binding transcription factor.
2. A maximum concentration of 500 mM imidazole is recommended for the final elution. DTT-, EDTA-, and EGTA-containing buffers must not be used for the preparation of His-tagged proteins.
3. An appropriate gel concentration should be used for each purified protein. Histones (H2A, H2B, H3, and H4) are assessed using 15% SDS-polyacrylamide gels. NAP-1, Sox9, Smad3, and Smad4 are visualized in 10% gels. ACF complex (ISWI and Acf-1) and p300 are evaluated by 6% gels (10).
4. Baculovirus and recombinant protein of p300 degrade more easily than the other components used for chromatin assembly and in vitro transcription analyses. Careful handling and storage are necessary.
5. The DNA-binding activity of purified Sox9 should be assessed by electrophoretic mobility shift assay (EMSA) (15).
6. Preliminary experiments should be performed to obtain equal amounts of phosphorylated Smad3 and Smad4. Excess Smad4

interferes with production of Smad3 in Sf9 cells. Excess untagged Smad4 is not immunoprecipitated with FLAG-tagged Smad3 by ANTI-FLAG Affinity Gel purification.

7. Protein-protein interactions among purified transcriptional complexes should be evaluated by immunoprecipitation (IP) analysis (10, 15).
8. Nuclear extracts derived from HeLa cells can be used as a positive control. However, the amount of each component (Sox9, p300, and Smad3/4) should be adjusted in order to detect an optimum effect of the third molecule (Smad3/4) on chromatinized DNA template.
9. An appropriate balance among plasmid DNA, histones, NAP-1, and ACF complex should be determined in preliminary studies. The recommended ratio of DNAs and histones is 3:2. Recombinant NAP-1 (2–10-times volume of histones) should be added as a histone-shuttling protein in the reaction. The recommended ratio of Acf1 and histones is 1:150.
10. A reaction volume of 15–20 μL is needed for a single sample of in vitro transcription after chromatin assembly.
11. Incubation should be performed for 3–5 h. Longer incubation (>6 h) is detrimental.
12. The optimum amounts of each transcriptional component (i.e., Sox9, p300, and Smad3/4) should be examined in this in vitro transcription model. In our experience, the third stimulators (Smad3/4) should be added at five–tenfold volumes of DNA-binding transcription factor (Sox9). The HAT assay should be performed using radiolabeled ^{14}C-AcCoA to confirm the p300 activity and the balance between chromatinized template and purified proteins (10).
13. A pellet may be rarely visible. A rotary evaporator can be used.
14. The S1 nuclease assay detects the specific signals derived from mRNA products in this experiment more effectively than primer extension analysis. Nonspecific signals in primer extension obscure small increases in transcriptional activation caused by the third component (stimulator) on chromatin templates (22).

Acknowledgments

We thank Dr. Takashi Ito and Dr. Toshifumi Ozaki for their kind assistance. This work was supported by Okayama Medical Foundation, Japan Orthopaedics and Traumatology Foundation (No. 225), and JSPS Fujita Memorial Fund for Medical Research.

References

1. Jones PA, Takai D (2001) The role of DNA methylation in mammalian epigenetics. Science 293:1068–1070
2. Quina AS, Buschbeck M, Di Croce L (2006) Chromatin structure and epigenetics. Biochem Pharmacol 72:1563–1569
3. Wolffe AP, Hayes JJ (1999) Chromatin disruption and modification. Nucleic Acids Res 27:711–720
4. Jenuwein T, Allis CD (2001) Translating the histone code. Science 293:1074–1080
5. Furumatsu T, Ozaki T (2010) Epigenetic regulation in chondrogenesis. Acta Med Okayama 64:155–161
6. Akiyama H, Chaboissier MC, Martin JF et al (2002) The transcription factor Sox9 has essential roles in successive steps of the chondrocyte differentiation pathway and is required for expression of Sox5 and Sox6. Genes Dev 16:2813–2828
7. Shi Y, Massagué J (2003) Mechanisms of TGF-β signaling from cell membrane to the nucleus. Cell 113:685–700
8. Kamachi Y, Uchikawa M, Kondoh H (2000) Pairing SOX off: with partners in the regulation of embryonic development. Trends Genet 16:182–187
9. Tsuda M, Takahashi S, Takahashi Y et al (2003) Transcriptional co-activators CREB-binding protein and p300 regulate chondrocyte-specific gene expression via association with Sox9. J Biol Chem 278:27224–27229
10. Furumatsu T, Tsuda M, Yoshida K et al (2005) Sox9 and p300 cooperatively regulate chromatin-mediated transcription. J Biol Chem 280:35203–35208
11. Liu F (2003) Receptor-regulated Smads in TGF-β signaling. Front Biosci 8:s1280–s1303
12. Furumatsu T, Tsuda M, Taniguchi N et al (2005) Smad3 induces chondrogenesis through the activation of SOX9 via CREB-binding protein/p300 recruitment. J Biol Chem 280:8343–8350
13. Furumatsu T, Asahara H (2010) Histone acetylation influences the activity of Sox9-related transcriptional complex. Acta Med Okayama 64:351–357
14. Ferguson CM, Schwarz EM, Reynolds PR et al (2000) Smad2 and 3 mediate trans forming growth factor-β1-induced inhibition of chondrocyte maturation. Endocrinology 141:4728–4735
15. Furumatsu T, Ozaki T, Asahara H (2009) Smad3 activates the Sox9-dependent transcription on chromatin. Int J Biochem Cell Biol 41:1198–1204
16. Dignam JD, Lebovitz RM, Roeder RG (1983) Accurate transcription initiation by RNA polymerase II in a soluble extract from isolated mammalian nuclei. Nucleic Acids Res 11: 1475–1489
17. Ito T, Bulger M, Kobayashi R et al (1996) Drosophila NAP-1 is a core histone chaperone that functions in ATP-facilitated assembly of regularly spaced nucleosomal arrays. Mol Cell Biol 16:3112–3124
18. Ito T, Levenstein ME, Fyodorov DV et al (1999) ACF consists of two subunits, Acf1 and ISWI, that function cooperatively in the ATP-dependent catalysis of chromatin assembly. Genes Dev 13:1529–1539
19. Chan HM, La Thangue NB (2001) p300/CBP proteins: HATs for transcriptional bridges and scaffolds. J Cell Sci 114:2363–2373
20. Asahara H, Santoso B, Guzman E et al (2001) Chromatin-dependent cooperativity between constitutive and inducible activation domains in CREB. Mol Cell Biol 21:7892–7900
21. Fyodorov DV, Kadonaga JT (2003) Chromatin assembly in vitro with purified recombinant ACF and NAP-1. Methods Enzymol 371:499–515
22. Konesky KL, Laybourn PJ (2007) Biochemical analyses of transcriptional regulatory mechanisms in a chromatin context. Methods 41:259–270

Chapter 16

Using FRET to Monitor Protein-Induced DNA Bending: The TBP-TATA Complex as a Model System

Rebecca H. Blair, James A. Goodrich, and Jennifer F. Kugel

Abstract

Proteins that bind to DNA can elicit changes in DNA conformation, such as bending and looping, which are important signals for later events such as transcription. TATA-binding protein (TBP) is one example of a protein that elicits a conformational change in DNA; TBP binds and sharply bends its recognition sequence, which is thought to facilitate the recruitment of other protein factors. Here we describe the use of fluorescence resonance energy transfer (FRET) to evaluate DNA bending using TBP as a model system. FRET is a useful technique to measure changes in DNA conformation due to protein binding because small changes in the distance between two fluorophores (2–10 nm) translate into large changes in energy transfer.

Key words: FRET, DNA bending, Protein–DNA interaction, TBP, Fluorophore

1. Introduction

TBP recognizes and binds the TATA box in the promoter region of many genes. Upon binding, phenylalanines in TBP insert between specific base pairs in the TATA box, which causes the DNA to bend, producing two kinks in the DNA (1–6). The extent of DNA bending by TBP has been described using a specific two-kink model, and studies have found that the extent of bending changes based on TATA sequence and the organism from which TBP is purified (7–9). We have measured the extent of DNA bending due to human TBP by adding specific fluorophores to the ends of the target TATA DNA and measuring changes in the fluorescence resonance energy transfer (FRET) between the dyes (9, 10).

Minou Bina (ed.), *Gene Regulation: Methods and Protocols*, Methods in Molecular Biology, vol. 977,
DOI 10.1007/978-1-62703-284-1_16,

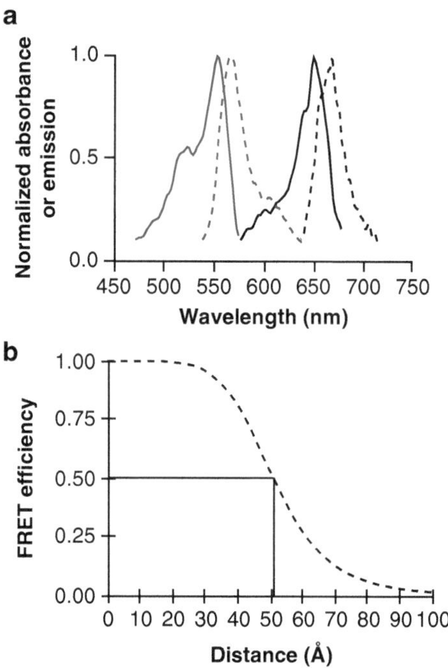

Fig. 1. (**a**) Spectral overlap of Alexa 555/647. The excitation spectra (*solid*) and emission spectra (*dashed*) of Alexa 555 (*grey*) and Alexa 647 (*black*) are shown. Peak excitation of Alexa 555 is 555 nm and peak emission is 565 nm. Peak excitation of Alexa 647 is 650 nm and peak emission is 670 nm (12). (**b**) FRET efficiency is plotted versus distance for the Alexa555/647 FRET pair, where the R_0 is 51 Å.

FRET is a useful tool for studying biological molecules because a large change in FRET can be observed for distance changes in the low nanometer range, which occur frequently in biology. FRET requires two fluorophores, a donor and an acceptor, that spectrally overlap. After the donor dye is excited, the energy is transferred to the acceptor dye and light is given off at the acceptor emission wavelength. The amount of energy transfer is dependent on the distance between the two dyes; when the donor and acceptor are closer, there is higher energy transfer. This efficiency is described by the following equation:

$$\text{FRET efficiency} = \frac{R_0^{\,6}}{R_0^{\,6} + r^6}$$

Where r is the distance between the donor and acceptor dyes and R_0 is the Förster radius, which is equal to the distance at which the FRET efficiency is 0.5. The value of R_0 is unique to each dye pair (11). The studies described here use Alexa 555 as the donor dye and Alexa 647 as the acceptor; the spectral overlap of these two dyes is shown in Fig. 1a. For Alexa 555 and Alexa 647 the Förster radius is 51 Å (12); which is illustrated in the plot of FRET versus distance shown in Fig. 1b. The FRET efficiency between two dyes can be experimentally determined using the following equation:

$$\text{FRET efficiency} = \frac{A}{A + D}$$

Where A is the acceptor emission signal after excitation at the donor wavelength and D is the donor emission signal (11).

There are a few important points to consider when using FRET to study protein-induced DNA bending. First, FRET measures the change in distance between the two dyes upon protein binding, but more information is needed to determine how the change in distance relates to the geometry and conformation of the DNA. For example, for the TBP/TATA DNA complex, we can convert distance to bend angle because a model of the complex exists that is based on crystal structures and extensive biochemical data (3–5, 7, 13). Moreover, FRET efficiency can change not only due to changes in distance, but changes in the conformation and orientation of the two dyes with respect to one another, so care must be taken to ensure that the change in FRET efficiency is attributable to a change in distance. Another important consideration is that DNA bending is measured in a solution of many DNA and TBP molecules. If every DNA molecule in solution is not bound by protein, then the measured FRET efficiency will both reflect the change in the conformation of the DNA upon protein binding, as well as the fraction of DNA that is bound, which is driven by the affinity of the interaction and the concentrations of protein and DNA.

Here we describe an experimental procedure for measuring DNA bending by TBP using a double-stranded TATA DNA with a donor dye on one 5′ end and an acceptor dye on the other 5′ end. After adding saturating amounts of TBP, the FRET efficiency is measured. An increase in FRET corresponds to a decrease in the distance between the dyes, which indicates DNA bending.

2. Materials

Prepare all solutions using ultrapure water (18 MΩ). Store all solutions in Subheading 2.1 in small aliquots at –20°C, unless otherwise stated. These are then used to make the buffers described in Subheadings 2.2 and 2.3, which should be prepared fresh prior to each experiment.

2.1. General Solutions

1. 2 M Tris, pH 7.9: Dissolve 4.84 g Tris base into 10 mL water. Add concentrated HCl until pH is approximately 7.9. Let the solution sit at room temperature for a few hours (since pH is sensitive to temperature for Tris). Check pH and add concentrated HCl until pH is 7.9. Add water to a final volume of 20 mL.

Table 1
The sequences and fluorescent modifications of oligos used in these studies

Oligo #	Oligo description	Sequence (5′ to 3′)
1	Donor-top strand	Alexa 555—CAGGCTATAAAAGGGACG
2	Unlabeled-top strand	CAGGCTATAAAAGGGACG
3	Acceptor-bottom strand	Alexa 647—CGTCCCTTTTATAGCCTG
4	Unlabeled-bottom strand	CGTCCCTTTTATAGCCTG

2. 100 mM $MgCl_2$: Dissolve 1.02 g $MgCl_2$•$6H_2O$ in water to a final volume of 50 mL.
3. 2 M KCl: Dissolve 7.5 g KCl in water to a final volume of 50 mL.
4. Bovine Serum Albumin (BSA, 20 mg/mL, from Roche)
5. 1 M DTT (1,4–dithiothreitol): Dissolve 15.4 g in water to a final volume of 50 mL.
6. 80% glycerol: Mix 10 mL 100% glycerol with 2.5 mL water (see Note 1).
7. 1 M HEPES, pH 7.9: Dissolve 4.76 g HEPES into 10 mL water. Check pH and add 4 M NaOH until the pH reaches 7.9. Add water to a final volume of 20 mL.
8. 0.5 M EDTA, pH 8.0: Combine 10 g NaOH and 400 mL water, and add 93 g Na_2EDTA while stirring. Increase the volume to 500 mL with water. Check pH to 8.0. Store at room temperature.
9. 20% Nonidet-P40 (NP-40): Combine 1 mL 100% NP-40 and 4 mL water. Store in dark at 4°C.

2.2. Solutions and Equipment Needed for Making and Analyzing Annealed TATA DNA

1. 10× Anneal Buffer: 200 mM Tris (pH 7.9), 20 mM $MgCl_2$, 500 mM KCl. Mix 5 μL 2 M Tris (pH 7.9), 10 μL 100 mM $MgCl_2$, 12.5 μL 2 M KCl, and 22.5 μL water.
2. Single-stranded TATA oligos: Order each oligo (top and bottom strands) with and without a fluorophore (Invitrogen); the sequences of the oligos and positions of the labels are shown in Table 1. Dilute the lyophilized oligos to a final concentration of 100 μM in 18 MΩ water. Store at −20°C.
3. Ammonium persulfate (APS): 10% solution in water. Store at 4°C.
4. *N′*,*N′*,*N′*,*N′*-Tetramethylethylenediamine (TEMED). Store at 4°C.

5. Equipment for running a native gel: Low fluorescence notched glass plate set (20 × 22 cm, CBS scientific), 1.5 mm thick × 22 cm spacer set (CBS scientific), 1.5 mm thick 20-well comb for 20 cm wide units (CBS scientific), and an Owl electrophoresis system (Thermo Scientific).
6. 5× TBE: 0.45 M Tris (pH 8.3), 0.44 M Boric Acid, 10 mM EDTA. Combine 54 g Tris Base, 27.5 g Boric Acid, 20 mL 0.5 M EDTA (pH 8.0), and water to a final volume of 1 L. The pH should be 8.3. Can store at room temperature.
7. Running buffer for native gel: 100 mL 5× TBE, 50 mL 100% glycerol, 850 mL water.
8. 30% acrylamide/Bis solution (37.5:1).
9. Buffer A: 1 mM DTT, 10 mM $MgCl_2$, 25 mM HEPES (pH 7.9). Combine 0.5 μL 1 M DTT, 50 μL 100 mM $MgCl_2$, 12.5 μL 1 M HEPES (pH 7.9), and 437 μL water.

2.3. Solutions and Equipment Used for TBP/DNA Complex Formation and FRET Measurements

1. Buffer A (described in Subheading 2.2).
2. Buffer B: 20% glycerol, 20 mM Tris (pH 7.9), 300 mM KCl, 1 mM DTT, 0.1 mg/mL BSA, 0.2% NP-40. Combine 125 μL 80% glycerol, 5 μL 2 M Tris (pH 7.9), 75 μL KCl, 0.5 μL 1 M DTT, 2.5 μL 20 mg/mL BSA, 5 μL 20% NP-40, and 287 μL water.
3. TBP: 2 μM solution in Buffer B. Human TBP can be purchased (ProteinOne) or expressed in E.coli and purified as described previously (14). Solutions of TBP should be stored in small aliquots at −80°C.
4. 384-well SensoPlate, PS, Glass Bottom, which has an optically clear bottom (Greiner).
5. Typhoon 9400 Scanner (GE Healthcare).

3. Methods

3.1. Annealing Reactions for the DNA Constructs

Three double-stranded DNAs need to be prepared: (1) doubly labeled TATA DNA (DL) containing both the donor and acceptor dyes, (2) donor-only TATA DNA (DO), and (3) acceptor-only TATA DNA (AO). For the DL DNA, every strand with a donor dye needs to have a strand with an acceptor dye annealed to it to obtain an accurate FRET measurement. For the AO and DO DNAs, every labeled strand needs to have an unlabeled strand annealed to it. To accomplish this the annealing reactions contain unequal amounts of the oligos. The annealing reactions are run on a native gel to assess whether the aforementioned parameters were obtained. Note: the oligo numbers referred to below are provided in Table 1.

1. Prepare the DL DNA: Combine 1 μM donor-labeled top oligo with 1.5 μM acceptor-labeled bottom oligo in 1× anneal buffer (see Note 2). To prepare a 50 μL annealing reaction, add 5 μL 10× anneal buffer, 0.5 μL 100 μM Oligo #1, 0.75 μM 100 μM Oligo #3, and 43.75 μL water.
2. Prepare the AO DNA: Combine 1.5 μM unlabeled top oligo with 1 μM acceptor-labeled bottom oligo in 1× anneal buffer. To prepare a 50 μL annealing reaction, add 5 μL 10× anneal buffer, 0.75 μL 100 μM Oligo #2, 0.5 μM 100 μM Oligo #3, and 43.75 μL water.
3. Prepare the DO DNA: Combine 1 μM donor-labeled top oligo with 1.5 μM unlabeled bottom oligo in 1× anneal buffer. To prepare a 50 μL annealing reaction, add 5 μL 10× anneal buffer, 0.5 μL 100 μM Oligo #1, 0.75 μM 100 μM Oligo #4, and 43.75 μL water.
4. In a thermal cycler, heat the annealing reactions to 95°C for 5 min, then drop the temperature to the approximate melting temperature of the two oligos (60°C for the TATA DNA constructs) for 45 min, followed by a slow cool of 0.1°C/s to 15°C. Store in the freezer at −20°C in small aliquots.
5. Pour a 7% native polyacrylamide gel: Combine 18.7 mL 30% acrylamide, 8 mL 5× TBE, 47.7 mL water, 5 mL 80% glycerol, and mix well. To polymerize, add 450 μL 10% APS and 110 μL TEMED. Immediately fill plates with acrylamide mix without getting bubbles in the gel. Insert comb and let sit until polymerized (~20 min).
6. Dilute annealed DNAs: Prepare dilutions of all three annealing reactions (DL, DO, AO) and the single-stranded labeled oligos (Oligo #1 and Oligo #3) to 5 nM in 20 μL buffer A.
7. After the gel is polymerized, place the gel in the gel apparatus and add native gel running buffer to upper and lower reservoirs of gel apparatus. Pre-run the gel at 150 V for 5 min before loading samples into the wells. Load 18 μL of each DNA sample into separate lanes and run the gel at 150 V for 2 h (see Note 3).
8. Scan the gel on a Typhoon 9400 scanner, for the donor (532 nm excitation, 580/30 nm emission), and acceptor (633 nm excitation, 670/30 nm emission) fluorophores. Make sure that the focal plane of the scan is at +3 mm (see Note 4).
9. Open the donor and acceptor output scans of the gel using the software of your choice, for example ImageJ or ImageQuant. Overlay your gels to determine the ratio of double-stranded versus single-stranded DNA (see Note 5). The native gel will reveal: (1) whether all the donor oligo in the DL DNA is double-stranded; (2) whether the DL DNA contains too much

excess single-stranded acceptor; (3) whether in the AO and DO constructs the fluorescent DNA is all double stranded; (4) whether excess free dye is present in the single stranded oligos.

10. If necessary, the annealing procedure can be repeated while adjusting the ratios of donor to acceptor oligo accordingly so that in the DL DNA, all the donor oligo is double-stranded with minimal single-stranded acceptor present (see Note 6).

3.2. Binding Reactions

Twenty-one reactions will be assembled according to Table 2. These include both the experimental and control reactions that are required to determine the extent to which a given protein bends a single DNA sequence.

1. Make 20 μL of 10 nM DNA solutions by diluting a portion of the DNA annealing reactions above (DL, DO, AO) in Buffer A.
2. Set up binding reactions on ice according to Table 2 in microcentrifuge tubes. Add the components in the order listed in the table (i.e., add component in first row to all tubes, then move on to next component down). Mix well prior to incubating at room temperature for 20 min in the dark (see Notes 7 and 8).
3. After 20 min, transfer 18 μL of each reaction to a well in the 384-well plate, being careful not to get bubbles in the wells. Make sure that all liquid is at the bottom of the well (i.e., not stuck to the sides of the well).
4. Scan the plate using the Typhoon scanner. Three fluorescence intensity measurements will be made: (1) I_{DD}; excitation and emission from the donor fluorophore (532 nm excitation, 580/30 nm emission), (2) I_{AA}; excitation and emission from the acceptor fluorophore (633 nm excitation, 670/30 nm emission), and (3) I_{DA}; excitation of the donor fluorophore with emission from the acceptor fluorophore (532 nm excitation, 670/30 nm emission), which provides a direct measurement of the FRET. Make sure that the focal plane of the scan is set at +3 mm and the PMT is 600 V (see Note 9).

3.3. Analysis of the Data to Determine the FRET Efficiency and the Distance Between the two Dyes

The calculations described below refer to the scans described in Subheading 3.2, step 4 (I_{AA}, I_{DD}, and I_{DA}).

1. Quantitate the fluorescence in each well using the software program of your choice. We use ImageJ (NIH, http://rsbweb.nih.gov/ij/) to obtain a signal intensity from each well of each scan (see Note 10).
2. Copy all measurements into a spreadsheet program such as Excel to perform the following mathematical manipulations.
3. Use the buffer only reactions to determine average background (reactions 19–21) for each scan (I_{AA}, I_{DD}, and I_{DA}). Subtract

Table 2
The amount of each solution (in μL) that should be added to reactions

Reaction component (μL)	Reaction number																				
	1	2	3	4	5	6	7	8	9	10	11	12	13	14	15	16	17	18	19	20	21
Buffer A	8	8	8	8	8	8	8	8	8	8	8	8	8	8	8	8	8	8	10	10	10
10 nM DL DNA	2	2	2	2	2	2	–	–	–	–	–	–	–	–	–	–	–	–	–	–	–
10 nM DO DNA	–	–	–	–	–	–	2	2	2	2	2	2	–	–	–	–	–	–	–	–	–
10 nM AO DNA	–	–	–	–	–	–	–	–	–	–	–	–	2	2	2	2	2	2	–	–	–
Buffer B	10	10	10	8.8	8.8	8.8	10	10	10	8.8	8.8	8.8	10	10	10	8.8	8.8	8.8	10	10	10
2 μM TBP	–	–	–	1.2	1.2	1.2	–	–	–	1.2	1.2	1.2	–	–	–	1.2	1.2	1.2	–	–	–

the average intensity of the background from all other intensity values within the scan. Use these background-subtracted intensities for all subsequent calculations.

4. Calculate the fluorescence of the donor fluorophore at the acceptor emission wavelength (B_D, donor bleed-through), using the average I_{DA} and I_{DD} obtained from the DO DNA in the absence of TBP (reactions 7–9):

$$B_D = \frac{\text{avg} I_{DA}}{\text{avg} I_{DD}}$$

In the example experiment shown here, B_D was 0.07.

5. Calculate the amount of fluorescence emitted by the acceptor fluorophore when it is excited at the donor wavelength (B_A, direct excitation of the acceptor), using the average I_{AA} and I_{DA} obtained from the AO DNA in the absence of TBP (reactions 13–15):

$$B_A = \frac{\text{avg} I_{DA}}{\text{avg} I_{AA}}$$

In the example experiment shown here, B_A was 0.03.

6. For reactions containing DL DNA, determine the corrected FRET values (I_{DA}*) using the intensities obtained from the I_{DA} scan, and the correction factors determined in steps 4 and 5. For reactions 1–6, apply the following equation:

$$I_{DA}^* = (I_{DA} - (B_D \times I_{DD}) - (B_A \times I_{AA}))$$

7. Calculate the effect of TBP on the donor fluorophore (P_D) using the average intensities of I_{DD} for the DO reactions, in the absence of TBP (reactions 7–9) and the presence of TBP (reactions 10–12):

$$P_D = \frac{\text{avg} I_{DD} \text{ with TBP}}{\text{avg} I_{DD} \text{ without TBP}}$$

In the example experiment shown here, P_D was 1 (see Note 11).

8. Calculate the effect of TBP on the acceptor fluorophore (P_A) using averages of I_{AA} measured for the AO reactions, in the absence of TBP (reactions 13–15) and the presence of TBP (reactions 16–18):

$$P_A = \frac{\text{avg} I_{AA} \text{ with TBP}}{\text{avg} I_{AA} \text{ without TBP}}$$

In the example experiment shown here, P_A was 1 (see Note 11).

Table 3
Experimentally determined FRET efficiency and distance (in Å) for the unbound and TBP-bound DL DNA

	FRET Efficiency	Distance (Å)
Unbound DNA (reactions 1–3)	0.18 ± 0.01	65.6 ± 0.04
Bound/bent DNA (reactions 4–6)	0.36 ± 0.02	55.9 ± 0.04

9. For the reactions containing DL DNA and TBP (reactions 4–6), correct the donor signal for the effect of the protein on the donor fluorophore to obtain the corrected donor intensity (D_{corr}):

$$D_{corr}\frac{I_{DD}}{P_D}$$

Determine D_{corr} for reactions 4–6 separately.

10. For reactions containing DL DNA and TBP (reactions 4–6), correct the intensities from the I_{DA} scan for the effect of the protein on the acceptor fluorophore. Note that the following equation uses the I_{DA}* values previously determined.

$$A_{corr}\frac{I_{DA}^*}{P_A}$$

Determine A_{corr} for reactions 4–6 separately.

11. Determine the FRET efficiency for each sample containing DL DNA (reactions 1–6).

$$\text{FRET} = \frac{A_{corr}}{A_{corr} + D_{corr}}$$

Note that A_{corr} and D_{corr} are only used for reactions 4–6 that contained TBP. For reactions 1–3, use the I_{DA}* and I_{DD} values, respectively. The calculated, corrected FRET efficiencies for a sample experiment are presented in Table 3.

12. Determine the distance between the dyes. Rearranging the equation in the introduction that describes the relationship between FRET efficiency and distance, gives the following:

$$r = \left(\frac{R_0^6}{\text{FRET}} - R_0^6\right)^{1/6}$$

Where R_0 is 51 Å for the Alexa 555 and Alexa 647 dyes. A protein that bends DNA will cause a decrease in r, the distance between the two dyes (see Table 3 and Notes 12 and 13).

4. Notes

1. 100% glycerol is very viscous and it is difficult to pipet; therefore, carefully pour 10 mL 100% glycerol into a 15 mL conical and then add water to 12.5 mL.
2. You may need to try several different ratios of donor: acceptor DNA. Ideally, the entire donor DNA will be annealed with as little single-stranded acceptor oligo in solution as possible.
3. Do not run the gel at too high a voltage or it will get hot and change the equilibrium of annealed versus single-stranded DNA.
4. When scanning the gel, place the thinner plate against the glass surface of the Typhoon scanner.
5. You can overlay the donor and acceptor channels in ImageJ by converting the image to 8-bit format and then using the Colocalization plug-in.
6. The most cautious way to ensure that you will use fully annealed donor and acceptor oligos is to gel purify the dsDNA away from single-stranded oligos.
7. During the 20 min incubation, scan the empty 384-well plate on the Typhoon scanner at the donor and acceptor wavelengths to identify wells with no/low background. Use these wells for your experiment.
8. Keep the TBP on dry ice prior to use. After all other reaction components are assembled and mixed well, thaw the TBP and add it to reaction tubes. Snap-freeze the TBP in liquid nitrogen afterwards. TBP that is handled in this manner can typically survive 4 freeze-thaw cycles without losing activity.
9. To ensure that the intensity of each fluorophore in the experiment is within the linear range of the Typhoon scanner, perform separate titrations of single-stranded acceptor DNA and single-stranded donor DNA. When the intensity of the fluorescence plateaus as you further increase the concentration of a fluorophore-labeled oligo, then you have exceeded the linear range of the Typhoon scanner under its current settings. We set the PMT to 600 V, but the gain (PMT voltage) can be adjusted to keep the signal for a given concentration of fluorophore within the linear range.
10. If using ImageJ on data collected with the Typhoon scanner, each measurement (signal) should be corrected using the following equation, prior to background subtraction:

    $$\text{Corrected signal} = \text{signal}^2$$

    This conversion is only applied if using the Typhoon scanner (a file with a ".gel" extension). This is due to the fact that the software running the Typhoon scanner saves the square root of signal intensity. If you analyze in ImageQuant, the program automatically squares the signal. Also note that the equation assumes that the intensities of all wells were determined using squares (or rectangles) of the same size. If this is not the case, the corrected signal of each well should be multiplied by the area of the rectangle used for quantitation of that well.

11. If the correction factor (P_D) equals one, then the protein had no effect on the fluorophore. The more the protein affects the fluorescent properties of the dye, the lower the correction factor. If the correction factor is significant, you may want to consider altering the design of your DNA construct to minimize the effect of the protein, which may require choosing a different dye pair.
12. To assess DNA saturation, perform a titration of TBP up to 200 nM. Complete binding will be apparent when the FRET efficiency reaches a plateau. Also, note that for a given protein it might be difficult to fully saturate DNA for a variety of reasons, including poor affinity of the protein for the DNA or incomplete oligo annealing. More information on estimating protein/DNA binding affinities can be found in (15).
13. If you have a model for binding and bending, the angle at which the DNA is bent can sometimes be calculated from the distance between the fluorophores, as described for the TBP/TATA complex (7, 9, 10).

Acknowledgments

This work was supported by grant MCB-0919935 from the National Science Foundation.

References

1. Starr DB, Hawley DK (1991) TFIID binds the minor groove of the TATA box. Cell 67: 1231–1240
2. Starr DB, Hoopes BC, Hawley DK (1995) DNA bending is an important component of site-specific recognition by the TATA binding protein. J Mol Biol 250:434–446
3. Kim JL, Nikolov DB, Burley SK (1993) Co-crystal structure of TBP recognizing the minor groove of a TATA element. Nature 365:520–527
4. Kim Y, Geiger JH, Hahn S et al (1993) Crystal structure of a yeast TBP/TATA-box complex. Nature 365:512–520
5. Nikolov DB, Chen H, Halay ED et al (1996) Crystal structure of a human TATA box-binding protein/TATA element complex. Proc Natl Acad Sci USA 93:4862–4867
6. Lee DK, Horikoshi M, Roeder RG (1991) Interaction of TFIID in the minor groove of the TATA element. Cell 67:1241–1250
7. Wu J, Parkhurst KM, Powell RM et al (2001) DNA bends in TATA-binding protein-TATA complexes in solution are DNA sequence-dependent. J Biol Chem 276:14614–14622
8. Whittington JE, Delgadillo RF, Attebury TJ et al (2008) TATA-binding protein recognition and bending of a consensus promoter are protein species dependent. Biochemistry 47: 7264–7273
9. Hieb AR, Halsey WA, Betterton MD et al (2007) TFIIA changes the conformation of

the DNA in TBP/TATA complexes and increases their kinetic stability. J Mol Biol 372:619–632

10. Kugel JF (2008) Using FRET to Measure the Angle at Which a Protein Bends DNA. Biochem Mol Biol Edu 36:341–346
11. Lakowicz JR (2006) Principles of fluorescence spectroscopy. Springer, New York
12. Johnson I, Spence MTZ (2010) Molecular Probes Handbook, A Guide to Fluorescent Probes and Labeling Technologies, 11th edn. Life Technologies, Oregon
13. Wu J, Parkhurst KM, Powell RM et al (2001) DNA sequence-dependent differences in TATA-binding protein-induced DNA bending in solution are highly sensitive to osmolytes. J Biol Chem 276:14623–14627
14. Weaver JR, Kugel JF, Goodrich JA (2005) The sequence at specific positions in the early transcribed region sets the rate of transcript synthesis by RNA polymerase II in vitro. J Biol Chem 280:39860–39869
15. Goodrich JA, Kugel JF (2007) Binding and kinetics for molecular biologists. Cold Spring Harbor Laboratory Press, New York

Chapter 17

Promoter Independent Abortive Transcription Assays Unravel Functional Interactions Between TFIIB and RNA Polymerase

Simone C. Wiesler, Finn Werner, and Robert O.J. Weinzierl

Abstract

TFIIB-like general transcription factors are required for transcription initiation by all eukaryotic and archaeal RNA polymerases (RNAPs). TFIIB facilitates both recruitment and post-recruitment steps of initiation; in particular, TFIIB stimulates abortive initiation. X-ray crystallography of TFIIB-RNAP II complexes shows that the TFIIB linker region penetrates the RNAP active center, yet the impact of this arrangement on RNAP activity and underlying mechanisms remains elusive. Promoter-independent abortive initiation assays exploit the intrinsic ability of RNAP enzymes to initiate transcription from nicked DNA templates and record the formation of the first phosphodiester bonds. These assays can be used to measure the effect of transcription factors such as TFIIB and RNAP mutations on abortive transcription.

Key words: RNA polymerase, Abortive transcription, Biochemical analysis, Saturation mutagenesis, High-throughput, Luciferase

1. Introduction

The central dogma of biology renders transcription a fundamental process that is indispensable in all three domains of life (i.e. bacteria, eukaryotes, and archaea). RNA polymerases use rNTPs to synthesize single-stranded RNA from a double-stranded DNA template. RNAPs cannot function autonomously but require the aid of basal transcription factors for their progression through the transcription cycle.

The transcription cycle is a multistep process which can be broadly divided into three stages, namely, (1) initiation, (2) elongation, and (3) termination (reviewed in ref. 1). Transcription initiation is the most highly regulated of the three stages with a number of events required to occur prior to full length transcript production.

Minou Bina (ed.), *Gene Regulation: Methods and Protocols*, Methods in Molecular Biology, vol. 977,
DOI 10.1007/978-1-62703-284-1_17,

First, a preinitiation complex (PIC) consisting of a defined set of basal transcription factors assembles at the promoter of a gene. The basal transcription factors that constitute the PIC locate the promoter core elements of a gene (TBP), bridge between TBP/DNA and RNAP II (TFIIB), stabilize the complex and counteract the negative effect of repressive factors such as NC-1 (TFIIA), aid RNAP recruitment and change DNA topology (TFIIF), or assist in promoter melting and concomitant isomerization of the preinitiation complex (TFIIE, TFIIH) (2–4). Following the recruitment of RNAP, the "closed complex" is formed (3, 5, 6). The initiation complex, as well as RNAP itself, must undergo a series of structural rearrangements which occurs simultaneously with DNA strand separation and formation of the transcription bubble (7). This complex is referred to as the open complex (3, 5, 6). During the initial phase of transcription initiation RNAP synthesizes repeatedly short transcripts between 3 and 9 nucleotides in length; these are referred to as abortive transcripts and can represent up to 95% of the total transcription output (8–14). The underlying mechanisms are unclear but in all likelihood involve tensions in the initiation complexes due to DNA "scrunching" effects dependent on interactions between RNAP and the promoter DNA. Subsequent structural rearrangements result in promoter escape and processive transcription elongation (15). The mechanistic details of transcription initiation (i.e., formation of a promoter complexes and structural rearrangements required to establish the transcription bubble and to make RNAP competent for transcription) are fundamental and follow the same molecular principles in bacteria, archaea and eukaryotes (16, 17).

1.1. Interactions Between TFIIB and RNAP

The investigation of protein-protein interactions within the PIC relies to a large extent on X-ray crystallography (18–20) and biochemical cross-linking assays (21, 22). Structural snapshots of different stages of the transcription cycle provide us with some insight into functional relationships (11, 23–28). Biochemical analysis strategies complement these structural data and promote our understanding of the molecular mechanisms operating in this system. We were particularly interested in the interface of RNAP II and TFIIB which—together with TBP—is indispensable for promoter-directed transcription both in eukaryotic RNAPI, II, and III and archaeal RNAPs (29–36).

TFIIB comprises a tripartite structure, (1) a C-terminal core domain to interact with TBP and promoter DNA, (2) an N-terminal zinc-ribbon domain to interact with and recruit RNAP, and (3) a flexible linker domain in between that has been demonstrated to stimulate the catalytic activity of RNAP (18–20, 37). TFIIB-RNAP II co-crystals show glimpses of the interactions between the TFIIB linker and the RNAP II surface, suggesting that the TFIIB linker penetrates the active center cleft (18–20). Biochemical analyses demonstrated that the linker domain of TFIIB actively contributes to the catalytic activity of RNAP, which in its presence is substantially higher than in its absence (35, 38, 39). The resolution of these structures is, however, relatively poor and lacks the nucleic acid substrates present in preinitiation complexes.

By applying a saturation mutagenesis approach (i.e., by substituting every amino acid residue of a given sequence by all 19 other amino acids) we were able to investigate the influence that TFIIB has on the abortive transcription activity of RNAP (35, 39). This approach was facilitated by adopting a model system that has proven its high degree of accessibility on several occasions. The RNAP of the hyperthermophilic archaeon *Methanocaldococcus jannaschii* (*mj* RNAP) is highly orthologous to eukaryotic RNAP II in subunit composition, structure and function. A lifestyle in extreme environments has required the adaptation of the organism's proteins to high temperatures and pressures which has given them an exceptional degree of robustness. The experimental handling of these proteins is therefore much easier and has facilitated the successful in vitro assembly of the *mj* RNAP from recombinantly produced subunits (40). The archaeal counterparts of TBP (*mj* TBP), TFIIB (*mj* TFIIB), and TFIIE (*mj* TFIIE) can be studied in this system as well (38). This provides us with a powerful tool that is readily accessible to saturation mutagenesis approaches during which we can study protein-protein interactions in an unbiased manner and on a single-residue level. Saturation mutagenesis has already given us extensive insight into the protein dynamics accompanying the nucleotide addition cycle catalyzed by *mj* RNAP that structural data have failed to uncover (41, 42). Given the success in this system, we recently included the interface of the *mj* TFIIB linker and *mj* RNAP into our studies to further investigate the stimulation effect of the linker on *mj* RNAP activity (39).

1.2. Promoter-Dependent and -Independent Transcription Assays

The catalytic activity of RNAP enzymes is usually assessed by measuring the production of RNA transcripts under defined reaction conditions. Different types of such transcription assays accommodate the need to investigate different aspects and stages of the transcription cycle.

Promoter-dependent assays make use of a well-defined biologically relevant DNA template, which can be linear or supercoiled. They measure the rate of transcription that occurs from a particular promoter and rely on transcription factors, and on a PIC forming at that promoter in a sequence-dependent manner. These assays have the caveat of being biased by the promoter sequence and its strength. The efficiency of RNAP recruitment, promoter opening, promoter escape, etc., are dependent on several transcription factors. Therefore, masking effects and redundancies occur within a PIC and make it difficult to quantitate the contribution of individual factors.

In contrast, promoter-independent assays constitute a minimal in vitro system and do not rely on any sequence properties and they can test purely for the catalytic activity of RNAP without the interference of other transcription factor activities. Transcription is randomly initiated at 3′ overhangs or nicks in DNA templates.

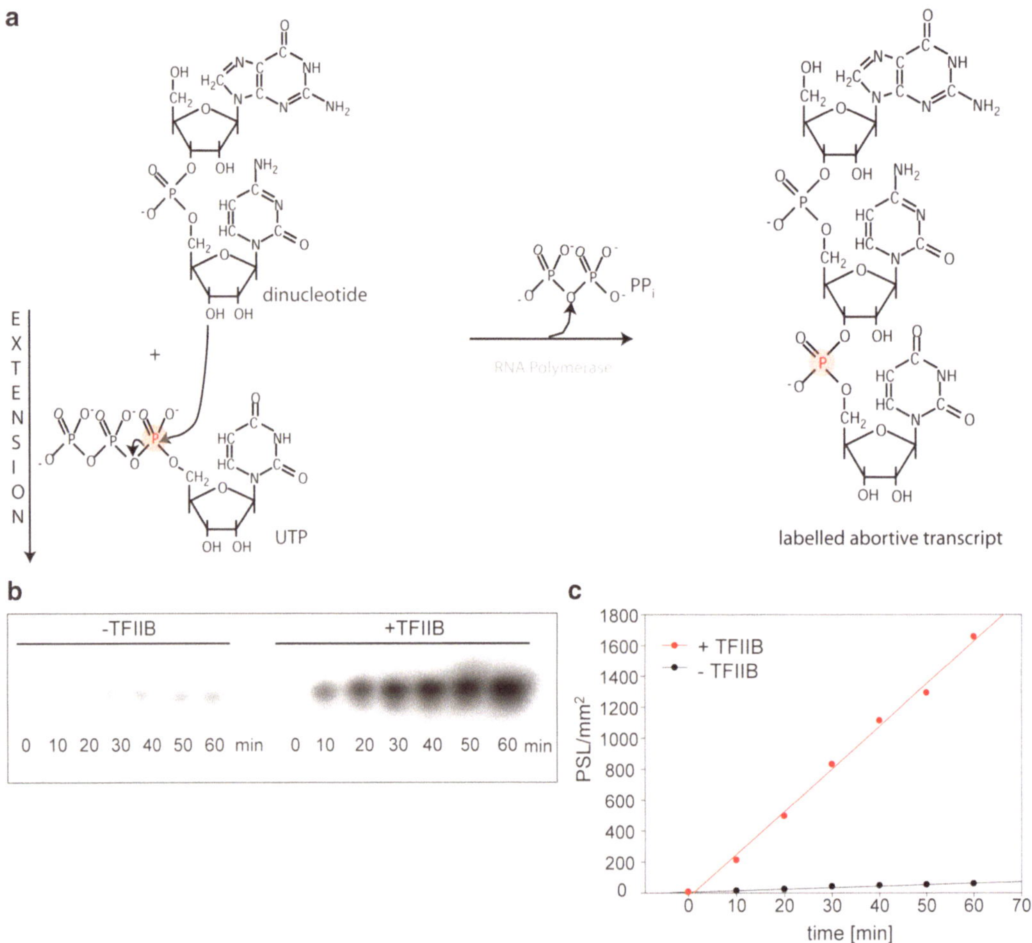

Fig. 1. Promoter-independent abortive initiation assays. (**a**) The assay exploits the intrinsic ability of RNAP to initiate from nicked DNA templates in a nonspecific manner. In the presence of a single type of (labelled) rNTP, RNAP catalyzes the extension of a dinucleotide primer to form a short, labelled, abortive transcript which can be visualized on a gel. Pyrophosphate is released as a by-product of the reaction. (**b**) Abortive transcripts can be visualized on a gel. In the presence of *mj*TFIIB, transcripts accumulate at a much higher rate. (**c**) The rate at which abortive transcripts are produced is time-dependent.

RNAPs do have a basal affinity to DNA and the availability of such nicks abolishes the need for precise promoter positioning. The single-stranded DNA can enter the catalytic site of RNAP without the requirements of transcription. Such assays are particularly suitable to study the catalytic mechanism of RNAP. This type of assays also sets a clear standard for RNAP activity by measuring it under particular experimental conditions and can be used to assess how the activity of RNAP is affected upon changing experimental conditions e.g., the presence of transcription factors.

In a promoter-independent transcription assay, *mj*TFIIB stimulates RNAP activity significantly (38). We found that this effect was even more pronounced in a promoter-independent abortive initiation assay (Fig. 1b, c).

2. Materials

2.1. Gel-Electrophoresis Components

1. TBE/acrylamide/urea mix (UBS).
2. 40% acrylamide (19:1) (Bio-Rad).
3. Glycerol.
4. *N*, *N*, *N*, *N*2-tetramethyl-ethylenediamine (TEMED).
5. 10% APS solution: dissolve 1 g ammonium persulfate in 10 ml distilled H_2O.
6. Commercial 10× Tris–borate/EDTA running buffer.
7. Vertical gel unit, 100×200 mm (Fisher Scientific, FB69602).

2.2. Dinucleotide Extension Assay

1. 5× transcription buffers
   (a) *M. jannaschii* transcription buffer: 250 mM Tris–Cl [pH7.5], 375 mM KCl, 125 mM $MgCl_2$.
   (b) *E. coli* transcription buffer: 200 mM Tris–Cl [pH7.5], 750 mM KCl, 50 mM $MgCl_2$, 0.05% Triton-X100.
   (c) T7 transcription buffer: 400 mM Tris–Cl [pH7.9], 60 mM $MgCl_2$, 20 mM spermidine.
2. Working stock solution of 0.3 μg/μl activated calf thymus DNA in TE buffer.
3. Working stock of 100 mM DTT in distilled H_2O.
4. Working stock of 10 mM GpC in distilled H_2O.
5. Working stocks of 0.25 mM and 10 mM unlabeled rUTP in distilled H_2O.
6. α^{32}P-UTP (3,000 Ci/mmol; New England Nuclear).
7. Recombinant *mj*RNAP (40) or use commercially available *E. coli* RNAP or T7 RNAP.
8. Recombinant *mj*TFIIB (39).
9. Kodak IP screen.

2.3. Luciferase Assay

1. ATP Assay mix: dissolve ATP-Assay mix (Sigma) in 5 ml sterile H_2O, incubate on ice for at least 1 h, freeze in 500 μl aliquots.
2. Working stock of 20 mM adenosine-5′-phosphosulfate in sterile H_2O.
3. Working stock of 10U/ml ATP sulfurylase in 50% glycerol.
4. ATP sulfurylase reaction mix: mix 0.1 μl ATP sulfurylase stock solution with 0.05 μl APS stock solution in a total volume of 7.5 μl sterile H_2O.
5. White square-bottom 384-well microtiter plate.

3. Methods

3.1. Promoter-Independent Abortive Initiation Assays

1. Carry out the reactions in a total volume of 25 μl containing 1× transcription buffer, 10 mM DTT, 600 ng activated calf thymus DNA, 400 μM GpC, 10 μM unlabelled rUTP, and 2.5 μCi α^{32}P-UTP (3000 Ci/mmol). We typically use 250–500 ng *mj*RNAP or 1U *E. coli* RNAP or 50U T7 RNAP and 750–1,750 ng *mj*TFIIB per reaction.
2. Incubate the samples at 65°C—the temperature optimum for *mj*RNAP in this assay (40). For mesophilic RNAPs such as *E. coli* RNAP or T7 RNAP, incubate the reactions at 37°C. The accumulation of abortive products is time-dependent and strictly linear over a course of 60 min (Fig. 17.1c) and we routinely incubate the reactions for 30 min (see Note 1).
3. To prevent evaporation at high temperatures, add a drop of mineral oil to seal the reaction surface.
4. Mix 10 ml TBE/acrylamide/urea mix with 4 ml 40% acrylamide–bisacrylamide (19:1) and 0.7 ml glycerol. Add 14 μl TEMED and 140 μl 10% APS to start the polymerization reaction. Cast the gel between two 200 × 100 mm glass plates using 1 mm spacers and a 24-well comb and allow the mixture to polymerize.
5. Assemble the gel system and wash the wells with running buffer using a disposable syringe. Pre-run the gel in 1× TBE running buffer for 20–30 min at 224 V.
6. Wash the wells again. Load 10 μl of each sample and run the gel for 70 min at 224 V. Remove one of the glass plates and cover the gel with cling film. Expose the gel to a Kodak IP screen for 90 min. Read the screen with a PhosphoImager and analyze the data.

3.2. A Luciferase-Based Abortive Initiation Assay

1. Prepare the abortive initiation reactions as described above (Subheading 3.1, step 1), but use only half of all the volumes stated and include 10 mM (instead of 0.25 mM) stock of unlabelled rUTP (see Notes 2 and 3).
2. Transfer 10 μl of undiluted ATP assay mix to each well of a white, square bottom 384-well plate and incubate it in the dark at room temperature for 10 min.
3. Use 2.5 μl of the transcription reaction and mix it with 7.5 μl of the ATP sulfurylase reaction mix.
4. Add the entire mixture to the ATP assay mix and read the plate in a luminescence plate reader without further delay (see Note 4).

4. Notes

1. Promoter-independent abortive initiation assays use "activated" calf thymus DNA as template, (i.e., genomic DNA that has been treated with DNaseI (43) to introduce nicks). A dinucleotide (e.g., GpC) serves as initiating nucleotide or priming agent and is elongated by a single nucleotide (e.g., UTP) in a step mimicking abortive transcript synthesis (Fig. 1a). We applied this assay to characterize the effect of TFIIB on abortive transcription of RNAP in the *M. jannaschii* system (Fig. 1b, c, Fig. 2). With small modifications, the protocol can also be used for bacterial and bacteriophage RNAPs such as *E. coli* or T7 RNAP (Fig. 3d).
2. Abortive initiation assays rely on the separation of transcripts employing denaturing gel electrophoresis (Fig. 1b). This is time-consuming and not amenable to high-throughput screenings. Attempts to automate this type of assay have been hampered by

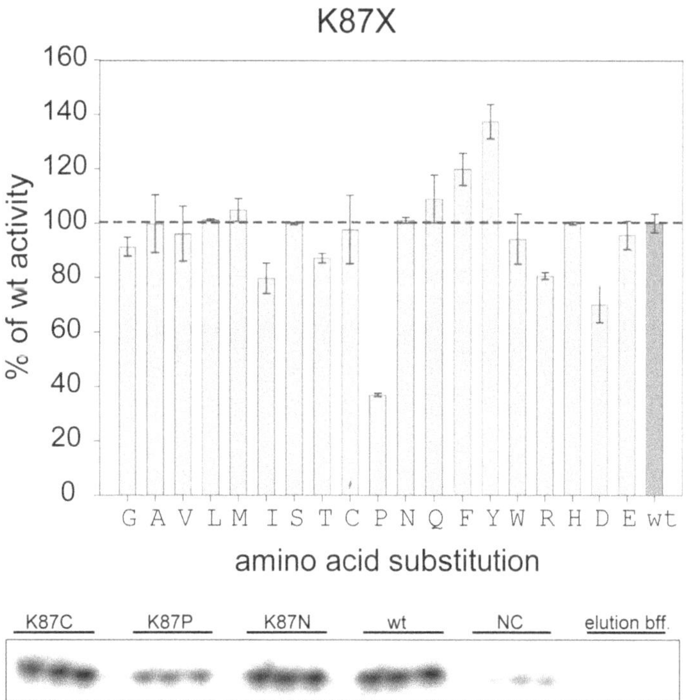

Fig. 2. Abortive initiation assays uncover a stimulatory effect of *mj*TFIIB on *mj*RNAP. The histogram shows the altered stimulation activity of a full library of point mutations in *mj*TFIIB residue K87 on mjRNAP relative to the stimulation activity obtained with wild type (wt) TFIIB. Each mutant was tested in triplicate. The error bars (representing standard deviations) as well as the sample gel underneath illustrate the reproducibility of the results.

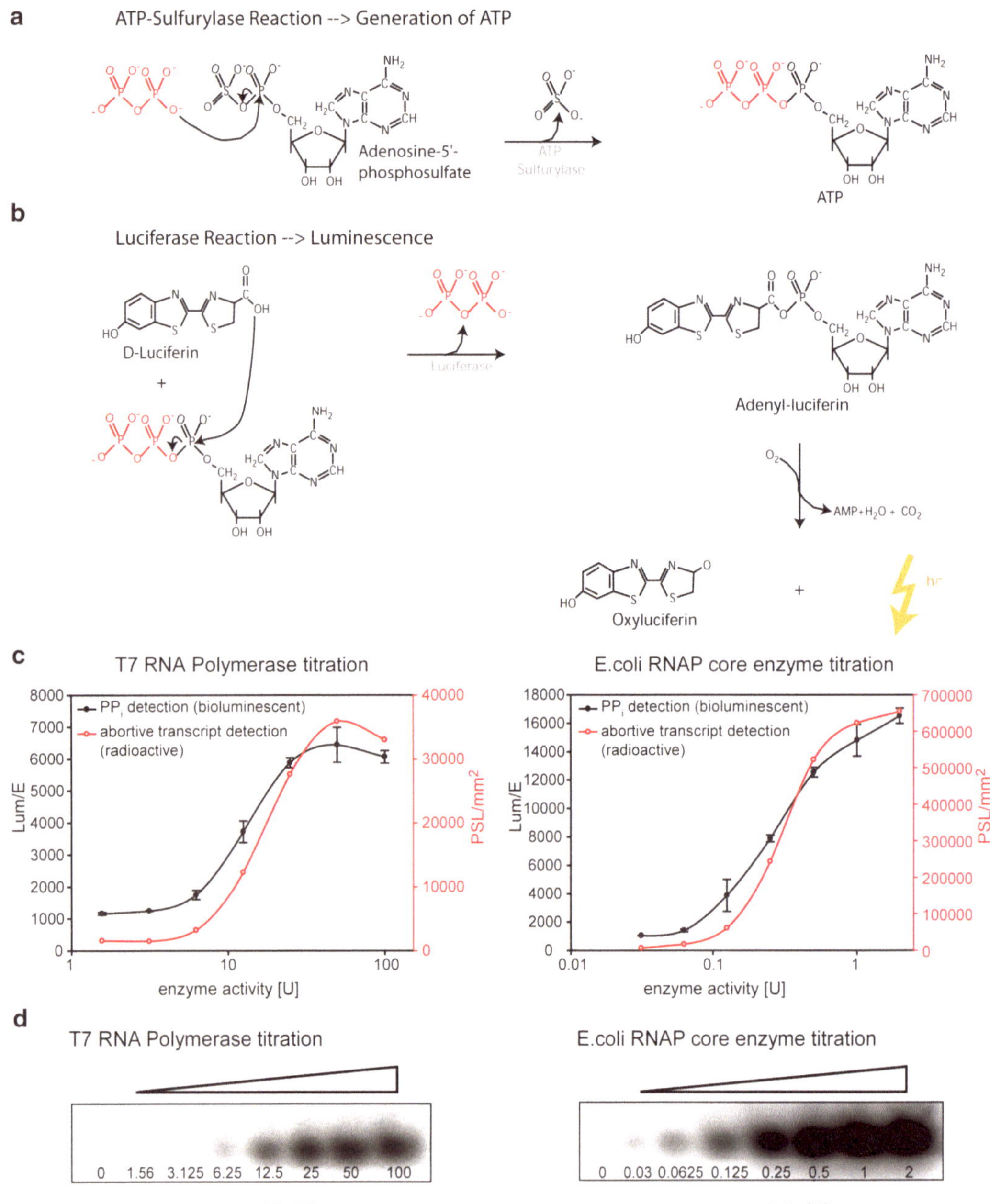

Fig. 3. A coupled luciferase-based abortive transcription assay. (**a**) PPi which is released as a by-product during the transcription reaction and APS are converted to ATP by ATP-sulfurylase. The amount of PPi present is proportional to the amount of ATP being produced. (**b**) Luciferase converts ATP and D-luciferin to adenyl-luciferin which—in a light-producing reaction—is oxidized to oxyluciferin. The amount of ATP present is proportional to the intensity of the light being produced. (**c**) Comparison of RNAP titrations evaluated by the radioactive (*red*) or the bioluminescent (*red*) read-out methods. The titration curves show comparable sensitivity for both methods when tested with *E. coli* or *T7* RNAP. (**d**) Gel images of the RNAP titrations analyzed in C.

the short size of abortive transcripts that cannot be separated sufficiently from free nucleotides chromatographically. We have overcome this limitation by adopting an approach that relies on the determination of the amount of inorganic pyrophosphate (PP_i) released as a by-product during the transcription reaction. In a series of enzymatic reactions, PP_i together with adenosine-5′-phosphosulfate (APS) can be converted to ATP by ATP-sulfurylase (Fig. 3a). ATP is subsequently used by luciferase to activate D-luciferin which, in this activated form, can be oxidized to oxyluciferin in a light producing reaction (44).

3. The promoter-independent abortive initiation assay described here measures phosphodiester bond formation and is thus a useful tool to assess the catalytic activity of RNAPs during the transcription initiation steps, either on their own or in the presence of stimulating factors (such as TFIIB). The gel electrophoresis-based variant of this assay has been successfully and extensively used to analyze the effect of the *mj* TFIIB linker on *mj* RNAP activity. It has not only complemented structural studies but has also given us novel insight into the functional aspects of these interactions. The sensitivity of the luciferase-based method is comparable to the isotope-based protocol. The former has a major experimental advantage over the latter, since it can be adapted to a high-throughput format.

4. Luminescence was measured immediately using a multimode microplate reader (such as the BioTek Synergy HT). The intensity of the emitted light is directly proportional to the concentration of PP_i and thus correlates stoichiometrically with the number of transcripts (Fig. 3b). Such an approach has also been applied for sequencing purposes (45–47), as well as for assaying the activity of RNA-dependent RNAP (48). We modified this assay to detect abortive transcription activity of DNA-dependent RNAPs and obtained reliable results for both *E. coli* RNAP and T7 RNAP. The degree of sensitivity obtained was comparable to gel-based assays (Fig. 3 c, d). We are currently in the process of adapting this assay for the archaeal system. This is technically more challenging because a heat-stable pyrophosphatase activity co-purifies with one of the *mj* RNAP subunits expressed in *E. coli* (Rpo12[P]).

Acknowledgments

This work was supported by a Wellcome Project Grant (078043/Z/05/Z) to R.O.J.W.

References

1. Borukhov S, Nudler E (2008) RNA polymerase: the vehicle of transcription. Trends Microbiol 16:126–134
2. Buratowski S, Hahn S, Guarente L, Sharp PA (1989) Five intermediate complexes in transcription initiation by RNA polymerase II. Cell 56:549–561
3. Buratowski S, Sopta M, Greenblatt J, Sharp PA (1991) RNA polymerase II-associated proteins are required for a DNA conformation change in the transcription initiation complex. Proc Natl Acad Sci USA 88:7509–7513
4. Reese JC (2003) Basal transcription factors. Curr Opin Genet Dev 13:114–118
5. McClure WR (1985) Mechanism and control of transcription initiation in prokaryotes. Annu Rev Biochem 54:171–204
6. Walter G, Zillig W, Palm P, Fuchs E (1967) Initiation of DNA-dependent RNA synthesis and the effect of heparin on RNA polymerase. Eur J Biochem 3:194–201
7. Chen BS, Hampsey M (2004) Functional interaction between TFIIB and the Rpb2 subunit of RNA polymerase II: implications for the mechanism of transcription initiation. Mol Cell Biol 24:3983–3991
8. Artsimovitch I, Vassylyev DG (2006) Is it easy to stop RNA polymerase? Cell Cycle 5:399–404
9. Carpousis AJ, Gralla JD (1980) Cycling of ribonucleic acid polymerase to produce oligonucleotides during initiation in vitro at the lac UV5 promoter. Biochemistry 19:3245–3253
10. Cheetham GM, Jeruzalmi D, Steitz TA (1999) Structural basis for initiation of transcription from an RNA polymerase-promoter complex. Nature 399:80–83
11. Gnatt AL, Cramer P, Fu J, Bushnell DA, Kornberg RD (2001) Structural basis of transcription: an RNA polymerase II elongation complex at 3.3 A resolution. Science 292:1876–1882
12. Kapanidis AN, Margeat E, Ho SO, Kortkhonjia E, Weiss S, Ebright RH (2006) Initial transcription by RNA polymerase proceeds through a DNA-scrunching mechanism. Science 314:1144–1147
13. Krummel B, Chamberlin MJ (1989) RNA chain initiation by Escherichia coli RNA polymerase. Structural transitions of the enzyme in early ternary complexes. Biochemistry 28:7829–7842
14. Revyakin A, Liu C, Ebright RH, Strick TR (2006) Abortive initiation and productive initiation by RNA polymerase involve DNA scrunching. Science 314:1139–1143
15. Orphanides G, Reinberg D (2002) A unified theory of gene expression. Cell 108:439–451
16. Ebright RH (2000) RNA polymerase: structural similarities between bacterial RNA polymerase and eukaryotic RNA polymerase II. J Mol Biol 304:687–698
17. Cheung AC, Sainsbury S, Cramer P (2011) Structural basis of initial RNA polymerase II transcription. EMBO J 30(23):4755–4763
18. Kostrewa D, Zeller ME, Armache KJ, Seizl M, Leike K, Thomm M, Cramer P (2009) RNA polymerase II-TFIIB structure and mechanism of transcription initiation. Nature 462(7271):323–330
19. Bushnell DA, Westover KD, Davis RE, Kornberg RD (2004) Structural basis of transcription: an RNA polymerase II-TFIIB cocrystal at 4.5 Angstroms. Science 303:983–988
20. Liu X, Bushnell DA, Wang D, Calero G, Kornberg RD (2010) Structure of an RNA Polymerase II-TFIIB Complex and the Transcription Initiation Mechanism. Science 327(5962):206–209
21. Chen HT, Hahn S (2004) Mapping the location of TFIIB within the RNA polymerase II transcription preinitiation complex: a model for the structure of the PIC. Cell 119:169–180
22. Chen HT, Hahn S (2003) Binding of TFIIB to RNA polymerase II: mapping the binding site for the TFIIB zinc ribbon domain within the preinitiation complex. Mol Cell 12:437–447
23. Wang D, Bushnell DA, Westover KD, Kaplan CD, Kornberg RD (2006) Structural basis of transcription: role of the trigger loop in substrate specificity and catalysis. Cell 127:941–954
24. Boeger H, Bushnell DA, Davis R, Griesenbeck J, Lorch Y, Strattan JS, Westover KD, Kornberg RD (2005) Structural basis of eukaryotic gene transcription. FEBS Lett 579:899–903
25. Westover KD, Bushnell DA, Kornberg RD (2004) Structural basis of transcription: nucleotide selection by rotation in the RNA polymerase II active center. Cell 119:481–489
26. Westover KD, Bushnell DA, Kornberg RD (2004) Structural basis of transcription: separation of RNA from DNA by RNA polymerase II. Science 303:1014–1016
27. Bushnell DA, Cramer P, Kornberg RD (2002) Structural basis of transcription: alpha-amanitin-RNA polymerase II cocrystal at 2.8 A resolution. Proc Natl Acad Sci USA 99:1218–1222
28. Cramer P, Bushnell DA, Kornberg RD (2001) Structural basis of transcription: RNA polymerase II at 2.8 angstrom resolution. Science 292:1863–1876

29. Bell SD, Magill CP, Jackson SP (2001) Basal and regulated transcription in Archaea. Biochem Soc Trans 29:392–395
30. Parvin JD, Sharp PA (1993) DNA topology and a minimal set of basal factors for transcription by RNA polymerase II. Cell 73:533–540
31. Tyree CM, George CP, Lira-DeVito LM, Wampler SL, Dahmus ME, Zawel L, Kadonaga JT (1993) Identification of a minimal set of proteins that is sufficient for accurate initiation of transcription by RNA polymerase II. Genes Dev 7:1254–1265
32. Naidu S, Friedrich JK, Russell J, Zomerdijk JC (2011) TAF1B is a TFIIB-like component of the basal transcription machinery for RNA polymerase I. Science 333:1640–1642
33. Kassavetis GA, Geiduschek EP (2006) Transcription factor TFIIIB and transcription by RNA polymerase III. Biochem Soc Trans 34:1082–1087
34. Schramm L, Hernandez N (2002) Recruitment of RNA polymerase III to its target promoters. Genes Dev 16:2593–2620
35. Weinzierl RO, Wiesler SC (2011) Revealing the functions of TFIIB. Transcription 2: 254–257
36. Knutson BA, Hahn S (2011) Yeast Rrn7 and human TAF1B are TFIIB-related RNA polymerase I general transcription factors. Science 333:1637–1640
37. Buratowski S, Zhou H (1993) Functional domains of transcription factor TFIIB. Proc Natl Acad Sci USA 90:5633–5637
38. Werner F, Weinzierl RO (2005) Direct modulation of RNA polymerase core functions by basal transcription factors. Mol Cell Biol 25:8344–8355
39. Wiesler SC, Weinzierl RO (2011) The linker domain of basal transcription factor TFIIB controls distinct recruitment and transcription stimulation functions. Nucleic Acids Res 39:464–474
40. Werner F, Weinzierl RO (2002) A recombinant RNA polymerase II-like enzyme capable of promoter-specific transcription. Mol Cell 10:635–646
41. Tan L, Wiesler S, Trzaska D, Carney HC, Weinzierl RO (2008) Bridge helix and trigger loop perturbations generate superactive RNA polymerases. J Biol 7:40
42. Weinzierl RO (2010) The nucleotide addition cycle of RNA polymerase is controlled by two molecular hinges in the Bridge Helix domain. BMC Biol 8:134
43. Aposhian HV, Kornberg A (1962) Enzymatic synthesis of deoxyribonucleic acid. IX. The polymerase formed after T2 bacteriophage infection of Escherichia coli: a new enzyme. J Biol Chem 237:519–525
44. Sun Y, Jacobson KB, Golovlev V (2007) A multienzyme bioluminescent time-resolved pyrophosphate assay. Anal Biochem 367:201–209
45. Ronaghi M, Karamohamed S, Pettersson B, Uhlen M, Nyren P (1996) Real-time DNA sequencing using detection of pyrophosphate release. Anal Biochem 242:84–89
46. Nyren P, Karamohamed S, Ronaghi M (1997) Detection of single-base changes using a bioluminometric primer extension assay. Anal Biochem 244:367–373
47. Karamohamed S, Nyren P (1999) Real-time detection and quantification of adenosine triphosphate sulfurylase activity by a bioluminometric approach. Anal Biochem 271: 81–85
48. Lahser FC, Malcolm BA (2004) A continuous nonradioactive assay for RNA-dependent RNA polymerase activity. Anal Biochem 325:247–254

Chapter 18

Fluorescence Cross-correlation Spectroscopy (FCCS) to Observe Dimerization of Transcription Factors in Living Cells

Hisayo Sadamoto and Hideki Muto

Abstract

Fluorescence cross-correlation spectroscopy (FCCS) is an established spectroscopic method to observe the interaction between the different fluorescent molecules. Using FCCS, researchers can assess the interaction of target molecules in the aqueous condition, and can apply the technique in cultured cells. Here, we describe the method of FCCS to demonstrate direct observation of dimerization between transcription factors in a living cell.

Key words: Fluorescence cross-correlation spectroscopy (FCCS), Transcription factor, Dimerization, Living cells, CREB

1. Introduction

Transcription factors play critical roles in regulating numerous biological systems, and function in part through protein interactions. Recently, we succeeded to determine strong interaction between different isoforms of cAMP responsive element binding protein (CREB) by fluorescence cross-correlation spectroscopy (FCCS) (1). In this chapter, we introduce the method of FCCS to demonstrate direct observation of dimerization between transcription factors in a living cell.

FCCS is a technique to examine the direct interaction between two distinct fluorophores, such as EGFP-(Green) or RFP(red)-labeled molecules, in a small detection volume defined by the optics system of confocal microscopy (2, 3). We can study the motions and interactions of fluorescent molecules with analyzing fluorescent intensity fluctuations, arising from a single molecules diffusing in and out of sub-femto-liter confocal volume (Fig. 1).

Minou Bina (ed.), *Gene Regulation: Methods and Protocols*, Methods in Molecular Biology, vol. 977,
DOI 10.1007/978-1-62703-284-1_18, © Springer Science+Business Media, LLC 2013

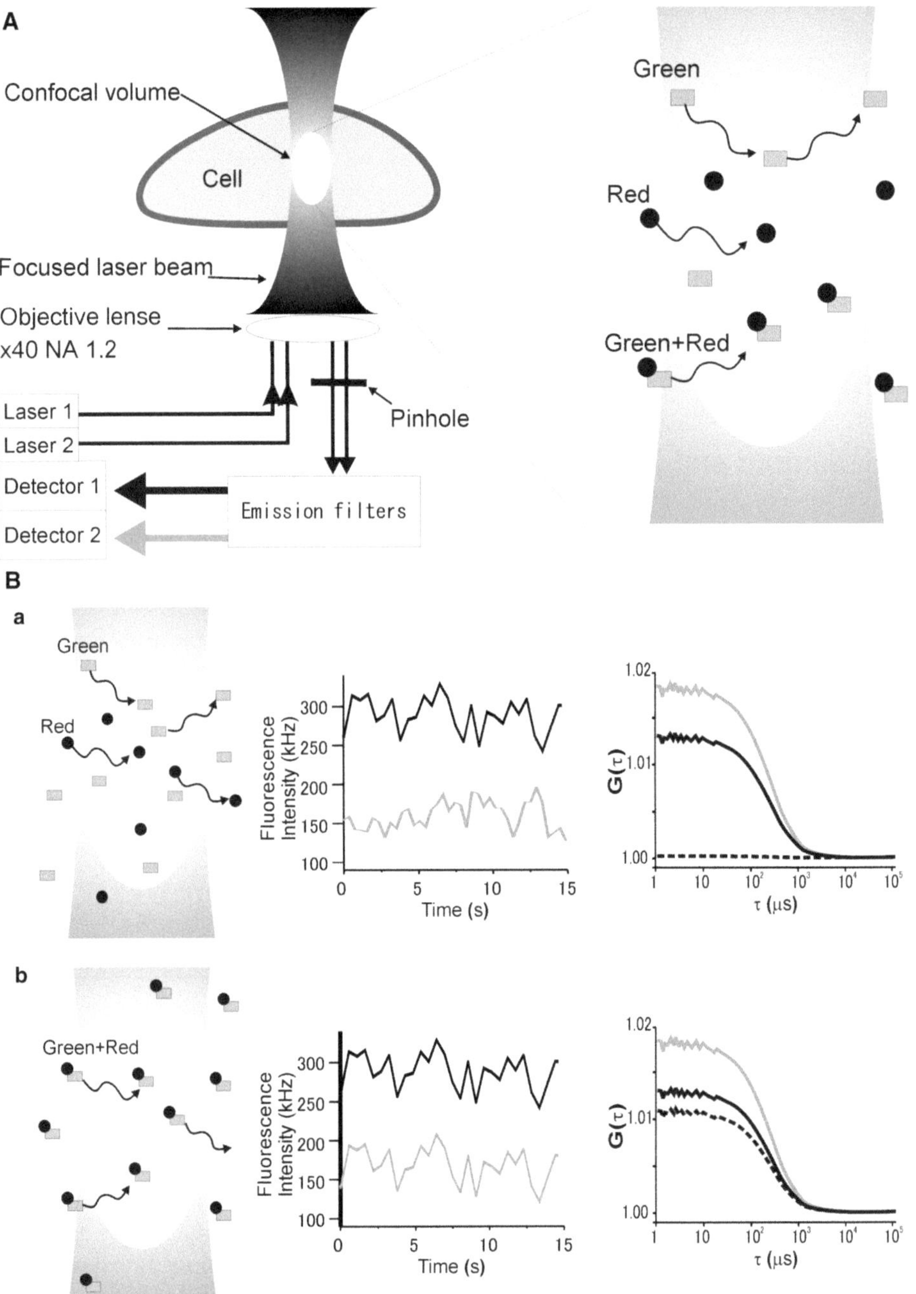

Fig. 1. The fluorescence cross-correlation spectroscopy (FCCS) system. (**A**) In FCCS analysis, a small detection volume is defined by the optics system of confocal microscopy. Two different fluorescent, EGFP-(*green*) and RFP(*red*)-labeled molecules are diffusing in and out of a confocal volume. Each fluorescent signal is simultaneously detected by each detector which carries out single-photon counting. (**B**) Schematic drawing of non-interacting (row a) and interacting molecules (row b) observed by FCCS. The two auto-correlation functions and a cross-correlation function of the signal are calculated by a built-in correlator. The middle columns represent fluorescence intensity expressed in count rate (Hz, or number of photon per second). The *grey* and *black lines* represent the fluorescence of *green* and *red*, respectively. The fluorescence intensity fluctuates because the fluorescent molecules are diffusing in and out of a confocal volume. The right columns represent the auto- and cross-correlation curves calculated using the fluctuation data of fluorescence intensity. The *grey* and *black lines* denote the auto-correlation of the green channel [$G_g(\tau)$] and the red channel [$G_r(\tau)$], respectively. The dots represent the cross-correlation between the two channels [$G_c(\tau)$]. No amplitude of cross-correlation is observed when the two molecules are diffusing independently and do not interact with each other (a). On the other hand, high amplitude of cross-correlation is observed when the two molecules are diffusing together and interact with each other (b).

The correlated motion between two fluorescent molecules is demonstrated as the cross-correlation amplitude. This method is suitable for detecting relatively strong interactions (4) and has been applied to characterize molecular interactions in the intracellular environment (5–8).

Several reports have further revealed FCCS is applicable for direct examination of dimerization between transcription factors in living cells. For example, the transcription activator proteins Fos and Jun were also examined for dimerization (9), for which in vitro studies had shown heterodimer formation during the course of their action. In line with the previous reports, the FCCS method will facilitate the future investigation that accurately reflects the molecular dynamics between transcription factors in vivo.

2. Materials

2.1. Cell Culture and Transfection

1. Culture medium: Dulbecco's Modified Eagle's Medium (DMEM, Sigma-Aldrich, St. Louis, MO) supplemented with heat inactivated 10% fetal bovine serum (Gibco/Invitrogen, Carlsbad, CA), 100 U/ml penicillin, and 0.1 mg/ml streptomycin (Penicillin-Streptomycin, 100×; Sigma) stored at 4°C.
2. 0.25% trypsin-EDTA solution (Sigma) stored at −20°C.
3. Effectene transfection reagent (Qiagen, Inc., Valencia, CA) stored at 4°C (see Note 1).
4. Opti-MEM I Reduced-Serum Medium (Gibco/Invitrogen) stored at 4°C.
5. CO_2 incubators.
6. Phosphate-buffer saline (PBS): 137 mM NaCl, 2.7 mM KCl, 4.3 mM Na_2HPO_4, 1.47 mM KH_2PO_4, pH 7.4. Store at room temperature after sterilization by autoclaving.
7. TE: 10 mM Tris–HCl pH 7.5, 1 mM EDTA. Store at room temperature after sterilization by autoclaving.
8. One hundred-millimeter diameter tissue culture dishes.
9. Chambered coverslips: Lab-Teck #1.0 borosilicate cover glass system, 8 wells (Nalge Nunc, Rochester, NY).
10. Sterile 15-ml centrifuge tubes, 1.5-ml microtubes and tips.

2.2. Equipment and Solutions for FCCS Analysis

1. LSM510-ConfoCor 3 system, Objective C-Apochromat, 40× 1.2 NA W Corr. (Carl Zeiss, Jena, Germany, see Note 2).
2. Rhodamine 6 G (Molecular Probe, OR) diluted to 1×10^{-7} M and stored in the dark at room temperature.
3. Alexa Fluor 594 (Molecular Probe) diluted to 1×10^{-9} M and stored in the dark at room temperature.

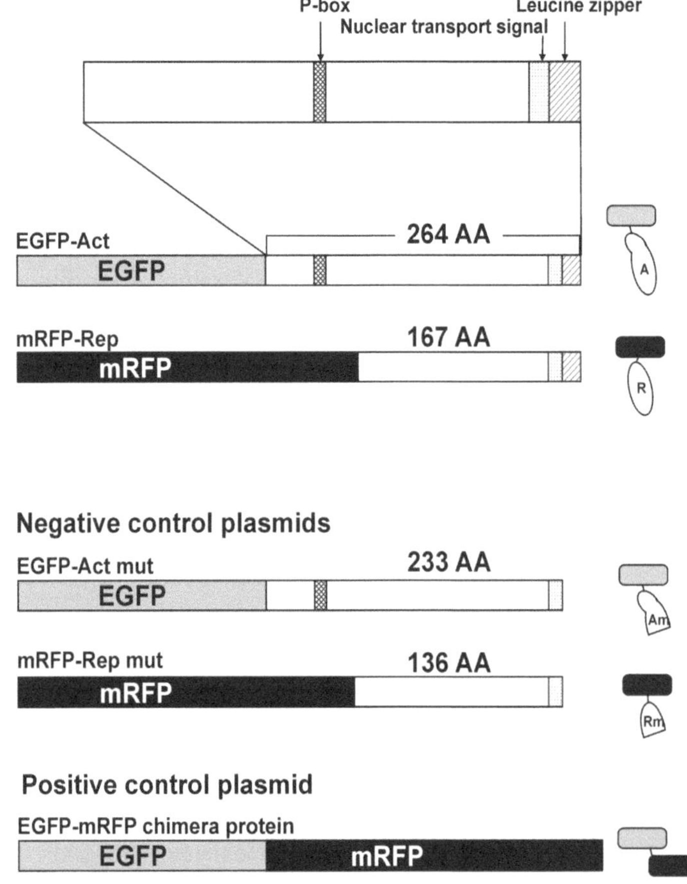

Fig. 2. Details of plasmid constructions for FCCS analysis. Expression plasmids were constructed for differently fluorescent-labeled CREB1 activator (Act) and CREB1 repressor (Rep). The fluorescent proteins were fused to the N-terminus of CREB1 proteins. For negative control experiments, truncated CREB1 activator (Act mut) and repressor (Rep mut) that lack dimerization abilities were also constructed. EGFP (*grey box*) was fused to Act or Act mut protein, and mRFP (*black box*) was fused to Rep or Rep mut proteins. For positive control experiment, plasmid encodes EGFP- mRFP chimera protein was also prepared. Schematic diagram of the fluorescent-labeled CREB1 isoform proteins are shown on the right side (reproduced from ref. 1 with permission).

2.3. Plasmids

1. Plasmids encoding EGFP tagged CREB1 activator (EGFP-Act) and tandem mRFP tagged CREB1 repressor (mRFP-Rep) (see Note 3, Fig. 2)
2. Plasmid encoding both of EGFP and mRFP (EGFP-mRFP chimera) for positive control experiment (see Note 4, Fig. 2)
3. Plasmids encoding fluorescent-labeled truncated CREB1 proteins (EGFP-Act mut or mRFP-Rep mut) for negative control experiments. Truncated proteins were designed to lack the C-terminal regions for CREB1 dimerization (see Note 5, Fig. 2)

3. Methods

3.1. Cell Culture

1. HeLa cells were seeded into 100-mm culture dishes and grown in DMEM at 37°C in a CO_2 incubator until 70–80% confluent (see Note 6).
2. Discard culture medium using an aspirator.
3. Wash cells, by adding 5 ml of PBS, swirling, and discarding PBS with an aspirator. Repeat washing twice.
4. Add 0.5 ml of 0.25% trypsin and swirl it over the monolayer of cells. Incubate for a few minutes at 37°C.
5. Stop trypsin activity with 5 ml DMEM addition and transfer cell suspension to a 15-ml tube.
6. Centrifuge cell suspension at 180 × *g* for 1 min.
7. Suspend the cell pellet with 5 ml of culture medium, and centrifuge again as above.
8. Resuspend the cell pellet with 4 ml of culture medium.
9. Transfer 300 μl to 3 ml of culture medium in a new 15-mL tube, and carefully pipette and down without making bubbles.
10. Seed 400 μl of the HeLa cell suspension to each well of an 8-chambered coverslip.
11. Incubate for 6–8 h at 37°C in a CO_2 incubator.

3.2. Transfection

1. Aspirate medium from each well and gently replace with 400 μl of fresh culture medium.
2. Prepare new 1.5-ml microtubes. Add 100 ng of plasmid DNA (minimum DNA concentration: 0.1 μg/μl) and 15 μl of EC buffer (see Note 7).
3. Add 0.4 μl of Enhancer, vortex for 1–3 s to mix, and incubate tubes at room temperature for 2–5 min.
4. Add 1.25 μl Effectene solution to each tube, vortex for 10 s to mix, and incubate tubes at room temperature for 5–10 min.
5. Add all the solution to each well.
6. Incubate at 37°C overnight in a CO_2 incubator (see Note 8).

3.3. Confocal Setting

1. Place 15–20 μl of solution containing 10^{-7} M Rhodamine 6 G and 10^{-9} M Alexa Fluor 594 on a well of an 8-chambered coverslip.
2. Adjust correction ring to 0.14 or appropriate position. And adjust *xy*-axis at the fluorophore solution and *z*-axis at 200 μm above the upper surface of the coverslip.
3. Set the intensity of 2 mW He-Ne laser (594 mm) and 25 mW Ar + laser (488 nm) to 5%, the beam path to BP615–680 for red channel and BP505–540 for green channel; set the pinhole diameter to 70 nm (see Note 9).

4. Adjust pinhole position by auto-adjustment.
5. Measure fluorescence intensity for 15 s.
6. Auto-correlation curves for red and green channels and a cross-correlation curve are obtained by using the algorithm described below in the software package for ConfoCor3 (Carl Zeiss). The fluorescence auto-correlation functions of green and red channel, $G_g(\tau)$ and $G_r(\tau)$, and the fluorescence cross-correlation function, $G_c(\tau)$, are calculated by

$$Gx(\tau) = \frac{< \delta I_i(t) \cdot \delta I_j(t+\tau) >}{< I_i(t) >< I_j(t) >} + 1 \qquad \text{(i)}$$

   where τ denotes the time delay; I_i is the fluorescence intensity of the green fluorescence (i=g) or red fluorescence (i=r); and $G_g(\tau)$, $G_r(\tau)$, and $G_c(\tau)$ denote the auto-correlation functions of green (i=j=x=g), red (i=j=x=r), and cross (i=r, j=g and x=c), respectively.
7. The $G(\tau)$ curves are fitted with the FCS fit program. Set component at 1, triplet fraction and structural parameter as free. The detection volume element is defined by the structure parameter (s) representing the ratio of the beam waist w_0 and the axial radius z_0, $s=z_0/w_0$. The structure parameter for green and red channel is calibrated at this step.

3.4. Laser Scanning Microscope Imaging and Measurement of FCCS

1. Discard culture medium of the transfected HeLa cells in an 8-chambered coverslip using an aspirator.
2. Add 400 μl of Opti-MEM I to each well of an 8-chambered coverslip.
3. Observe the cells under a confocal fluorescent microscopy (LSM 510), and select the cells that express both green and red fluorescence with an appropriate intensity for the FCCS measurement (see Note 10, Fig. 3).
4. Scan the cells using a water immersion objective (C-Apochromat, 40× 1.2 NA W Corr.; Zeiss).
5. Select measuring positions in a scanned image (see Note 11, Fig. 3).
6. Set the intensity of excitation lasers to 0.3% for 488 nm and 0.5% for 594 nm, respectively.
7. Set the values of bleach time to 0 s, measuring time to 30 s, and repeat count to 3.
8. Measure the fluorescent intensity for each measuring position. Auto-correlation curves for red and green channels and a cross-correlation curve are obtained as described in Subheading 3.5 (see Note 12, Figs. 1 and 4).

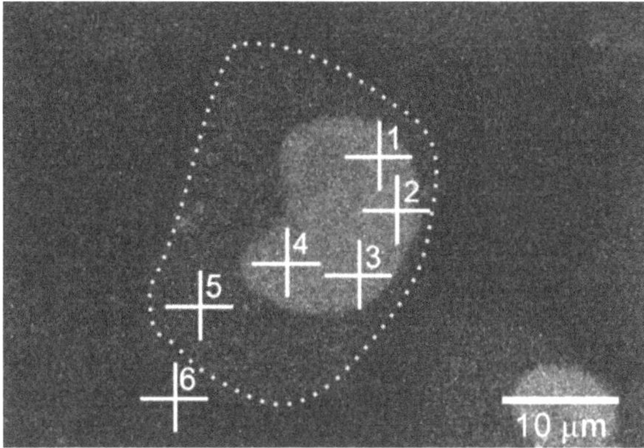

Fig. 3. LSM image of a HeLa cell expressing fluorescent-labeled CREB1 isoforms. For FCCS analysis, we generally select the cells that show the fluorescent intensity at rather low levels and with 1.5–2 times higher intensity of red fluorescence than that of green (see Notes 10 and 15). The crosses indicate the positions where FCCS measurements were performed, and the dotted line represents the outline of cell. Because the transcription factor CREB1 localizes in the nucleus, the measuring positions were set mainly in the nucleus (1–4), and also set in the cytoplasm (5) and out of the cell (6) as controls.

3.5. Curve Fitting

1. The correlation curves $G(\tau)$ are fitted with the FCS fit program by a one- or two-component model as follows.

$$G(\tau) = \frac{1 - F_{\text{triplet}} + F_{\text{triplet}} \exp(-\tau / \tau_{\text{triplet}})}{N(1 - F_{\text{triplet}})} \times \sum_i F_i \left(1 + \frac{\tau}{\tau_i}\right)^{-1} \left(1 + \frac{\tau}{s^2 \tau_i}\right)^{-1/2} + 1 \qquad \text{(ii)}$$

where F_{triplet} is the average fraction of triplet state molecules, τ_{triplet} is the triplet relaxation time, F_i and τ_i are the fraction and diffusion time of component i, respectively. N is the average numbers of green fluorescent particles (N_g) and red fluorescent particles (N_r) in the confocal volume, which can be calculated respectively, by

$$N_g = \frac{1}{G_g(0) - 1}, \qquad \text{(iii)}$$

$$N_r = \frac{1}{G_r(0) - 1} \qquad \text{(iv)}$$

N for cross-correlation (N_c) can be calculated by

$$N_c = \frac{1}{G_c(0) - 1} = \frac{N_g N_r}{N_{gr}} \qquad \text{(v)}$$

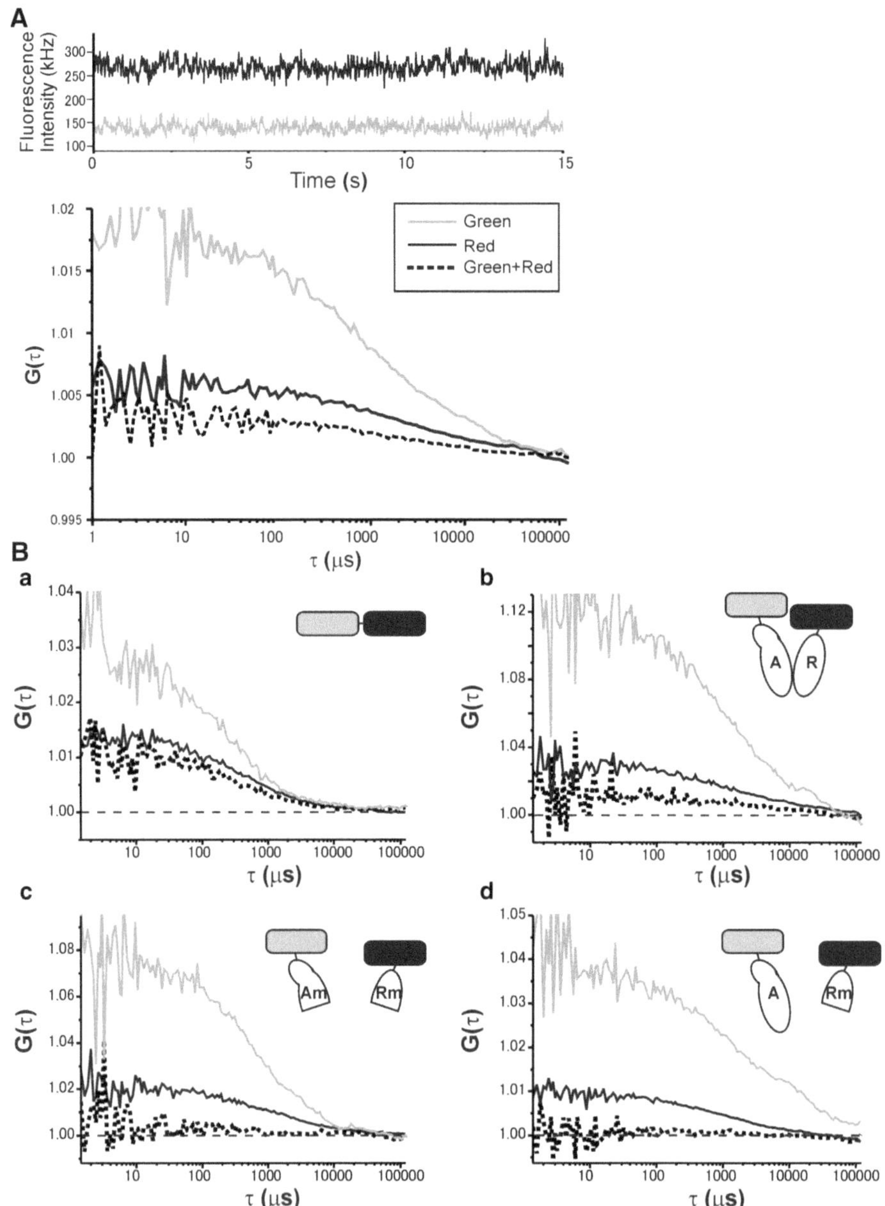

Fig. 4. FCCS measurement of differently labeled CREB1 isoforms in the nuclei of living cells. (**A**) The example data of FCCS analysis. Intensity of each fluorescence is measured by single-photon counting and expressed in count rate (Hz). The *grey* and *black lines* represent *green* and *red* fluorescence, respectively. Using the fluctuation data of fluorescence intensity, the auto- and cross-correlation functions are calculated [auto-correlation: $G_g(\tau)$ and $G_r(\tau)$; cross-correlation: $G_c(\tau)$]. The *grey* and *black lines* denote the auto-correlation of the green channel [$G_g(\tau)$] and the red channel [$G_r(\tau)$], respectively. The *dots* represent the cross-correlation between the two channels [$G_c(\tau)$]. (**B**) Auto- and cross-correlation curves of fluorescent-labeled proteins were shown for EGFP-mRFP chimera protein (a), EGFP-Act and mRFP-Rep (b), EGFP-Act mut and mRFP-Rep mut (c), and EGFP-Act and mRFP-Rep mut (d), respectively (reproduced from ref. 1). The inset is a schematic diagram showing fluorescent-labeled proteins used with each experiment as Fig. 2.

where N_{gr} is number of particles that have both green and red fluorescence. Set the value of fit limits longer than 10 μs (see Note 13). Fit the auto-correlation curves by two-component model with triplet fraction at 0%, and with structural parameters in the calibration (Subheading. 3.3). For cross-correlation, fit the curve by one-component model with triplet fraction at 0%, and with structural parameter as 5 (see Note 14).

2. For the quantitative evaluation of cross-correlations among various samples, the cross-correlation amplitude $[G_c(0)-1]$ is normalized by $[G_r(0)-1]$, i.e., the relative cross-correlation amplitude, $RCA=[G_c(0)-1]/[G_r(0)-1]$, which corresponds to the fraction of the associated molecules in total number of green molecules (N_{gr}/N_g) given by the equations (iv) and (v). The negative control experiments using truncated or mutant proteins can help to determine functional domains necessary for the interaction. In this case, negative control data using truncated proteins that lacked bZIP domains clearly evidenced CREB1 dimerization via bZIP domains (see Notes 15 and 16, Fig. 5).

4. Notes

1. Chemicals for transfection can be changed as long as the transfection efficiency and the biological activities of HeLa cells are not affected critically. For an example, FuGENE 6 Transfection Reagent (Roche, Basel, Switzerland) can be used as well (8).
2. Here we present a protocol using ConfoCor3 system as the most up-to-date system version. All figures in this chapter are based on the previous results (1) using ConfoCor2, however, we have confirmed the same results using ConfoCor3.
3. For measurement of protein–protein interactions, each target protein must be tagged by a different fluorophore, such as enhanced GFP (EGFP) with pEGFP-Cl vector (Clontech Laboratories, Mountain View, CA, USA) for the green fluorescence, monomeric RFP (mRFP) tandem dimer (7, 8) and mCherry (10) for the red fluorescence. The constructs should be designed to protect the interaction ability of target proteins. In the case of CREB proteins (1, 11), the expression constructs code for the N-terminal fusions of CREB isoform proteins with EGFP or tandem mRFP respectively, because the dimerization domain (leucine zipper motifs) exists at the C-terminal part.
4. To measure the maximum cross-correlation, a plasmid that encodes tandem dimer of two different fluorophores is used as a positive control.

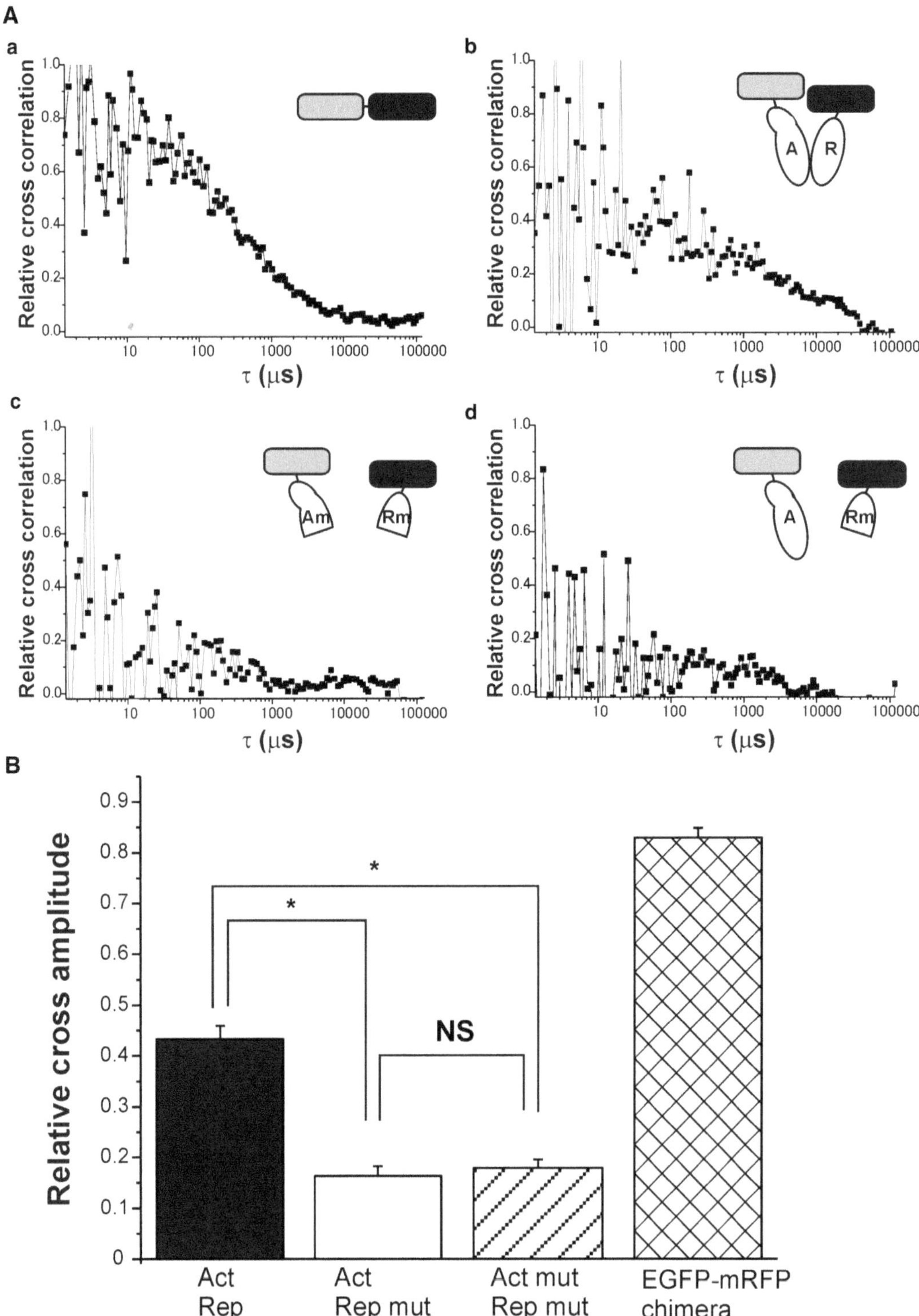

Fig. 5. The quantitative evaluation of interaction between CREB1 isoforms. (**A**) Relative cross-correlation curves $[(G_c(\tau) - 1)/(G_r(0) - 1)]$ are calculated for EGFP-mRFP chimera protein (a), EGFP-Act and mRFP-Rep (b), EGFP-Act mut and mRFP-Rep mut (c), and EGFP-Act and mRFP-Rep mut (d), respectively. (**B**) Summary of relative cross amplitudes. The interaction between different CREB1 isoforms was observed as significant cross-correlation between *green* and *red* fluorescence. The negative control experiments also showed that bZIP domain at the C-terminus of each isomer is necessary for dimerization (reproduced from ref. 1 with permission).

5. Plasmids that encode truncated target proteins, lacking a protein-interaction region, can be also constructed for the negative control experiment. In the case of CREB proteins, the dimerization domains of CREB were selectively truncated without affecting the nucleus translocation signal and DNA binding domains.
6. In this case, we successfully analyzed CREB1 dimerization using HeLa cells and observed no cell damage after transfection, such as cell death or a significant number of changes. If necessary, select the cell line depending on the effects which one intends to study.
7. The amount of plasmid necessary for transfection may be different from each plasmid sample and affected by purification methods.
8. Before starting FCCS analysis, observe the target protein expression and the intracellular localization using a fluorescent microscopy. For example, expressed CREB proteins should localize in the nucleus because of their nucleus translocation signal.
9. The optimal pinhole diameter is primarily dependent on the numerical aperture (NA) number of an objective lens and the wavelength of each excitation laser used in FCCS. In our system using a water immersion objective (C-Apochromat, 40×, 1.2 NA; Zeiss), the theoretical pinhole size is 70 nm for 488 nm, which correspond to one Airy unit.
10. Select the cells that show the fluorescent intensity at rather low levels because the optimal intensity of fluorescence is 100–500 kHz in count rate. And select the cells with 1.5–2 times higher intensity of red fluorescence than that of green, because EGFP-tandem mRFP chimera proteins emit the two fluorescent lights with this ratio of intensity.
11. Set the measuring positions mainly in the nucleus, and also in the cytoplasm and out of the cell as controls.
12. As a result of molecule diffusion, auto-correlation functions of green and red fluorescence should have significant high amplitudes at near 0 s, and should converge on 1 as shown in Fig. 4. Typically, the auto-correlation from inappropriate experimental data does not converge on 1 at a longer time point. The fluctuation of fluorescence intensity with longer periods than 1 s is resulted from movement of HeLa cells or vibration of the microscopic equipment. Thus, it does not reflect the free diffusion of target molecules and not derive accurate correlation data. Photobleaching is another serious problem for the analysis and it often occurs when the intensity of red fluorescence is higher than 500 kHz. If strange data are obtained in two of the three repeated measurements, change the measuring positions or select another cell. If two of three measurements are successful, use the two as appropriate experimental data.
13. As shown in the fitting equation (ii), a correlation function is also affected by the fluorescence fluctuation occurred in

the fluorophore transition from a singlet to a triplet excited state. This fluctuation occurs within a few μs, which is shorter than the diffusion time of measured molecules. Thus, curve fitting should be done in a longer period than 10 μs with triplet fraction as 0%. By this fitting, triplet fraction is eliminated as a residue.

14. The structural parameter (also called the axial ratio) represents a ratio between the radius to the half height of the detection volume which has a cylinder-like shape. It is determined for each green and red channel by measuring Rhodamine 6 G and Alexa 594 (see Subheading. 3.3, step 7). However, the detection volume of cross-correlation cannot be determined because detection volumes of green and red channels overlap each other. Thus, the structural parameter for cross-correlation should be fixed simply to 5.
15. The minimum value of RCA obtained from negative control experiment is generally above 0 because of cross-talk between green and red channels. Because EGFP has an emission spectrum longer than 610 nm, the EGFP fluorescence is also detected in red channel. If only EGFP is measured, both green and red fluorescent signals come from EGFP and a cross-correlation showed the high amplitude. To minimize the effect of cross talk, the signal intensity in red channel should be kept higher than that of green channel. On the other hand, the maximum value of RCA is usually under 1 even though all molecules have both green and red fluorophores, because of non-ideal overlap of two detection (confocal) volumes. In our ConfoCor2 system, for an example, the minimum and maximum values were 0.18 ± 0.02 and 0.43 ± 0.03, respectively (1).
16. FCCS evaluates interaction between GFP- and mRFP-labeled proteins. However, homodimer of the same fluorescent labeled proteins is also existed in the same FCCS measurement. This may cause some complications to quantitative interpretation of dimer formation from the result. Thus, to compare the strength of interaction between different dimers (e.g., Act-Act homodimer vs. Act-Rep heterodimer), another experiment using pairs of differently labeled isomers are necessary. For example, homodimer formation can be analyzed using the same proteins with different fluorescent tags (e.g., EGFP-Act and mRFP-Act), and the RCA is comparable to that of heterodimer (e.g., EGFP-Act and mRFP-Rep).

Acknowledgments

This work was supported by grants from the Japan Society for the Promotion of Science (Nos. 10765, 19770059, and 21770081 to H.S.) and by an OM Award from the Zoological Society of Japan (to H.S.). The authors would like to thank Prof. Masataka Kinjo (Hokkaido University) and Prof. Etsuro Ito (Tokushima Bunri University) for providing the devices and helping with much useful discussion.

References

1. Sadamoto H, Saito K, Muto H et al (2011) Direct observation of dimerization between different CREB1 isoforms in a living cell. PLoS One 6:e20285
2. Kogure T, Karasawa S, Araki T et al (2006) A fluorescent variant of a protein from the stony coral Montipora facilitates dual-color single-laser fluorescence cross-correlation spectroscopy. Nat Biotechnol 24:577–581
3. Muto H, Kinjo M, Yamamoto KT (2009) Fluorescence cross-correlation spectroscopy of plant proteins. Methods Mol Biol 479:203–215
4. Bacia K, Schwille P (2007) Practical guidelines for dual-color fluorescence cross-correlation spectroscopy. Nat Protoc 2:2842–2856
5. Bacia K, Kim SA, Schwille P (2006) Fluorescence cross-correlation spectroscopy in living cells. Nat Methods 3:83–89
6. Park H, Pack C, Kinjo M et al (2008) In vivo quantitative analysis of PKA subunit interaction and cAMP level by dual color fluorescence cross correlation spectroscopy. Mol Cells 26:87–92
7. Saito K, Wada I, Tamura M et al (2004) Direct detection of caspase-3 activation in single live cells by cross-correlation analysis. Biochem Biophys Res Commun 324:849–854
8. Muto H, Nagao I, Demura T et al (2006) Fluorescence cross-correlation analyses of the molecular interaction between an Aux/IAA protein, MSG2/IAA19, and protein-protein interaction domains of auxin response factors of Arabidopsis expressed in HeLa cells. Plant Cell Physiol 47:1095–1101
9. Baudendistel N, Muller G, Waldeck W et al (2005) Two-hybrid fluorescence cross-correlation spectroscopy detects protein-protein interactions in vivo. Chemphyschem 6:984–990
10. Shu X, Shaner NC, Yarbrough CA et al (2006) Novel chromophores and buried charges control color in mFruits. Biochemistry 45:9639–9647
11. Sadamoto H, Kitahashi T, Fujito Y et al (2010) Learning-Dependent Gene Expression of CREB1 Isoforms in the Molluscan Brain. Front Behav Neurosci 4:25

Chapter 19

Nuclear Recruitment Assay as a Tool to Validate Transcription Factor Interactions in Mammalian Cells

C.J.J. Boogerd, V.M. Christoffels, and P. Barnett

Abstract

Identification and verification of novel transcription factor interactions is an inherent step in the discovery of molecular mechanisms driving gene transcription and regulation. Co-immunoprecipitation and GST-pull down are often key techniques in the verification process. Despite wide applicability, their use may sometimes be restricted. We provide a detailed protocol for an intracellular immunofluorescence technique that may be used as an alternative or complimentary study for transcription factor interaction verification.

Key words: Transcription factor, Transfection, Nuclear import, Interaction, Immunofluorescence

1. Introduction

Identification of novel interactions among transcription factors requires implementation of secondary techniques to confirm the validity of the findings. Several tried and well-published techniques, such as co-immunoprecipitation (CoIP), GST-pull down, and 2-hybrid assays are available to this end. Of these, the first two represent perhaps the most commonly reported in vitro assays used to assess and explore singular interactions of eukaryotic transcription factors (1–3). Such techniques are choice in their ease of use and application to aspects of interaction such as binding domains and mutational studies. However, these techniques have in common that the interactions are shown in an artificially buffered solution or a yeast cell, which are different from the physiological microenvironment of the living mammalian cell (4).

Minou Bina (ed.), *Gene Regulation: Methods and Protocols*, Methods in Molecular Biology, vol. 977, DOI 10.1007/978-1-62703-284-1_19,

The protocol we present here describes the nuclear recruitment assay, a fast and easy method to demonstrate direct interactions between proteins inside living mammalian cells. This assay is based on the idea that interacting proteins will co-localize and it involves the active recruitment of one protein to the cellular compartment in which the other protein is localized. We have used this method in a recent publication describing a novel interaction between a developmental T-box transcription factor (Tbx3) and an SRY-box transcription factor (Sox4) (5) in order to complement methods demonstrating in vitro interaction between these proteins. In transfected Hek293 cells, Tbx3 localizes to the cytoplasm, whereas Sox4 is in the nucleus. In most other cell lines, Tbx3 is known to localize to the nucleus. Upon co-transfection of Sox4 and Tbx3 in Hek293, Tbx3 was efficiently recruited to the nucleus, a feature that was not observed upon co-transfection of a non-interacting nuclear localized protein. Another direct and elegant example of this method made use of the localization of Tbx transcription factors with disrupted nuclear localization signals (6). In the assays, cytoplasmic mislocalization of Tbx15 or Tbx18 with disrupted NLS sequence could be rescued by co-expression of the proteins with intact NLS, indicating the direct formation of homo- and heterodimers in living mammalian cells.

The protocol we provide offers details of how to carry out visualization of the interaction and does not deal with actual molecular characteristics of the interactions under study. Prior knowledge of the NLS would make this approach applicable over a wider spectrum of interactions of transcription factors with other proteins. To this end, we would propose first demonstrating nuclear import of at least one of the transcription factor interaction partners and then generation of a subsequent mutation within the NLS that disturbs its nuclear import. The protocol, as set out below, can then be implemented, the expectation being that loss of import via the NLS system can be restored by co-transfection of a nuclear targeted interacting transcription factor.

2. Materials

Although we provide a simple transfection protocol, this method may of course be substituted for another preferred protocol for transfection.

2.1. Materials and Chemicals

1. Sterile Glass Coverslips (round) diameter 14 mm (see Note 1).
2. Standard culture medium.
3. Serum free standard culture medium.
4. Gelatin from bovine skin.

5. PEI (Polyethylenimine).
6. Paraformaldehyde.
7. BSA (Bovine serum albumin).
8. Sytox green (Molecular probes, cat#S7020).
9. DAPI (Molecular probes, cat# D1306).
10. Vectashield (Vector Labs, cat#H1000).
11. Glycerol.
12. PBS (Phosphate buffered saline) tablets (Invitrogen).
13. Fluorescence Mounting medium (Dako, cat#S3923).
14. Primary antibodies to target proteins (see Note 2).
15. Secondary fluorescent dye labelled antibodies (see Note 3).

2.2. Working Solutions

2.2.1. Cell Culture and Transfections

1. Coating solution: 0.1% gelatin dissolved in cell culture grade water (see Note 4).
2. 1× PEI (see Note 5). Dissolve PEI to 1 mg/ml in ultrapure water, heating to 80°C for 10 min. Filter sterilize. This working solution may be stored for up to a year at −20°C. A 10 mg/ml stock may also be made and similarly stored at −20°C.

2.2.2. Immunofluorescence

1. 2% PFA: Dissolve 2 g paraformaldehyde in 100 ml PBS, stir and heat to (maximum) 60°C in a fume hood. The solution may be stored at −20°C for up to 2 months.
2. Permeabilization buffer: 0.3% Triton X-100 in PBS.
3. Blocking buffer: 1% BSA in PBS (see Note 6).
4. Nuclear counterstain: Dilute 5 μl Sytox green in 195 μl PBS. This may be stored in smaller aliquots at −20°C. Discard after freeze-thawing three times.
5. Mounting medium: Vectashield, fluorescence mounting medium or 50% glycerol in PBS (freshly made before use).

3. Methods

Figure 1 provides an overview of some of the key steps.

3.1. Cell Seeding

1. Place sterile coverslip in well (see Note 1).
2. Add 800 μl 0.1% gelatin.
3. Incubate for 10 min then aspirate.
4. Add 1 ml of standard cell culture medium and seed cells to about 75% density (see Note 7).
5. Incubate for 12–24 h using optimal cell growth conditions, allowing the cells to attach and grow to about 80% confluency.

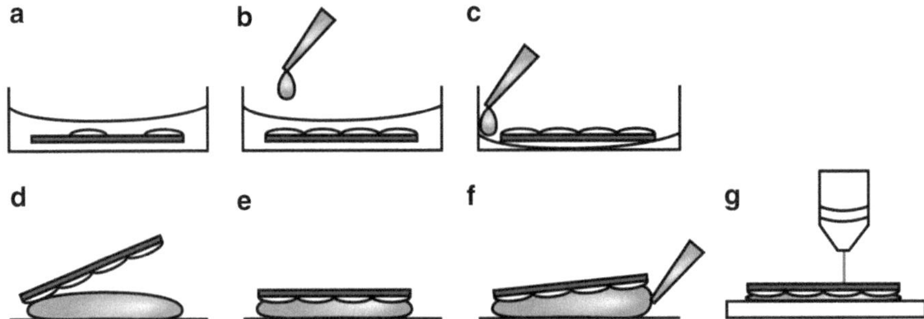

Fig. 1. Critical steps in the protocol. (**a**) Seed cells on coverslip. (**b**) Transfect cells by adding DNA:PEI mix dropwise on top of the cells. (**c**) Prevent detachment of cells when changing media or buffers by adding fluids at the side of the dish. (**d**, **e**) Place coverslip cell side down on primary antibody mix. (**f**) Add some PBS to the side of the coverslip to lift it and take it off with forceps. (**g**) Image stained cells with fluorescence microscope.

3.2. Transfection Using PEI

This stage may be substituted with a lab's preferred transfection routine.

1. Refresh the culture medium 1–3 h before transfection (see Note 8).
2. Pipette plasmid DNA(s) in to a sterile microfuge tube (see Note 9).
3. Add 1/20 of end volume of serum free culture medium (see Note 10) and mix well.
4. In a separate tube dilute the PEI with the same amount of medium (see Note 11) and mix well.
5. Add the PEI medium mix to the DNA medium mix and mix well. Incubate for 20 min at room temperature (see Note 12).
6. Gently pipette the mixture a couple of times up and down before adding dropwise to the cell culture well (Fig. 1b).
7. Optional step (see Note 13). 4–6 h after transfection, refresh the cell culture medium.
8. Continue growth of the cells for 24–48 h.

3.3. Immuno-fluorescence

1. Remove medium and wash cells twice with PBS (see Note 14).
2. Fix cells by adding 1 ml 2% PFA. Incubate at room temperature for 10 min.
3. Wash cells 3×, with a 5 min incubation each time, with PBS.
4. Add 1 ml permeabilization buffer and incubate for 10 min at room temperature.
5. Wash cells 3×, with a 5 min incubation each time, with PBS.
6. Add 1 ml blocking buffer and incubate for 1 h at room temperature (see Note 6).

   From this step onwards cells should be protected as much as possible from the light.

7. Dilute the primary antibody of choice in blocking buffer, gently mix and place a drop of approximately 30 μl on a clean piece of parafilm (Fig. 1d). Carefully remove the coverslip and place it, cell side down, on to the drop. Incubate for 1 h at room temperature, or overnight at 4°C in a moisture-sealed container to prevent drying out.
8. Add 100 μl PBS to the side of the coverslip so that the floating coverslip is raised and take it off with forceps (Fig. 1f). Transfer to a fresh 12 wells plate (cell side up).
9. Wash cells briefly twice with PBS.
10. Wash cells 3×, with a 10 min incubation each time, with PBS.
11. Dilute secondary antibody in 1 ml blocking buffer and add 1 μl nuclear counterstain.
12. Add this mix to the coverslip containing well (see Note 15). Incubate for 1 h at room temperature.
13. Wash cells 3×, with a 10 min incubation each time, with PBS.

3.4. Mount Slides

1. Place a drop of mounting medium (25 μl) on a clean glass slide.
2. Place coverslip cell side down on the drop and lower slowly, being careful not to trap any air bubbles.
3. Allow medium to set fully, especially when using oil-based immersion lenses or inverted microscopes.
4. Image slides on fluorescence microscope or confocal microscope.

4. Notes

1. The size of the coverslip is chosen on the basis of a 12 well scale culture. Sterilize by autoclaving.
2. Alternatively, GFP and/or RFP tagged target proteins or suitably antigen tagged versions of target proteins could be an alternative if no suitable primary antibody is available to your proteins of interest. Use of GFP and/or RFP will circumvent the necessity of fixation and immunohistochemistry, though their implementation may interfere with interactions and/or localization.
3. A full range of Alexa fluorescent dye labelled secondary antibodies is available from Invitrogen.
4. Alternatively, use an experimentally determined optimal coating for the cell type being used.
5. We have found polyethylenimine (PEI) to be a cheap but very efficient alternative transfection reagent for many other commercially available reagents.

6. The addition of serum to the blocking buffer can generally be omitted when staining cultured cells. When experiencing background problems, add 1% serum of the species in which secondary antibodies were generated.
7. We have found that the best efficiencies of transfection when using PEI as a reagent, are achieved when cells are seeded to a relatively high density, covering approximately 75% of the well's surface.
8. We have found that this step is best omitted when using Hek293 cells, which are easily disrupted during washing.
9. Good transfection efficiencies are obtained using 1–2.
10. For a single well of a 12 wells plate, this is $1/20 \times 1$ ml = 50 μl.
11. Although the amount of PEI to be used should be optimized for a particular cell line, we have found a ratio (PEI:DNA) in the range of 2.5–3.0:1 to be an effective range for many cell lines.
12. Usually the whole process, including incubation, is carried out in a sterile lamina flow cupboard.
13. PEI is not toxic to many cell lines and may be left overnight in the growth medium.
14. To reduce the chance of dislodging of the cells, add buffers to the side of the well, and perform incubations without agitation (Fig. 1c).
15. Alternatively, an approach similar to that in step 7 may be taken.

Acknowledgments

This work was supported by European Community's Sixth Framework Programme contract HeartRepair LSHM-CT-2005-018630.

References

1. Miller J, Stagljar I (2004) Using the yeast two-hyrbid system to identify interacting proteins. Methods Mol Biol 261:247–262
2. Vikis HG, Guan KL (2004) Glutathione-S-transferase-fusion based assays for studying protein-protein interactions. Methods Mol Biol 261:175–186
3. Cekan SZ (2002) Genes and transcription factors, including nuclear receptors: methods of studying their interactions. J Lab Clin Med 140:215–227
4. Mackay JP, Sunde M, Lowry JA, Crossley M, Matthews JM (2007) Protein interactions: is seeing believing? Trends Biochem Sci 32: 530–531
5. Boogerd CJ, Wong LY, van den Boogaard M, Bakker ML, Tessadori F, Bakkers J, 't Hoen PA, Moorman AF, Christoffels VM, Barnett P (2011) Sox4 mediates Tbx3 transcriptional regulation of the gap junction protein Cx43. Cell Mol Life Sci 68:3949–3961
6. Farin HF, Bussen M, Schmidt MK, Singh MK, Schuster-Gossler K, Kispert A (2007) Transcriptional repression by the T-box proteins Tbx18 and Tbx15 depends on Groucho corepressors. J Biol Chem 282:25748–25759

Chapter 20

Preparation of Cell Lines for Single-Cell Analysis of Transcriptional Activation Dynamics

Ilona U. Rafalska-Metcalf and Susan M. Janicki

Abstract

Imaging molecularly defined regions of chromatin in single living cells during transcriptional activation has the potential to provide new insight into gene regulatory mechanisms. Here, we describe a method for isolating cell lines with multi-copy arrays of reporter transgenes, which can be used for real-time high-resolution imaging of transcriptional activation dynamics in single cells.

Key words: Live-cell imaging, Transgene array, Stable-cell lines, Transcription, Chromatin

1. Introduction

For a gene to be expressed, the functions of multiple regulatory factors must be coordinated at the chromatin/DNA interface. Therefore, it is essential that techniques and tools be developed to directly visualize molecularly defined regions of chromatin in single living cells. Additionally, the ability to experimentally regulate the transcriptional activity of these sites provides the opportunity to define the temporal and spatial dynamics of regulatory factor interactions with chromatin during activation.

For the purpose of directly visualizing transcriptional activation in single cells, transgenes, which allow DNA, RNA and protein to be visualized in living cells, have been engineered and stably introduced into eukaryotic cells (1–3). The inclusion of sequence elements, such as the lac operator and tetracycline operator repeats from bacteria, allow the integration site to be identified when their binding proteins fused to auto-fluorescent proteins are expressed (4). It is also possible to visualize the RNA transcribed from these sites when

Minou Bina (ed.), *Gene Regulation: Methods and Protocols*, Methods in Molecular Biology, vol. 977,
DOI 10.1007/978-1-62703-284-1_20, © Springer Science+Business Media, LLC 2013

repeats of the MS2 bacteriophage RNA stem-loop are included and the auto-fluorescently tagged MS2 coat protein is expressed (5). The recently developed RNA aptemer/fluorophore complex, Spinach, is a new sequence element that can potentially be incorporated into these types of reporter transgenes to visualize RNA (6).

Here we describe a method for isolating cell lines with multi-copy arrays of these types of transgenes. This technique produces homogenous cell lines, which can be used for single cell imaging studies.

2. Materials

2.1. Plasmids

1. Reporter transgenes: Plasmids containing the transcription unit, *cis*-regulatory elements and the sequence elements that allow DNA and RNA to be visualized in living cells. See the following references for examples (1–3, 7, 8).
2. Drug resistance plasmids: Examples include, pTK-Hyg, (Clontech), which confers resistance to Hygromycin B, and, pPUR (Clontech), which confers resistance to Puromycin.
3. Plasmids expressing auto-fluorescent proteins fused to DNA and RNA binding proteins: Examples include, the lac repressor protein, the Tet-Off construct (Clontech) fused to the estrogen receptor hormone binding domain (9), the tet repressor (10), and the MS2 coat protein (11).

2.2. Cell Lines

1. Cell lines into which the reporter plasmids will be stably introduced (see Note 1).

2.3. Cell Culture Reagents

1. Media recommended for the cell line being grown.
2. Drugs for selection: examples include, Hygromycin B and Puromycin.
3. Conditioned media from the parental cell line: Media that cells, at ~80% confluency, have been cultured in for ~24 h. Store at −20°C in 10 ml aliquots (see Note 2).
4. Trypsin EDTA.
5. 1× PBS.
6. Cell culture dishes: 5 and 10 cm dishes and 96, 24 and 6-well plates.
7. 15 and 50 ml conical tubes.
8. 50 and 500 ml 0.22 μm filter units.
9. Hemocytometer.
10. Cryovials for freezing cell lines.
11. Freezing media recommended for the cell line being grown.

2.4. Calcium Phosphate Transfection Reagents

1. 2 M CaCl2: Weigh out 8.76 g $CaCl_2 \cdot 6H_2O$ and dissolve in 15 ml of molecular biology grade water (Fisher Bioreagents). Make up volume to 20 ml. Sterilize by filtration through a 0.22 μm filter. Store in 1 ml aliquots at −20°C
2. 2× HBS: Weigh out 1.6 g NaCl, 0.074 g KCl, 0.027 g $Na_2HPO4 \cdot 2H_2O$, 0.2 g dextrose, and 1 g HEPES. Add 90 ml distilled water. Adjust to pH 7.05 with NaOH. Make up volume to 100 ml. Sterilize the solution by filtration through a 0.22 μm filter. Store in 5 ml aliquots at −20°C.
3. 10×HEPES buffered saline (HBS): Weigh out 40 g NaCl, 1.85 g KCL, 0.65 g Na_2HPO_4 ($2H_2O$), 5 g dextrose and 25 g HEPES. Add 400 ml of distilled water. Adjust pH to 7.05. Adjust volume to 500 ml with distilled water. Filter-sterilize through a 22 μm filter and store at 4°C.
4. 1× HBS: prepare 500 ml by adding 50 ml of the 10× solution to a cylinder, with 400 ml of distilled water. Adjust pH to 7.05. Adjust volume to 500 ml with distilled water. Filter sterilize through a 0.22 μm filter. Store at 4°C.
5. 20% Glycerol in 1× HBS: Add 10 ml of Glycerol (Ultrapure, Invitrogen) and 5 ml 10× HBS to a 50 ml tube. Adjust volume to 50 ml with distilled water. Filter sterilize through a 0.22 μm filter. Use fresh.

2.5. Microscopy Reagents

1. Fluorescence microscope with a 63× oil objective and filters for GFP, YFP, and/or RFP or Cherry.
2. Coverslips: 12 mm round 0.13–0.17 mm thick (Fisherbrand 12-545-80 12CIR-1). These coverslips fit into the wells of a 24-well plate (see Note 3).
3. Glass slides.
4. 3% formaldehyde in 1× PBS: Weigh out 3 g of paraformaldehyde in a hood and add it to a heat-proof bottle with a stir bar containing 100 ml of 1× PBS (see Note 4). Stir the solution with heat (~130°C) until the powder has dissolved which will take ~2 h. Allow the solution to cool before filtering it through a 0.22 μm filter. Store in 5 or 10 ml aliquots at −20°C.
5. Anti-fade slide mounting solution.

3. Methods

Do all cell culture in biosafety cabinets and grow cells in humidified CO_2 incubators at percentage concentrations appropriate to the cell lines being cultured.

3.1. Determination of Drug Concentration for Stable Cell Line Selection

1. Seed seven dishes (5 cm) of cells in a total of 5 ml of media such that they will be ~60% confluent the next day.
2. Add incrementally increasing and decreasing concentrations of the selection drug to each dish based on typical concentrations used for the cell line being cultured (see Note 5).
3. Monitor cell density over 10 days. Change media and add fresh drug as needed to remove dead cells. The drug concentration to be used for stable cell line selection is the one that kills all of the cells by 10 days (see Note 6).

3.2. Calcium Phosphate Transfection

1. Seed cells on a 10 cm dish in 10 ml of media such that they will be ~60% confluent the next day (Fig. 1).
2. The next day, remove the media from the cells and replace it with fresh media at least 3–4 h before the transfection.
3. To a 15 ml conical tube, add 36 μl of 2 M $CaCl_2$, 20 μg of DNA (e.g., 18 μg of the reporter transgene, 2 μg of the drug resistance plasmid) (see Note 7), and water to a final volume of 300 μl. Vortex to mix well and spin down.
4. To another 15 ml conical tube, add 300 μl of 2× HBS.
5. Add the DNA/$CaCl_2$ solution to the 2× HBS dropwise mixing well after each addition by flicking the tube.
6. Incubate the solution at room temperature for 30 min.
7. Add the solution dropwise to the plate of cells.
8. Return cells to the incubator for 4 h (see Note 8).
9. To glycerol shock the cells, remove the media from the dish, and gently add 4 ml of the freshly prepared 20% glycerol 1× HBS solution. Leave it on the cells for exactly 1.5 min.
10. After removing the 20% glycerol 1× HBS solution, gently wash the cells 2× with 1× HBS. Add 10 ml of fresh media to the dish and return it to the CO_2 incubator.
11. The next day, rinse cells in 1× PBS and remove cells from the dish by adding 1–2 ml of Trypsin EDTA. Place the dish in the CO_2 incubator for ~2 min or until the cells detach from the dish. Collect the cells in 5 ml of media and spin down in a 15 ml conical tube.
12. Aspirate the media from the cell pellet and resuspend it in 3 ml of media. Add 1 ml of the cell suspension to each of three 10 cm dishes containing 9 ml of media. Gently swirl the dishes to distribute the cells evenly. Place the dishes in the CO_2 incubator and leave overnight.
13. The next day, add the predetermined concentration of the selection drug to each dish.
14. Remove the media and add fresh drug-supplemented media to the cells, as needed, in order to eliminate dead cells. Discrete colonies should be visible in ~14–20 days (see Note 9).

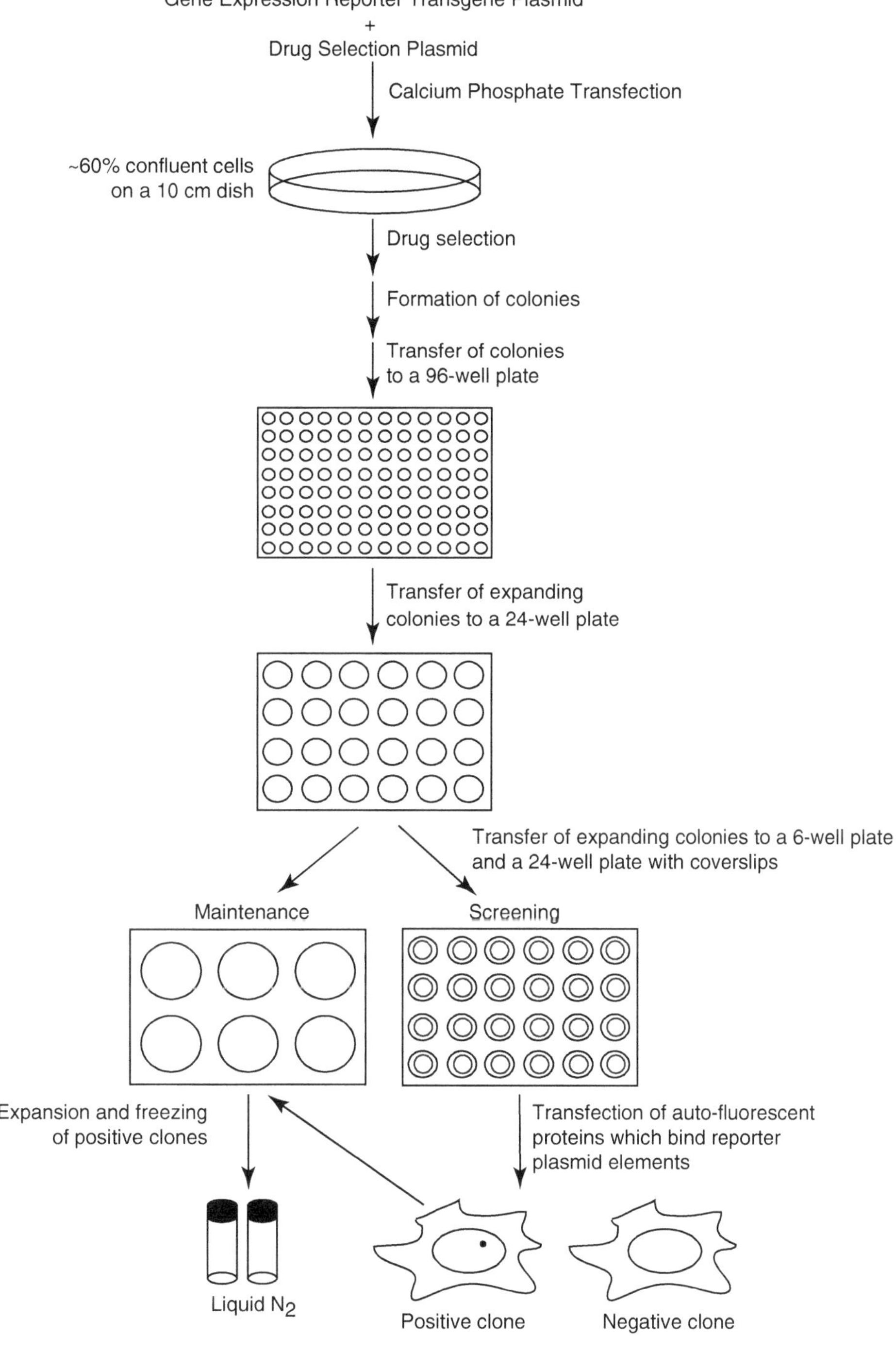

Fig. 1. A flowchart of the steps of cell line isolation.

3.3. Stable Cell Line Isolation

1. Use a permanent marker to put dots on bottom of the dish in the location of the cell colonies.
2. Using a p1000 pipettor set to ~600 μl, partially depress the button and firmly and perpendicularly place the tip over a marker spot on the dish. Partially release the button and suck up ~200 μl of media and add it to a well in a 96-well plate. The cells from the colony will be transferred with the media (Fig. 1).
3. Repeat step 2 with the remaining colonies. Add several drops of drug-supplemented media to each well and place the plate in the CO_2 incubator.
4. In ~5 days when adherent growing cells can be seen in the wells, remove the media from each well and replace with fresh drug-supplemented media (see Note 10).

3.4. Stable Cell Line Screening

1. In ~7–10 days after colony transfer, scan the wells in the 96-well plate and mark those in which the cells are expanding and filling the well.
2. To transfer the cells from the 96-well plate to a 24-well plate, remove the media from the marked wells, rinse each with 1× PBS, add a drop of Trypsin EDTA, and place the plate in the CO_2 incubator for ~2 min or until the cells detach from the wells. Add several drops of drug-supplemented media to each well. Using a 200 μl pipettor, transfer the cells with the media directly to wells in a 24-well plate. Add 500 μl of drug-supplemented media to each well and place the plate in the CO_2 incubator (Fig. 1).
3. Scan the wells in the 24-well plate in 5–7 days and mark those in which the cells have expanded to ~70–80% confluency. Rinse each with 1× PBS, add several drops of Trypsin EDTA and place in the incubator for ~2 min or until cells have detached from the well. Add 500 μl of media to each well.
4. Add a couple of drops of the cell suspension to a well in the 24-well plate containing the coverslips (see Note 11) (this will be used for screening) and transfer the rest of the cell suspension to a well in a correspondingly numbered 6-well plate (this will be used for maintenance and expansion) (Fig. 1).
5. When the cells have attached to the coverslips and begun to divide (1–2 days), transfect them with an auto-fluorescent protein that binds to elements in the reporter transgene, which will allow the integration site to be visualized (see Note 12).
6. 18–24 h later, fix the cells on the coverslips in 3% formaldehyde/1× PBS for 15 min. Rinse the coverslips, in 1× PBS and mount them on slides using an anti-fade solution. Scan the coverslips, using a fluorescent microscope with a 63× oil objective, for cells, which contain a dot. This indicates that the transgene has been stably incorporated into the genome (see Note 13) (Fig. 1).

7. When cells lines positive for transgene integration have been identified, expand the corresponding cells in the 6-well plates, until they have reached ~70% confluency. To transfer to a 5 cm dish, rinse the cells with 1× PBS, add 4–5 drops of Trypsin EDTA and place in CO_2 incubator for ~2 min or until cells begin to detach from the dish. Add 1 ml of drug-supplemented media and transfer the cell suspension directly to a 5 cm dish with 4 ml of drug-supplemented media (Fig. 1).
8. When the cells in the 5 cm dish have reached ~70–80% confluency, freeze them using the method specified for the parental cell line. Split the cells between two cryovials (see Note 14) (Fig. 1).

3.5. Single Cell Sorting the Cell Lines

1. When a cell line positive for transgene integration is identified it may be necessary to expand it from a single cell in order to have a homogeneous population (see Note 15). To do this, use a hemocytometer to count the number of cells in a suspension, determine the volume, which contains ~100 cells, and add it to a 10 ml aliquot of drug-supplemented conditioned medium.
2. Mix the cells in the conditioned media gently but thoroughly and pipet 100 μl into each well of a 96-well plate (see Note 16).
3. Place the plate in the CO_2 incubator and wait for colonies to form (~2 weeks) (see Note 17).
4. Repeat the steps in Subheading 3.4. Expand only cells in the wells in the 96-well plate which have a single colony.

4. Notes

1. Criteria for selecting a cell line to stably introduce a reporter plasmid depends on a variety of factors, which could include, genetic background, adherence properties, pluripotency, etc. It is also important to determine whether the cell line of choice can be transfected by the calcium phosphate technique, which is described in this protocol. Other transfection methods may result in lower copy number insertions.
2. Cells release metabolites and growth factors into the media and this media is called conditioned media. In order to expand a cell population from a single cell, it is necessary to dilute the cells into conditioned media before plating. It improves the rate of cell division and prevents them from dying.
3. Although not absolutely necessary, acid washing coverslips will facilitate cell attachment and growth because it removes dirt and debris. To do this, place coverslips in a glass beaker in a 2:1 mixture of HNO_3 to HCl for 2 h. Dispose of the acid mixture

in accordance with chemical waste rules. To remove the acid residue, run distilled water into the beaker for ~1 h or until pH is in the range of 5.5–6.0. Drain coverslips and store in 100% EtOH.

4. Paraformaldahyde is a suspected carcinogen and, therefore, should be weighed in a hood to prevent inhalation. It requires heating for it to go into solution. The heated stir plate should also be kept in a hood to prevent breathing the vapor that is produced. Keep the lid on the heat-proof bottle loose during the mixing to prevent the bottle from exploding.
5. The concentration of drug to be used for stable cell line selection has to be determined for each cell type being used because it can vary widely between cell lines. Working concentrations can range from 50–1,000 μg/ml for Hygromycin B and 1–10 μg/ml for Puromycin.
6. If the cells are totally dead before day 10 or more than 10% of the cells have survived by day 10, it is better to set up another set of plates and move the drug concentration up or down the scale rather than to proceed with an inaccurate drug concentration. If fewer than 10% of cells are still alive by day 10, the working drug concentration can be estimated at a value between that concentration and the one above it.
7. The drug resistance plasmids integrate into the same genomic site as the reporter transgene plasmids. Therefore, the copy number of the reporter plasmid can be crudely controlled by varying the ratio of the reporter plasmid to the drug resistance plasmid in the calcium phosphate precipitate. To recover cell lines with copy numbers greater than 100, combine the reporter plasmid and the drug resistance plasmid in a 9:1 ratio (e.g., 18 μg of the reporter plasmid and 2 μg of the drug resistance plasmid). A lower ratio (e.g., 5:1) will result in cell lines with smaller inserts. High copy number integrations make it easier to temporally and spatially resolve the events at the transcription site.
8. The appearance of a fine precipitate on the bottom of the dish, as seen using a light microscope, after the 4 h incubation is a good sign that the DNA precipitate has formed properly.
9. Colonies can be seen on the bottom of the dish when it is held up above the head against the light. They look like cloudy dots. When they are large enough to mark with a dot using a permanent marker, then they are ready to be transferred to a 96-well plate.
10. This is done in order to remove the dead cells, which were transferred with the colony.

11. Add the needed number of coverslips for the colonies to be screened to a 24-well plate before beginning to trypsinize the cells. To do this, remove individual coverslips from the EtOH using metal forceps and burn the EtOH off by waving them through a flame.
12. Some transfection reagents, such as Fugene (Promega) are inhibited by the presence of drugs in the media. Therefore it is better to plate the cells on the coverslips in the 24-well plates in media without the selection drugs. Changing the media before the transfection can also help to improve efficiency. Alternatively, viral preparations of the auto-fluorescently tagged binding proteins can be used and will result in a high percentage of expressing cells, which will make the screening process easier.
13. The visual screening of the cells for transgene integration is most efficiently done using a 63× oil lens because the field of view is large enough to both locate transfected cells and observe variations in the population. The magnification is also high enough to examine the size and structure of the transgene array. The use of an oil lens requires that the coverslips be firmly attached to the slide. Sealing the edges with nail polish will ensure this if a liquid mounting solution is used.
14. It is important to freeze the isolated cell lines at a low passage because if they are a mixture of positive and negative cells, it is possible that the negative ones will take over the culture before the positive ones can be purified by single-cell plating. Additionally, division between two cryovials ensures redundancy of the stock in case it is necessary to return to the original culture or in case cells from the first vial are lost.
15. Sometimes isolated cell lines contain a mixture of positive and negative cells. In order to acquire a homogenous population, it is necessary to expand them from single cells and then to repeat the screening procedures in Subheading 3.4. This is something that can be started at the time when the cell stocks are being frozen or immediately after the stocks have recovered from thawing.
16. Transfer the conditioned-media-cell suspension from the 15 ml tube to a 5 cm plate because it is easier to pipet from a flat dish into the 96-well plate.
17. Leaving the plate in the incubator for a long time with a small volume of media often results in evaporation, particularly in the wells on the outside of the plate. Check the plate at ~1 week to see if the media loss has been significant and consider adding a few drops of drug-supplemented media to each well before returning the plate to the incubator for the colonies to continue expanding.

References

1. Rafalska-Metcalf IU, Janicki SM (2007) Show and tell: visualizing gene expression in living cells. J Cell Sci 120:2301–2307
2. Wu B, Piatkevich KD, Lionnet T, Singer RH, Verkhusha VV (2011) Modern fluorescent proteins and imaging technologies to study gene expression, nuclear localization, and dynamics. Curr Opin Cell Biol 23: 310–317
3. Darzacq X, Yao J, Larson DR, Causse SZ, Bosanac L, de Turris V, Ruda VM, Lionnet T, Zenklusen D, Guglielmi B, Tjian R, Singer RH (2009) Imaging transcription in living cells. Annu Rev Biophys 38:173–196
4. Belmont AS (2001) Visualizing chromosome dynamics with GFP. Trends Cell Biol 11:250–257
5. Keryer-Bibens C, Barreau C, Osborne HB (2008) Tethering of proteins to RNAs by bacteriophage proteins. Biol Cell 100:125–138
6. Paige JS, Wu KY, Jaffrey SR (2011) RNA mimics of green fluorescent protein. Science 333:642–646
7. Tsukamoto T, Hashiguchi N, Janicki SM, Tumbar T, Belmont AS, Spector DL (2000) Visualization of gene activity in living cells. Nat Cell Biol 2:871–878
8. Janicki SM, Tsukamoto T, Salghetti SE, Tansey WP, Sachidanandam R, Prasanth KV, Ried T, Shav-Tal Y, Bertrand E, Singer RH, Spector DL (2004) From silencing to gene expression: real-time analysis in single cells. Cell 116:683–698
9. Rafalska-Metcalf IU, Powers SL, Joo LM, LeRoy G, Janicki SM (2010) Single cell analysis of transcriptional activation dynamics. PLoS One 5:e10272
10. Masui O, Bonnet I, Le Baccon P, Brito I, Pollex T, Murphy N, Hupe P, Barillot E, Belmont AS, Heard E (2011) Live-cell chromosome dynamics and outcome of X chromosome pairing events during ES cell differentiation. Cell 145:447–458
11. Shav-Tal Y, Darzacq X, Shenoy SM, Fusco D, Janicki SM, Spector DL, Singer RH (2004) Dynamics of single mRNPs in nuclei of living cells. Science 304:1797–1800

Chapter 21

Peptide Microarrays for Profiling of Serine/Threonine Kinase Activity of Recombinant Kinases and Lysates of Cells and Tissue Samples

Riet Hilhorst, Liesbeth Houkes, Monique Mommersteeg, Joyce Musch, Adriënne van den Berg, and Rob Ruijtenbeek

Abstract

Peptide microarray technology can be used to identify substrates for recombinant kinases, to measure kinase activity and changes thereof in cell lysates and lysates from fresh frozen (tumor) tissue. The effect of kinase inhibitors on the kinase activities in relevant tissues can be investigated as well. The method for performing experiments on dynamic peptide microarrays with real-time readout is described, as well as the influence of assay parameters and suggestions for optimization of experiments.

Key words: Tyrosine kinase, Serine/threonine kinase, Kinase activity, Peptide microarray, Multiplex assay, Kinase substrate identification, Kinase activity profiling, Kinase kinetics, Kinase inhibition

1. Introduction

Over 30% of the human proteome is estimated to be phosphorylated for the dynamic regulation of cellular processes. This illustrates the importance of the kinases responsible for the act of phosphorylation and the interplay between those important regulators. Measurement of their fluctuating activities provides more insight in the dynamics than just establishing their presence or phosphorylation status. Since many kinases are known to be promiscuous in their substrate preference and therefore phosphorylate a multitude of substrates, multiplex methods are required to monitor their activity in complex mixtures like cell or tissue lysates. Peptide microarrays are used for this purpose. The method is rapid, since apart from lysis, no preprocessing is required, and sensitive, because of amplification of signals by endogenous enzyme activity: a single kinase molecule will phosphorylate multiple peptide

Minou Bina (ed.), *Gene Regulation: Methods and Protocols*, Methods in Molecular Biology, vol. 977, DOI 10.1007/978-1-62703-284-1_21, © Springer Science+Business Media, LLC 2013

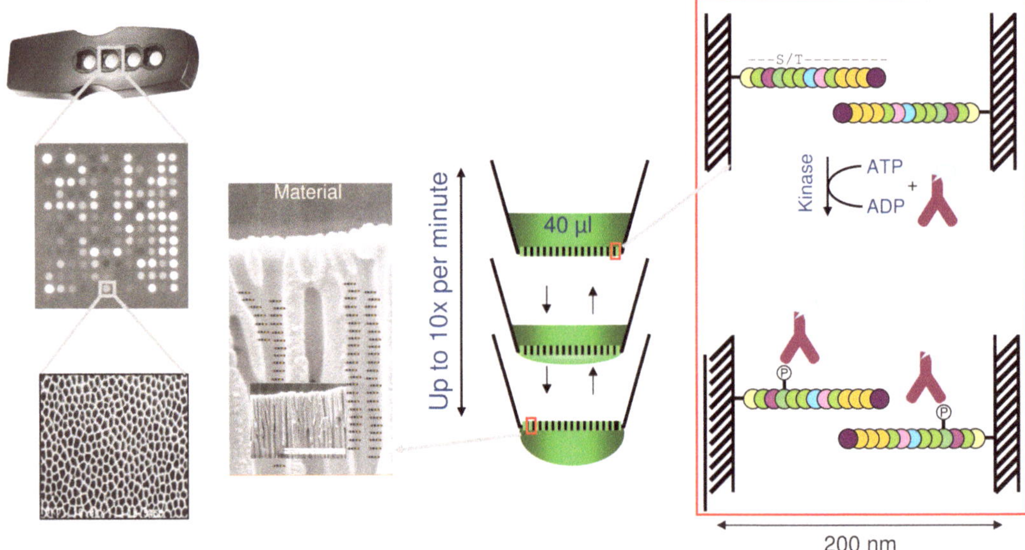

Fig. 1. Schematic overview of the reaction occurring on a PamChip® peptide microarray. Each PamChip® disposable contains four arrays with 144 peptides that are covalently attached to a porous material. Samples are pumped up and down through the porous array and kinases in the sample phosphorylate the peptides. Phosphorylation of peptides is detected by fluorescently labeled antibodies.

molecules, facilitating the detection of its activity. The first types of peptide microarrays were two-dimensional, and later three-dimensional formats were developed with additional features and benefits. The different types of peptide microarrays have recently been reviewed by Thiele et al. (1) and Arsenault et al. (2). Here, we focus on PamChip® arrays, the next generation of peptide microarrays. This is a flow-through microarray that allows real time monitoring of reactions (Fig. 1). The porous 3D surface permits immobilization of high concentrations of peptide and allows rapid peptide phosphorylation by active transport of the solutes. Up to 256 peptides of 13 amino acids long are covalently coupled to the porous internal surface of each array. Sequences are derived from known human phosphorylation sites. The phosphorylation activity in the sample is detected with fluorescently labeled antibodies. Whereas for detection of phosphotyrosines a generic antibody is available, for detection of serine/threonine phosphorylation a combination of antibodies is required. During the assay, images are taken of the developing signals on the microarray by a fluorescence imaging system being part of an integrated instrument, named a PamStation®12, resulting in registration of the kinetics of the reaction. This PamStation®12 is specifically intended for the fully automated processing including incubation and imaging of one up to three PamChip®4 disposables, each containing four microarrays. This allows the analysis of up to 12 samples in a single experimental run under the same experimental conditions. Experimental protocols for assays for tyrosine kinases on the platform have been described elsewhere (3, 4). Here we focus on the determination of serine/threonine kinase activity. Although most features are similar for both types of assay, sample input is about ten times lower (1,000–10,000 cells or 0.001–0.01 mm^3 of tissue per array) than for the tyrosine kinase assay on the same platform.

Dynamic peptide microarrays have been shown to be useful for substrate identification for recombinant kinases (5–7), for development of inhibitors (7, 8), for identification of kinases and pathways involved in disease (9–15), and for prediction of response to treatment with kinase inhibitors (16–18).

Since one and the same platform can be used both for the measurement of kinase activity of recombinant kinases and of tissue lysates in a variety of species ranging from zebra fish (3, 19) to humans, translational research is facilitated. An additional advantage is that inhibitors can be added to the (tumor) tissue of interest to study their effect ex vivo in the tissue of interest.

2. Materials

2.1. Reagents

Prepare all dilutions with ultrapure water. Use analytical grade chemicals to prepare solutions that are not provided in the reagent kit.

1. Mammalian extraction buffer (M-PER) (Pierce).
2. Halt Protease Inhibitor Cocktail, EDTA free (Pierce).
3. Halt Phosphatase Inhibitor Cocktail (Pierce).
4. Phosphate Buffered Saline.
5. PamChip®4 STK disposable peptide microarrays (PamGene International BV, 's-Hertogenbosch, The Netherlands).
6. PamStation® 12 (PamGene International BV, 's-Hertogenbosch, The Netherlands).
7. STK reagent kit (PamGene International BV, 's-Hertogenbosch, The Netherlands). The kit contains the following:
   - Blocking buffer.
   - 10× PK buffer.
   - 100× BSA.
   - 100 mM ATP.
   - STK Primary antibody mixture.
   - STK FITC labeled secondary antibody.

All materials should be stored as indicated in the kit information.

2.2. Protocol for Lysis of Cells or Tissues

Lysates of cells or of fresh frozen tissue can be used as source of kinase activity.

1. Lyse no more than four samples in one lysis run.
2. Prepare Mammalian Extraction Buffer (M-PER) with Halt Phosphatase Inhibitor Cocktail and Halt Protease inhibitor Cocktail (diluted 1:100) and store on ice (see Note 1). A 1:50 dilution of the inhibitors can be used for tissues expected to have a high protease or phosphatase activity.
3. Label at least four tubes for every sample and cool tubes in dry ice. This will cause the lysate to freeze immediately. Alternatively, lysates can be snap-frozen in liquid nitrogen.
4. Tissue: fine needle biopsies and endoscope biopsies can be lysed without cutting of sections. For core biopsies, cut a number of sections that gives about 1 mm^3 of tissue. For tissue blocks, cut six 10 μm sections, (total of 60 μm of material) of 5×5 mm surface from the fresh frozen specimen. For most tissues single sections of 60 μm can be used as well (see Note 2). Make sure that the sections remain frozen during the cutting process. The integrity of the sections is less important than in histology, since protein will be extracted from the sections anyway. Use first and last or middle sections for histological investigations if possible.
5. Place sections at the bottom of a precooled screw cap vial.
6. The material can be lysed immediately or stored in liquid nitrogen or at −80 °C for later use.
7. For lysis of tissues, add cold lysis buffer to the first vial with frozen tissue to start the lysis procedure (lyse at most four samples simultaneously). Use 100 μl lysis buffer per mm^3 tissue. Keep on ice.
8. Cells: remove culture medium from cells and wash twice with ice cold phosphate buffered saline. Add 100 μl of cold lysis buffer per 1×10^6 cells. Adhering cells must be scraped, not treated with trypsin since this might induce kinase activity. Keep on ice.
9. Promote the lysis by pipetting the mixture up and down (approximately ten times) in the vial until the solution is clear (no lumps should be visible). Expel the fluid gently from the pipette tip in order to prevent foaming and denaturation of the proteins.
10. Keep the lysate on ice for 30 min. Check the lysis process visually. Promote lysis by pipetting the liquid up and down a few times at intervals of about ten min.
11. Centrifuge the lysate for 15 min at $10{,}000 \times g$ at 4 °C in a precooled centrifuge.
12. The supernatant of the lysed sample should be collected and aliquotted in precooled vials (for example four times 5 μl/vial

Table 1
Composition of assay master mix

Solution	Volume (μl)
Water	To a final volume of 40 μl
10× PK buffer	4
100× BSA solution	0.4
STK primary antibody mix	Depends on batch
STK FITC labeled secondary antibody	Depends on batch
(Inhibitor in 10% DMSO)	(2)
Sample	Max 10
ATP	4

and four times 20 μl/vial). Lysates can be placed at the bottom of vials precooled on dry ice. Alternatively, vials can be snap-frozen in liquid nitrogen before storage. Store vials immediately at −80 °C. Keep a 5 μl aliquot for protein quantification purposes.

13. Repeat the procedure described above for each sample. A maximum of four samples can be lysed in parallel.
14. Determine protein concentration with a standard protein quantitation assay.

3. Methods

Before starting an experiment, a pipetting scheme must be prepared based on the volumes per array shown in Table 1. Multiply the volumes by the number of arrays and add at least 10% additional volume to account for pipetting loss during transfer of solutions (e.g. for four arrays, multiply volumes with at least a factor 4.4). Keep all solutions on ice, unless indicated otherwise.

1. Allow the PamChip®4 disposable(s) to come to room temperature.
2. Cool some ultrapure water on ice.
3. Dilute 10× PK buffer tenfold in ultrapure water to prepare 1× PK buffer for washing.
4. Dilute 100 mM ATP 25 fold in ultrapure water to prepare, e.g., 4 mM ATP (see Note 3).
5. When performing experiments with kinase inhibitors, dilute the inhibitors to 200× the final concentration in DMSO.

Just before addition of the inhibitor to the assay mixture, dilute tenfold in ultrapure water (see Notes 4–6).

6. Turn on the PamStation®12 following the instructions of the manufacturer and load the desired assay protocol into the Evolve software program. A more detailed protocol is given in the manual provided with the PamStation®12.
7. Place 1, 2, or 3 PamChip® STK disposables in the incubator.
8. Place a syringe with 1× PK wash buffer in the instrument at the appropriate position (indicated in the assay protocol).
9. Apply 30 μl of Blocking buffer to each array (see Note 7).
10. Start the blocking step in the assay protocol (see Note 8).
11. Prepare the assay master mix while the arrays are being blocked and washed. The assay master mix is always prepared just before the application onto the array and should not be stored. Mix gently, *do not vortex* and keep the assay mix on ice. Add components in the order indicated in Table 1. Refer to the instruction provided with the kit for amounts of antibody to be added (see Note 9). Prepare the assay master mix without sample and ATP. To reduce variation between arrays, prepare first an assay master mix containing all components common to all arrays; subsequently divide the mixture to tubes representing the different assay conditions.
12. Divide the desired amount of the assay master mix over pre-cooled tubes representing the different condition to be tested.
13. Add the sample to the assay mix just before application to the arrays. The amount of input material required depends on the sample type used (see Note 10).
14. For recombinant kinases, the amount can vary from 1 ng to 400 ng, depending on the activity of the recombinant kinase (see Notes 11–12).
15. For cell lysates ad tissue lysates, in general 0.2–2 μg protein per array will give a robust kinase activity (see Notes 13–17).
16. Start the reaction by adding ATP or water (No ATP control) to the tubes with assay mix and mix gently (see Note 18).
17. Apply 40 μl assay mix per array (see Note 19).
18. Run the assay protocol.
19. During the run, (a run takes 30 or 60 min in the standard assay protocol for recombinant kinases or lysates respectively), images will appear on the screen (Fig. 2). At the end of the run, an automated washing step is performed and images are taken at different exposure times.
20. When the running of the assay protocol is completed, remove the PamChip®4 disposable(s) and shut down the instrument according to the information in the PamStation®12 manual.

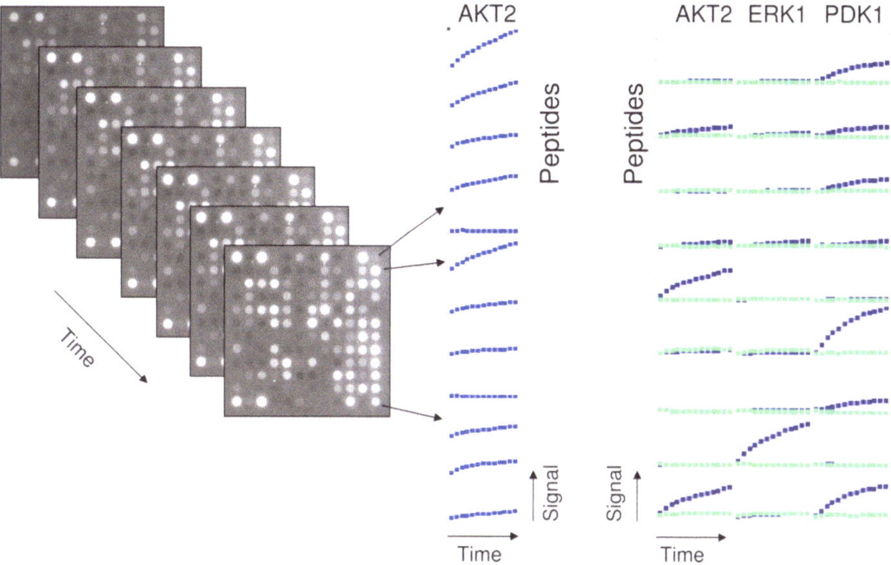

Fig. 2. *Left*: Time series of images of a PamChip® array phosphorylated by recombinant AKT2 kinase. *Middle*: Phosphorylation kinetics is shown for some peptides. *Right*: Kinetics of peptide phosphorylation for the recombinant kinases AKT2, ERK1 and PDK1 on some differentially phosphorylated peptides. *Dark curves* indicate the presence of ATP, *light curves* the absence of ATP.

21. Use BioNavigator software (PamGene International BV, 's-Hertogenbosch, the Netherlands) to quantitate the signal intensity of all peptide spots on all images (see Notes 20 and 21).

22. Inspect correct placement of each grid and spot by the software as well as the occurrence of irregularities, fluorescent particles and other phenomena affecting data quality (see Note 22). The software calculates the intensity in each spot, the local background around each spot and detects saturation in the pixels in the images. Subtraction of local background yields the value SmBg (signal minus background) for each spot. These values are used for further analysis.

23. The data points can be inspected as a time series of signal intensity per peptide (Fig. 2). The initial rate of reaction can be calculated and used as basis for data analysis. Alternatively, the SigmBg end levels either before or after the washing step can be used for further analysis. In general, these values are very similar. In case a high kinase activity leads to saturation of many spots, the values at an earlier time point can be used as input for data analysis.

24. The dynamic range of the assay can be increased by making use of a combination of different exposure times. Therefore the slope of the signal as function of exposure time is used. Saturated signals can be excluded from this analysis.

25. See Note 23 for additional tips for optimization various reactions.

4. Notes

1. MemPer and T-PER lysis buffers have been shown to be compatible with kinase activity.
2. Tissue-Tek® OCT™ (a wax-like substance used for embedding tissues before cryo-section) embedded material is compatible with PamChip® array analysis to a certain extent. If OCT™ is used, remove as much as possible. Tissue sections must contain less than 10% OCT™.
3. The 100 mM ATP provided in the kit is dissolved in water. ATP is an acidic molecule. Up to 2 mM ATP can be used in the assay without changing the pH of the PK buffer. This allows working at physiological concentrations (1–4 mM) of ATP.
4. When working with kinase inhibitors, usually a stock solution of 10 mM inhibitor in DMSO is made that is diluted in DSMO till 200× the desired final concentration. This solution is diluted tenfold. Per 40 μl assay mix, 2 μl of this solution is added, resulting in 0.5% DSMO in the final assay mix.
5. When dissolving and diluting inhibitors, the experimenter should check that all material dissolves. Depending on the inhibitor concentration, dilution in water may result in precipitation. When this happens, a less concentrated solution can be made. Up to 2% DMSO in the assay mixture is tolerated by kinases, but may result in lower activity.
6. Some inhibitors bind slowly to the kinase. To assure proper binding, add the inhibitor before adding ATP.
7. The material the arrays are made of is very brittle. Do not touch the arrays with the pipette tip to prevent breaking. Place the pipette tip in the dedicated area in the plastic housing adjacent to the array or let the fluid wet the array by touching the surface with the drop hanging from the pipette tip.
8. Complete drying of arrays after blocking affects assay performance negatively. When a long time passes between washing and application of the assay mixture, the arrays may dry completely (white appearance). When it is expected that the time between the end of the washing step and application of assay mix will be more than 15 min., the PamChip®4 disposable(s) may be removed from the instrument and stored in the original pouch to prevent further drying.
9. For the detection of phosphorylated serines or threonines no generic antibody is commercially available. Therefore a mixture of antibodies is used that detects a variety of phosphopeptides. These antibodies bind to some peptides on the array, even before phosphorylation has occurred. This nonspecific

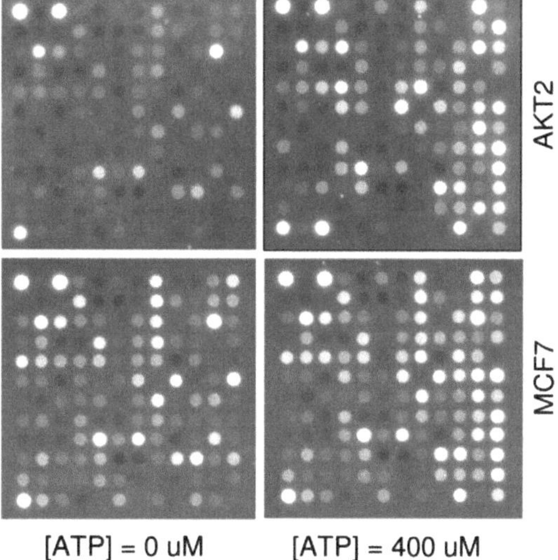

Fig. 3. Identification of ATP-independent and ATP-dependent binding of antibodies by comparison of assays performed in the absence and presence of ATP for the recombinant kinase AKT2 and an MCF7 cell lysate. The non-ATP dependent binding depends on the composition of the sample.

binding is counteracted by protein present during the assay. To correct for this binding effect, incubations without ATP are performed (Fig. 3).

10. For a recombinant kinase the optimal kinase input must be determined by testing a concentration series of the kinase. Since the sensitivity of the PamChip® assay is similar to a radioactive assay, the input suggested by the supplier is a good starting point. When such information is not available, 2, 20 and 200 ng per array can be tested to determine the optimal concentration. As control, an incubation without ATP must be performed to show ATP dependency of the reaction (Fig. 2).
11. When preparing a concentration series of a kinase, one assay master mix containing the highest concentration kinase is made and one without kinase. The latter is used to dilute the kinase containing assay master mixture. This method avoids errors due to variation in pipetting small volumes.
12. Since recombinant kinases may be sensitive to freezing–thawing cycles, the recombinant kinases must be treated according to the instruction of the manufacturer. In many cases, aliquotting of the recombinant kinases is suggested to avoid freezing–thawing cycles.
13. Lysates should be centrifuged shortly in a precooled centrifuge before addition to the assay mix to remove precipitates that could be present.

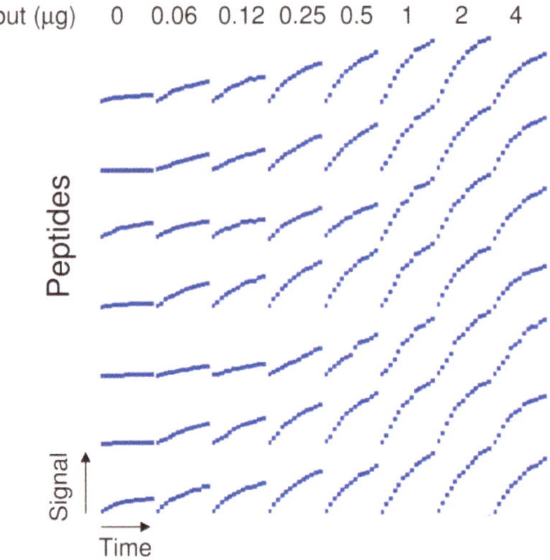

Fig. 4. Effect of increasing protein input per array from 0 to 4 μg protein per array. Phosphorylation kinetics for selected peptides is shown as function of protein input.

14. The optimal amount of lysate per array must be determined. In general, 0.2–1 μg of protein per array results in a robust signal (Fig. 4). For an optimization experiment, it is suggested to test three input concentrations, with the middle concentration repeated without ATP, to show ATP dependency of the signals. Signals increase with input concentration till an optimum is reached.
15. When using different amounts of sample it is advised to add M-PER lysis buffer with HALT Protease and Phosphatase Inhibitor Cocktails to the desired volume. Since the inhibitors are not resistant to freeze–thaw cycles, it is advised to make up the volume with a solution that has undergone a freeze–thaw cycle like the samples. No more than 10 μl of M-PER lysis buffer should be present per 40 μl assay mix.
16. The stability of kinase activity in lysates during storage on ice depends on the sample. When wanting to store a lysate for some hours, it is advised to leave it on ice rather than to subject it to a freeze–thaw cycle.
17. Although the kinase activity in some samples can withstand several freeze–thaw cycles, other lysates are more susceptible to freezing–thawing. Therefore, it is advised to aliquot samples after lysis so as to avoid freeze–thaw cycles.
18. To establish that *bona fide* kinase activity is measured controls should be performed to assure that activity depends on the concentration of ATP and input kinase/lysate and that it can be inhibited by a generic kinase inhibitor like staurosporine.
19. Broken arrays are detected automatically by the instrument. Broken arrays are not processed.

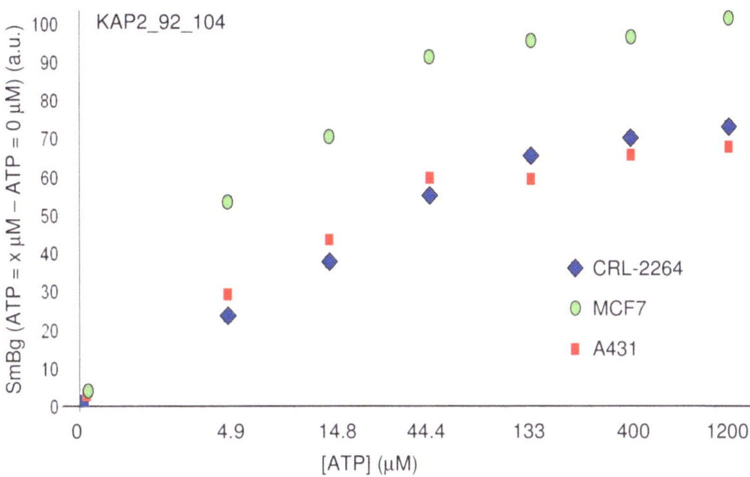

Fig. 5. Effect of variation of ATP concentration on signal intensity for three cell lysates.

20. Other programs can be used for signal quantitation, but are not designed to deal with a large number of arrays nor with the kinetic readouts.
21. The grid is placed on the last image of a series. When the camera position has slightly shifted during the run, grid placement may be inaccurate. In such cases separate analysis of images (see manual for more instructions) is helpful.
22. When fluorescent speckles are visible on many arrays, precipitates in the antibody solution may be the cause. Remove those by centrifugation of the antibody solutions and transfer to clean tubes. Since both primary and secondary antibodies are used in this assay, avoid long incubation of the primary and secondary antibodies as this may form insoluble aggregates.
23. Tips for optimization of various reactions:
    a. Kinase activity increases with increasing concentration of ATP till a maximum is reached. When a sample has a low activity, increasing the ATP concentration may increase the activity (Fig. 5).
    b. Many kinase inhibitors are ATP competitive inhibitors, i.e. they compete with ATP for the ATP binding site in the target kinase(s). In assays with recombinant kinases, inhibition is often measured at an ATP concentration equal to the Km of the recombinant kinase (for many kinases around 10 µM). In cell lysates, such a low ATP concentration results in low signals. Kinase activity profiles are usually measured at 100 or 400 µM ATP. For testing of inhibitors, 100 µM of ATP is advised. It should be noted that, given the higher concentration of ATP, higher concentrations of inhibitor are required. Inhibitor concentrations in the assay usually vary between 0.1 and 10 µM.

c. For detection of peptide phosphorylation on the PamChip® arrays, generic antibodies are used. If you wish to detect phosphorylation of a specific phosphosite, an antibody specific for that phosphorylation site can be used. Such an antibody (and its labeled detection antibody) should be checked for compatibility with the other antibodies, for nonspecific binding to peptides on the array and for the presence of contaminating kinases. The concentration of both primary and secondary antibody must be optimized to prevent high background signals. Staining by additional antibodies can be performed after completion of the standard assay. Antibodies may not only be labeled with FITC, but also with e.g. Cy3 or Cy5 to allow specific detection by Cy3 or Cy5 fluorescence.

d. Samples from zebra fish, as well as of murine, canine, porcine and human origin have shown kinase activity on PamChip® arrays.

References

1. Thiele A, Stangl GI, Schutkowski M (2011) Deciphering enzyme function using peptide arrays. Mol Biotechnol 49:283–305
2. Arsenault R, Griebel P, Napper S (2011) Peptide arrays for kinome analysis: new opportunities and remaining challenges. Proteomics 11:4595–4609
3. Lemeer S, Jopling C, Naji F et al (2007) Protein-tyrosine kinase activity profiling in knock down zebrafish embryos. PLoS One 2:e581
4. Versele M, Talloen W, Rockx C et al (2009) Response prediction to a multitargeted kinase inhibitor in cancer cell lines and xenograft tumors using high-content tyrosine peptide arrays with a kinetic readout. Mol Cancer Ther 8:1846–1855
5. Hilhorst R, Houkes L, van den Berg A et al (2009) Peptide microarrays for detailed, high-throughput substrate identification, kinetic characterization, and inhibition studies on protein kinase A. Anal Biochem 387:150–161
6. Sanz A, Ungureanu D, Pekkala T et al (2011) Analysis of Jak2 catalytic function by peptide microarrays: the role of the JH2 domain and V617F mutation. PLoS One 6:e18522
7. Poot AJ, van Ameijde J, Slijper M et al (2009) Development of selective bisubstrate-based inhibitors against protein kinase C (PKC) isozymes by using dynamic peptide microarrays. Chembiochem 10:2042–2051
8. Harmsen S, Dolman ME, Nemes Z et al (2011) Development of a cell-selective and intrinsically active multikinase inhibitor bioconjugate. Bioconjug Chem 22:540–545
9. Sikkema AH, Diks SH, den Dunnen WF et al (2009) Kinome profiling in pediatric brain tumors as a new approach for target discovery. Cancer Res 69:5987–5995
10. Bratland A, Boender PJ, Hoifodt HK et al (2009) Osteoblast-induced EGFR/ERBB2 signaling in androgen-sensitive prostate carcinoma cells characterized by multiplex kinase activity profiling. Clin Exp Metastasis 26: 485–496
11. Plaza-Menacho I, Morandi A, Mologni L et al (2011) Focal adhesion kinase (FAK) binds RET kinase via its FERM domain, priming a direct and reciprocal RET-FAK transactivation mechanism. J Biol Chem 286:17292–17302
12. Maat W, el Filali M, Dirks-Mulder A et al (2009) Episodic Src activation in uveal melanoma revealed by kinase activity profiling. Br J Cancer 101:312–319
13. Saelen MG, Flatmark K, Folkvord S et al (2011) Tumor kinase activity in locally advanced rectal cancer: angiogenic signaling and early systemic dissemination. Angiogenesis 14:481–489
14. Ter Elst A, Diks SH, Kampen KR et al (2011) Identification of new possible targets for leukemia treatment by kinase activity profiling. Leuk Lymphoma 52:122–130

15. Jinnin M, Medici D, Park L et al (2008) Suppressed NFAT-dependent VEGFR1 expression and constitutive VEGFR2 signaling in infantile hemangioma. Nat Med 14: 1236–1246
16. Folkvord S, Flatmark K, Dueland S et al (2010) Prediction of response to preoperative chemoradiotherapy in rectal cancer by multiplex kinase activity profiling. Int J Radiat Oncol Biol Phys 78:555–562
17. Hilhorst R, Schaake E, van Pel R et al (2011) Application of kinase activity profiles to predict response to erlotinib in a neoadjuvant setting in early stage non-small cell lung cancer (NSCLC). J Clin Oncol 28: suppl. abstract 10566
18. Hilhorst R, Schaake E., van Pel R et al (2011) Blind prediction of response to erlotinib in early stage non-small cell lung cancer (NSCLC) in a neoadjuvant settin based on kinase activity profiles. J Clin Oncol 29: suppl abstract 10521
19. Lemeer S, Ruijtenbeek R, Pinkse MW et al (2007) Endogenous phosphotyrosine signaling in zebrafish embryos. Mol Cell Proteomics 6:2088–2099

Chapter 22

Immunoaffinity Purification of Protein Complexes from Mammalian Cells

Chieri Tomomori-Sato, Shigeo Sato, Ronald C. Conaway, and Joan W. Conaway

Abstract

In this chapter, we describe a purification scheme designed to isolate multisubunit protein complexes gently and quickly from crude extracts of mammalian cells using immunoaffinity purification of epitope tagged proteins and the multisubunit complexes with which they associate. As an example we describe isolation of the mammalian Mediator complex from HeLa S3 cells.

Key words: Multisubunit protein complex, Immunoaffinity purification, Epitope tag, Stable cell line

1. Introduction

Many functions in gene regulation are carried out by multisubunit complexes that can be composed of anywhere from a few to dozens of different protein subunits. Experiments to define the composition and function of such complexes require the ability to isolate intact, functionally active complexes from tissues or cultured cells. Classical, conventional chromatography-based purification strategies separate proteins or multisubunit complexes from one another based on differences in physicochemical properties such as size, charge, or hydrophobicity. While conventional chromatography-based approaches have long been used for protein purification, they suffer from a number of disadvantages. First, many multiprotein complexes are quite fragile and are not stable to the extremes of ionic strength or other conditions encountered during ion exchange, hydrophobic interaction, gel filtration, or other

Minou Bina (ed.), *Gene Regulation: Methods and Protocols*, Methods in Molecular Biology, vol. 977,
DOI 10.1007/978-1-62703-284-1_22,

Table 1
Useful epitope tags and resins for immunoaffinity purification

Tag	Epitope peptide sequence	Affinity resin	Binding specificity
FLAG	DYKDDDDK	α-FLAG M2 agarose (SIGMA)	N, Met-N, Internal, C
HA	YPYDVPDYA	α-HA agarose (HA-7, SIGMA)	N, C
		α-HA agarose (HA.11, Covance)	N, Internal, C
cMyc	EQKLISEEDL	α-cMyc pAb agarose (SIGMA)	N, C
		α-cMyc agarose (9E10, Santa cruz)	N, C
V5	GKPIPNPLLGLDST	α-V5 agarose (V5-10, SIGMA)	N, C

Elution from antibody affinity resins is typically performed using peptides composed of one or three consecutive repeats of the epitope sequence

forms of conventional chromatography. Second, the degree of purification that can be obtained using any one separation method is typically limited, and it is almost always necessary to develop time-consuming and technically challenging strategies that combine multiple purification steps.

The use of immunoaffinity purification strategies can alleviate many of the problems associated with conventional chromatography. In an immunoaffinity purification, an antibody that recognizes a protein of interest is bound to a resin such as agarose or Sepharose beads. A cell extract or partially purified fraction is passed over the antibody-resin, unbound proteins are washed away, and specifically bound proteins are then eluted from the antibody with competing epitope peptides or by more harsh treatments that can result in complex dissociation or loss of activity, such as high salt or brief exposure to acidic pH. Using such methods, it is possible to achieve substantial purification in a single step; however, successful application of immunoaffinity approaches is dependent on the availability of antibodies with suitable affinity and specificity.

It is often not possible to obtain antibodies suitable for immunoaffinity purification for each individual protein that one wishes to study. An alternate strategy takes advantage of well-characterized antibodies that recognize short, defined peptide sequences with high specificity and affinity. These sequences, referred to as "epitope tags," are added to either the amino- or carboxyl-terminus of a protein of interest (1). When expressed in mammalian cells, the epitope tagged protein can be incorporated into a protein complex or complexes in place of its endogenous counterpart, allowing purification of the tagged protein and any proteins with which it is associated by immunoaffinity chromatography using anti-epitope antibodies (see Note 1). Table 1 shows a list of commonly used epitope tags for immunoaffinity purification (2–5).

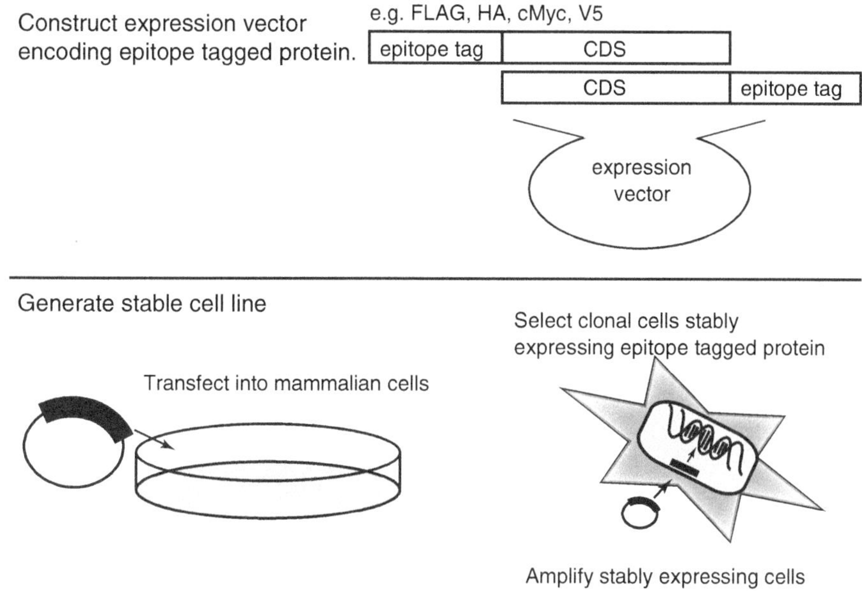

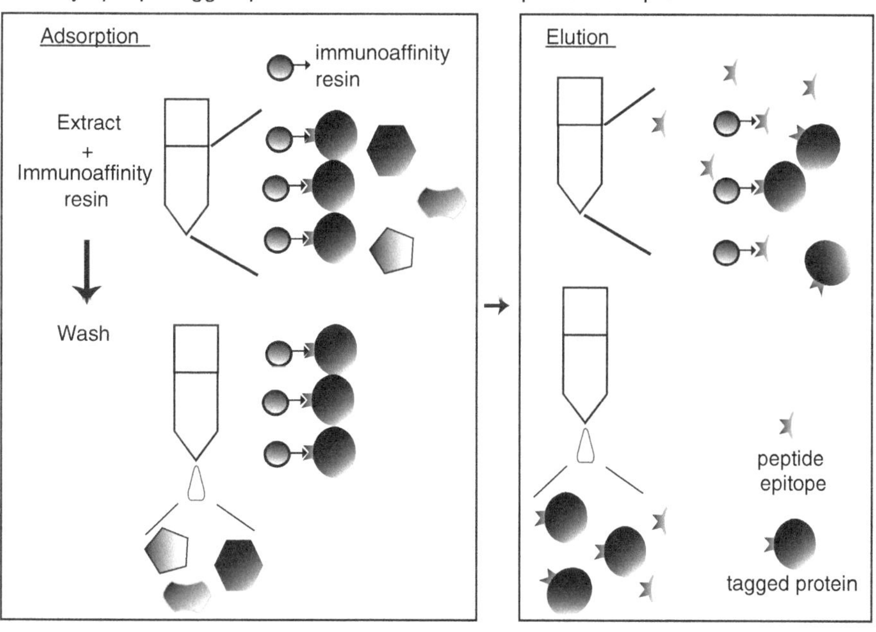

Fig. 1. Scheme for immunoaffinity purification of protein complexes.

A general strategy for the use of epitope-tagging and immunoaffinity purification of protein complexes is outlined in Fig. 1. The first step is to construct a suitable expression vector that encodes an epitope tagged protein that can be expressed in mammalian cells. The second step is to generate and amplify clonal cells stably

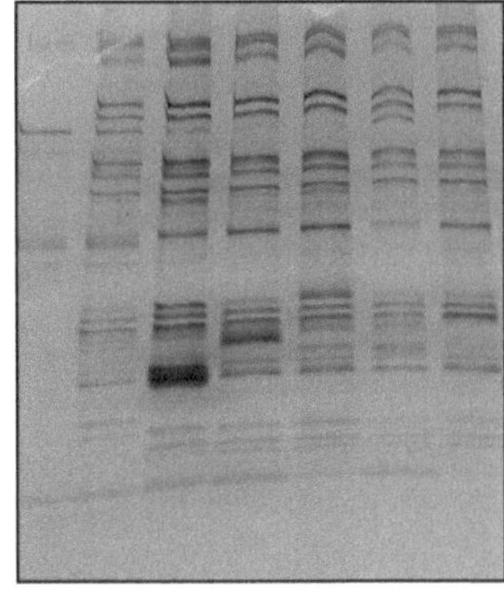

Fig. 2. Purified Mammalian Mediator complexes immunoaffinity purified from HeLa S3 nuclear extracts through FLAG-tagged mediator subunits.

expressing useful amounts of the epitope tagged protein. Finally, the protein of interest and any associated proteins can be purified from nuclear or cytoplasmic extracts by single-step immunoaffinity purification by binding to immobilized anti-epitope antibody and competitive elution with epitope peptides. Using this approach, we have successfully used anti-FLAG epitope immunoaffinity purification to purify the human Mediator of RNA polymerase II to near homogeneity from extracts of HeLa S3 cells stably expressing any of a large number of FLAG-epitope tagged Mediator subunits (6) (Fig. 2). Notably, using cell lines expressing FLAG-tagged versions of mutant Mediator subunits, we have been able to purify mutant Mediator complexes that have proven useful in functional studies (7).

2. Materials

2.1. Production of Mammalian Cell Lines

Host cells (e.g., HeLa S3 cells, HEK293/FRT cells).
Expression vector encoding epitope-tagged protein of interest.
Antibiotic needed for drug selection of stably transformed cells.

2.2. Cell Extract Preparation

0.4% (w/v) Trypan Blue Solution in PBS (#25-900-Cl, Mediatech).

Hypotonic buffer (10 mM HEPES (pH 7.9), 1.5 mM $MgCl_2$, 10 mM KCl, 0.5 mM Dithiothreitol).

Extraction buffer (20 mM HEPES (pH 7.9), 1.5 mM $MgCl_2$, 0.6 M KCl, 0.2 mM EDTA, 0.5 mM Dithiothreitol, 25% Glycerol).

Phosphate buffered saline.

Buffer C (20 mM HEPES (pH 7.9), 1.5 mM $MgCl_2$, 0.2 mM EDTA, 0.5 mM Dithiothreitol, 25% Glycerol).

Protease inhibitor (P8340, SIGMA), add to all buffers immediately before use.

Dounce homogenizer (40 ml, 15 ml, and/or 7 ml).

15 and 50 ml conical plastic tubes.

1.5 ml microcentrifuge tubes.

Beckman type JS4.2 or similar rotor.

Beckman J6 or similar centrifuge.

Beckman type 45Ti rotor and ultracentrifuge.

Polycarbonate ultracentrifuge bottles (#355622, Beckman).

Variable speed motorized tube rotator.

2.3. Immunoaffinity Purification

Equilibration buffer (50 mM HEPES (pH 7.9), 1.5 mM $MgCl_2$, 50 mM KCl, 0.3 M NaCl).

Washing buffer (50 mM HEPES (pH 7.9), 1.5 mM $MgCl_2$, 300 mM NaCl, 10 mM KCl, 0.2% TritonX-100).

Elution base buffer (50 mM HEPES (pH 7.9), 0.1 M NaCl, 1.5 mM $MgCl_2$, 0.05% TritonX-100).

Elution buffer (Elution base buffer with 200 or 500 μg/ml of epitope peptide).

Protease inhibitor cocktail (P8340, SIGMA); should be added to all buffers at 1:100 dilution immediately before use.

Beckman type 70.1Ti rotor and ultracentrifuge.

Polycarbonate ultracentrifuge tubes (#355630, Beckman).

15 and 50 ml conical plastic tubes.

1.7 ml low binding microcentrifuge tubes (#3207, Corning Costar).

Glass fiber pre-filter disk (Millex-AP, # SLAP02550, Millipore).

Sepharose 4B.

Micro Bio-Spin® chromatography columns (#732-6204, Bio-Rad).

Poly-Prep® chromatography columns (#731-1550, Bio-Rad).

Adams™ Nutator or similar mixer.

FLAG epitope: Anti-FLAG M2 affinity gel (A2220, SIGMA), FLAG peptide (F3290, SIGMA), 3× FLAG peptide (F4799, SIGMA).

HA epitope: Anti-HA agarose (clone HA-7, A2095, SIGMA), HA peptide (I2149, SIGMA).

cMyc epitope: Anti-cMyc agarose (sc-40AC, Santa Cruz), cMyc peptide (M2435, SIGMA).

V5 epitope: Anti-V5 agarose (A7345, SIGMA), V5 peptide (V7754, SIGMA).

3. Methods

3.1. Production of Cell Lines Expressing Epitope Tagged Proteins

To purify a multisubunit protein complex it is first necessary to establish a cell line that stably expresses an amino- or carboxyl-terminally epitope tagged protein. Detailed protocols for generating such cell lines have been provided elsewhere (8), so here we provide only a brief outline.

The choice of cell line and gene transfer method will depend on the specific experimental purpose; however, we often use HeLa S3 cells (CCL-2.2™, ATCC), which can be grown in suspension and are therefore advantageous when it is desirable to perform large-scale purifications. We have had good success using constitutive protein expression systems. Occasionally, however, we use a tetracycline inducible expression system, such as the Flp-In™ T-Rex™ Mammalian Expression System (Invitrogen), when the protein of interest is toxic and either kills cells or is expressed at particularly low levels when expressed constitutively. While it is possible to use mixed populations of stably transformed cells, we typically generate clonal cell lines because they can usually be grown for many passages without loss of protein expression.

Prior to attempting to generate stable cell lines, we first confirm that the desired protein can be expressed from the expression vector following transient transfection into HEK293T or other easily transfectable cells. Protein expression can be evaluated by western blotting of whole cell lysates using antibodies directed against the epitope and/or native protein. Lysates from transiently transfected cells can also be used in anti-epitope immunoprecipitations followed by western blotting if the level of expression is too low to be detected by western blotting of lysates and/or to confirm that the epitope is useful for immunoprecipitation.

Once the utility of the expression vector has been confirmed, it is introduced into the desired cell line using plasmid-based or virally-mediated gene transduction (Table 2), and stably transformed cells are selected and cloned using the appropriate drug selection (see Note 2) and screened for expression of the desired protein using western blotting and/or immunoprecipitation. Frozen stocks of cloned cell lines are prepared, and cells are amplified for further purification and analysis (see Note 3).

3.2. Preparation of Nuclear and Cytosolic (S100) Extracts

Below is the general procedure we use to prepare extracts from 12 L suspension cultures of HeLa S3 cells expressing FLAG-epitope tagged Mediator subunits. Additional protocols are provided elsewhere (9–11). Keep all buffers and equipment cold on ice, and perform the following procedures in a 4°C cold room if available.

Table 2
Gene transfer systems for generating stable mammalian expression cell lines

Gene transfer system	Advantage	Disadvantage
1. Plasmid based gene transduction		
(a) Transfection/Random integration (e.g., Calcium phosphate, Lipofection, Electroporation)	Simple procedure Wide range of cell types	Low efficiency Potential multiple integration
(b) Transfection/DNA recombinase mediated integration (e.g., Flp-In, cre-loxP, piggy-Bac transposon system)	Simple procedure Single copy of the gene of interest integrated at defined locus	Limited host cells
2. Virus mediated gene transduction		
Retrovirus or Lenti virus mediated gene integration (e.g., MMLV, MMSV, HIV)	Wide range of cell types	Virus handling technique is required Random and potential multiple integration
DNA virus infection (e.g., adenovirus, EB virus, Vaccinia virus)		Limited host cells DNA is not integrated into genome

1. Count cells using a hemocytometer, and evaluate cell viability using Trypan-blue staining; when observed under the microscope, dead or lysed cells will stain blue. Use cells with >95% viability.
2. Transfer cells from culture flasks to 1 L centrifuge bottles and spin 5 min at 1,500 rpm (500 × *g*) in a Beckman J6 centrifuge using a type JS4.2 rotor at 4°C.
3. Remove supernatants, gently resuspend each cell pellet in approximately 10 ml ice-cold PBS. Transfer cell suspensions from the twelve 1 L centrifuge bottles into a total of six 50 ml conical plastic tubes, and fill each tube to 50 ml with PBS.
4. Spin the cell suspensions 5 min at 1,500 rpm (500 × *g*) in the type JS4.2 rotor.
5. Remove supernatants, resuspend each pellet in ~10 ml PBS, transfer the resulting suspensions to two 50 ml conical tubes, and fill each tube to 50 ml with PBS.
6. Spin the cell suspensions 5 min at 1,500 rpm (500 × *g*) in the type JS4.2 rotor.
7. Wash each pellet one more time with ~50 ml PBS.
8. Add 30 ml hypotonic buffer to each of the two cell pellets. Gently resuspend the cells by blowing buffer onto them using a pipette. Do not draw the cells up into the pipet.

9. Incubate the cell suspensions for 15 min on ice.
10. Meanwhile, rinse the Dounce homogenizer and loose pestle with hypotonic buffer and keep cold on ice.
11. Transfer the cell suspension from one tube to prechilled 40 ml Dounce homogenizer.
12. Homogenize in the Dounce homogenizer with 15–20 gentle strokes of the loose pestle. Avoid generating foam.
13. Monitor cell disruption using Trypan-blue staining. Stop homogenizing when ~80% of cells have been disrupted.
14. Repeat steps 11–13 until all of the remaining cell suspension has been homogenized.
15. Centrifuge at 20,000 rpm (50,000 × *g*) for 20 min at 4°C in a Beckman type JA-20 rotor. Collect the supernatant, and add 0.11 volumes of Buffer C, then keep on ice. This supernatant is the cytosolic fraction.
16. To remove contaminating cytosol and lipids that are stuck to the walls of tubes, layer 15 ml of hypotonic buffer onto the nuclear pellets. Nuclei are very fragile at this point, so it is best to avoid resuspending them. Centrifuge at 5,000 rpm (3000 × *g*) for 8 min at 4°C in a Beckman type JA-20 rotor, and carefully remove the supernatants.
17. Add 5 ml extraction buffer to one of the nuclear pellets, and gently pipet up and down. Transfer the resulting suspension onto the second nuclear pellet, and gently resuspend. Finally, transfer the pooled suspension to a clean graduated 15 ml conical tube.
18. To estimate the volume of nuclear pellet contained in the suspension, measure the total volume of the suspension. Nuclear pellet volume = the total volume of the suspension–5 ml.
19. Transfer to a 70 ml polycarbonate bottle for a Beckman type 45Ti rotor. Add additional extraction buffer to bring the total volume of the suspension to twice the nuclear pellet volume. This will bring the final KCl concentration of the suspension to 0.3 M.
20. Pipet gently to generate a homogenous suspension. The suspension will become somewhat viscous at this stage due to salt-induced decondensation of chromatin. Decondensed chromatin can be easily sheared by over-vigorous mixing (i.e., vigorous pipetting, mixing, or vortexing). Avoid this, since the extract will contain a lot of contaminating DNA that can cause problems with subsequent purification steps if the chromatin becomes significantly sheared.
21. Place capped centrifuge bottle containing nuclear extract on a variable speed motorized tube rotator, and rotate end-over-end

for at least 30 min at 20 rpm at 4°C to extract nuclear proteins from chromatin.

22. Spin nuclear extract and S100 samples from step 14 at 35,000 rpm (145,000 × *g*) for 1 h at 4°C in a Beckman type 45Ti rotor.
23. Collect the cleared supernatants from the nuclear and cytosolic fractions; these are the nuclear extract and S100 fractions, respectively.

3.3. Immunoaffinity Purification of Epitope-tagged Proteins and Multi-protein Complexes

Carry out all procedures in a 4°C cold room or on ice unless otherwise specified.

For trial immunopurification, use whole cell lysates prepared from cells grown to ~90% confluency in 1 or 2 10 cm dishes with 20 μl packed bed volume of FLAG-M2 agarose immunoaffinity resin. For preparative immunopurification, use 5–6 ml of nuclear extract or 8–10 ml of cytosolic fraction (S100) prepared as described above (~30–50 mg total protein), with 100 μl FLAG-M2 agarose. In addition, the immunopurification method described below can be used with starting fraction prepared by other protocols (but see Note 4). We have also successfully used the methods described here to purify epitope-tagged proteins and complexes through additional epitopes using the immunoaffinity resins and elution peptides listed in Table 1.

3.3.1. Preparation of Immunoaffinity Beads

1. Transfer at least 1.5 packed volumes of the amount of immunoaffinity resin required for the purification to a clean 1.5 ml tube.
2. Add ten volumes of binding buffer, and gently invert tubes several times until resin is resuspended into an uniform slurry with no clumps.
3. Centrifuge at 200 × *g* for 30 s in a microcentrifuge. It is important not to spin too hard or beads will become difficult to resuspend.
4. Carefully remove the supernatant without disturbing the packed agarose bed.
5. Repeat steps 2–4 an additional three times.
6. Add enough binding buffer to make a 50% slurry of the immunoaffinity resin.
7. Beads should be stored on ice or at 4°C until use. Pre-equilibrated beads can be stored for up to 1 week.

3.3.2. Immunoaffinity Purification from Nuclear Extract (Batch Chromatography)

1. Thaw nuclear extract. To avoid proteolysis or protein denaturation, it is important to keep the extract cold during the thawing process. Place tube on bench top until frozen extract begins to thaw. Gently mix the frozen slurry by rocking the tube by hand until it is about two-thirds thawed, then lay it on a bed of ice or place it at 4°C until it is completely thawed.

2. Spin the thawed extract for 15 min at 35,000 rpm (100,000 × g) at 4°C in a Beckman type 70.1 Ti rotor in an ultracentrifuge. Save several 50 μl aliquots of the starting extract and keep on ice; these will be analyzed along with the final immunopurified material using western blotting or other assays to determine the efficiency of the purification.
3. Carefully transfer the cleared extract to a clean 15 ml conical tube.
4. Add 200 μl of the 50% slurry of pre-equilibrated immunoaffinity beads (equivalent to 100 μl of packed beads).
5. Incubate at 4°C for 4–12 h on a variable speed motorized tube rotator at <20 rpm (see Note 5).
6. Pellet the affinity matrix by centrifugation at 1,000 rpm (200 × g) for 5 min at 4°C in a Beckman type JS4.2 rotor and J6 centrifuge or other low speed centrifuge.
7. Remove the supernatant, and save it for a second round of immunoaffinity purification.
8. To wash residual unbound proteins away from the immunoaffinity matrix, add 5 ml of washing buffer to the beads, and incubate for 5 min at 4°C on a rocking mixer (such as an Adams™ Nutator Mixer).
9. Centrifuge at 1,000 rpm (200 × g) for 5 min at 4°C.
10. Carefully aspirate the supernatant without disturbing the resin.
11. Repeat steps 8–10 an additional four times, for a total of five washes.
12. Spin the washed resin one more time at 1,000 rpm (200 × g) for 5 min at 4°C to pack the beads and allow removal of remaining buffer.
13. Carefully remove the remaining buffer from the beads with a Pipetman or similar pipettor with a narrow tip.
14. Add 500 μl elution base buffer and transfer the immunoaffinity beads from the 15 ml tube to a new 1.5 ml low binding microcentrifuge tube.
15. To recover any beads that were left behind, rinse the 15 ml tube with an additional 500 μl elution base buffer and add to the 1.5 ml microcentrifuge tube from the previous step.
16. Centrifuge at 1,000 rpm (200 × g) for 1 min in a microcentrifuge.
17. Remove and discard the supernatant.
18. To ensure removal of as much buffer as possible, centrifuge the washed resin one more time at 1,000 rpm for 1 min in the microcentrifuge, and remove and discard any remaining liquid.

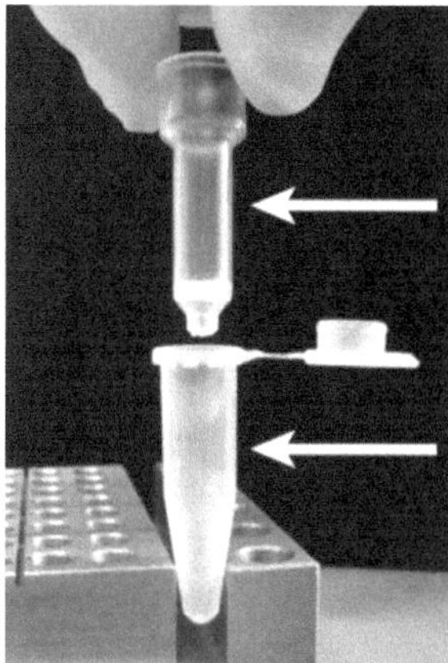

Fig. 3. Micro bio-spin chromatography column used to remove residual beads from immunoaffinity purified fractions.

To elute bound proteins, apply one packed bed volume (100 μl) of elution buffer containing 200–500 μg/ml FLAG peptide to the immunoaffinity resin (see Note 6).

19. Incubate at 4°C on the rocking mixer for at least 30 min.
20. Centrifuge at 1,000 rpm for 1 min. *IMPORTANT: The supernatant from this spin contains immunopurified protein or protein complexes that have been eluted from the resin.*
21. Remove the snap-off tip from an empty Micro Bio-Spin® chromatography column, and set the empty column on top of a new 1.5 ml low binding microcentrifuge tube (Fig. 3).
22. Carefully transfer the supernatant from step 21 into the empty Micro Bio-Spin® column on the microcentrifuge tube.
23. Centrifuge at 1,000 rpm for 1 min. The purified protein fraction will flow through the empty column into the microcentrifuge tube, and any remaining immunoaffinity resin will be filtered out of the purified protein fraction and will be retained in the column.
24. Repeat elution steps 18–23 an additional two times to elute additional protein from the immunoaffinity resin. Save the eluted fractions separately as "E1, E2, and E3."
25. Snap freeze immunopurified samples (E1, E2, and E3) in liquid nitrogen and store at −80°C for further analysis (see Note 7). At this time also freeze the aliquots of extract reserved at the beginning of the purification. Many proteins or protein complexes are unstable to repeated freeze-thaw, so it is a good idea to divide purified protein fractions into multiple small aliquots prior to freezing.

3.3.3. Purification of Epitope Tagged Proteins from the Cytosolic Fraction (S100)

1. Thaw the S100 fraction, and spin at 35,000 rpm (100,000 × *g*) at 4°C for 15 min in a Beckman type 70.1 Ti rotor.
2. Transfer to a 15 ml conical tube.
3. To minimize nonspecific protein binding to the immunoaffinity resin, preclear the S100 fraction by incubating it with Sepharose 4B (300 μl packed beads per 10 ml S100 fraction) for 10 min at 4°C with rotation on a variable speed motorized tube rotator at 20 rpm. Keep several 50 μl aliquots of the S100 fraction for further analysis.
4. Pellet Sepharose 4B resin by centrifugation at 1,000 rpm (200 × *g*) in the Beckman type JS4.2 rotor, and transfer the supernatant to a new 15 ml conical tube.
5. Repeat steps 3 and 4 an additional two times. The final supernatant is the precleared S100 fraction.
6. Spin the precleared S100 at 35,000 rpm (100,000 × *g*) in a Beckman type 70.1 Ti rotor for 15 min at 4°C.
7. Carefully collect the supernatant, and transfer to a 10 ml disposable syringe attached to a glass fiber pre-filter disk (Millex-AP).
8. Filter the S100 fraction through the glass fiber pre-filter disk into a new 15 ml conical bottom tube to capture contaminating lipids and minimize carryover of beads from the preclearing steps.
9. Proceed to immunoaffinity purification using the procedures described in Subheading 3.3.2, steps 4–26.

3.3.4. Immunoaffinity Purification Using Column Chromatography

Immunoaffinity purification using column chromatography can be used as an alternative to the batch purification method described above. Advantages of column purification methods include (1) ease of scale-up and (2) avoidance of nonspecific trapping of denatured or aggregated proteins present in the lysate.

1. Prepare nuclear extract samples for immunoaffinity column chromatography using steps 1 and 2 of Subheading 3.3.2. Prepare S100 fractions using steps 1–8 in Subheading 3.3.3.
2. Use 100 μl immunoaffinity agarose per approximately 30 mg of nuclear extract or S100 protein.
3. Pack immunoaffinity resin into a suitable column by gravity flow. We use Bio-Rad Poly-Prep® disposable chromatography columns for resin volumes up to ~1 ml. Do not allow the resin to become dry.
4. Float a glass fiber filter on top of the bed, and gently layer the cleared nuclear extract or S100 on top of the resin.

5. Allow the column to flow at 10–15 column volumes per hour until all of the nuclear extract or S100 has been loaded on to the column.
6. Collect the flow-through fractions and reload them on to the resin. Repeat this process until all of the nuclear extract or S100 has been passed through the column five times.
7. Wash the resin with 30 bed volumes of washing buffer at the maximum flow rate attainable with gravity flow (~50 μl/min for 100 μl resin).
8. Stop the flow. Gently apply 5–6 bed volumes of elution buffer on top of the resin. Incubate for 30 min at 4°C.
9. Allow the column to flow at ~10 column volumes per hour. Collect ten fractions of one bed volume each, and assay for the presence of the desired protein by western blotting or other suitable assays (see Note 7).

4. Notes

1. Epitope tags at one or the other end of a protein of interest can sometimes prevent complex assembly and/or be sterically occluded. Accordingly, it is advisable to generate both amino- and carboxyl-terminally tagged versions of the protein of interest.
2. Commonly used expression vectors carry genes encoding resistance to antibiotics (e.g., hygromycin, neomycin/G418, or puromycin); hence, the choice of antibiotic for selection of stably transformed cells is determined by the specific expression vector being used. Optimal antibiotic concentrations for selection of drug-resistant, stably transformed cells will vary with the specific cell line used. Before proceeding to stable cell line production, we recommend generating a "kill curve" to define the lowest concentration of antibiotic that will kill all non-transformed cells. To do so, culture parental cells in media containing various concentrations of the selection antibiotic. Observe the cells daily and determine the percent viable cells. We use the minimum concentration of selection antibiotic needed to kill all parental cells 6–8 days after placing cells under drug selection.
3. Even with initially clonal populations of cells it is possible to observe changes in expression level or even mutation of the exogenously expressed cDNA when cells are grown in culture for a long period of time. Accordingly, it is always good practice to prepare frozen stocks of newly established cell lines after as few passages as possible and to start large scale cultures for immunopurification from early passage cells.

4. Carefully examine the components of buffers in the starting fractions used for immunopurification, since binding of individual antibodies to their epitopes can be dramatically and differentially affected by detergents, salts, reducing agents, and other buffer components. Antibody manufacturers often provide information about reagent compatibility, and this should be carefully considered.
5. The optimal length of time for incubation of cell extracts with immunoaffinity beads needs to be determined. Shorter incubation times most often yield immunopurified material with fewer contaminants; however, with some epitope tagged proteins, a longer incubation can increase yield.
6. Optimal elution peptide, buffer composition, and the temperature at which the elution step is performed need to be determined empirically. Researchers often use elution peptides containing either one or three copies of the epitope sequence (1× or 3× epitope peptide, respectively). In our experience, 1× peptides are usually sufficient; in addition, they are less expensive and often give rise to immunopurified fractions with fewer contaminants. In initial trials, we typically test elution buffers containing 200 μg/ml or 500 μg/ml of 1× epitope peptide. If satisfactory yields are not obtained using 1× epitope peptide, we then try elution buffers containing 3× epitope peptide. We usually obtain good yields of FLAG-tagged proteins when elutions are performed at 4°C; however, yields of HA-, cMyc-, and V5-tagged proteins are usually better when elutions are performed between about 20°C and 37°C.
7. Following peptide elution from immunoaffinity beads, epitope-tagged proteins or protein complexes will be contaminated with a significant amount of the elution peptide. If necessary, desalting chromatography with resins such as Sephadex G-25 or G-50 (GE Healthcare Life Sciences) can be performed to remove peptide from immunopurified proteins.

Acknowledgments

Work in the authors' laboratory is supported by grant GM41628 from the National Institute of General Medicine, a grant to the Stowers Institute for Medical Research from the Helen Nelson Medical Research Fund at the Greater Kansas City Community Foundation, and the Stowers Institute for Medical Research.

References

1. Jarvik JW, Telmer CA (1998) Epitope tagging. Annu Rev Genet 32:601–618
2. Einhauer A, Jungbauer A (2001) The FLAG peptide, a versatile fusion tag for the purification of recombinant proteins. J Biochem Biophys Methods 49:455–465
3. Field J et al (1988) Purification of a RAS-responsive adenylyl cyclase complex from *Saccharomyces cerevisiae* by use of an epitope addition method. Mol Cell Biol 8:2159–2165
4. Ellison MJ, Hochstrasser M (1991) Epitope-tagged ubiquitin. A new probe for analyzing ubiquitin function. J Biol Chem 266:21150–21157
5. Southern JA et al (1991) Identification of an epitope on the P and V proteins of simian virus 5 that distinguishes between two isolates with different biological characteristics. J Gen Virol 72(Pt 7):1551–1557
6. Sato S et al (2004) A set of consensus mammalian mediator subunits identified by multidimensional protein identification technology. Mol Cell 14:685–691
7. Takahashi H et al (2011) Human mediator subunit MED26 functions as a docking site for transcription elongation factors. Cell 146:92–104
8. Mortensen RM, Kingston RE (2009) Selection of transfected mammalian cells. Curr Protoc Mol Biol 86:9.5.1–9.5.13
9. Manley JL et al (1980) DNA-dependent transcription of adenovirus genes in a soluble whole-cell extract. Proc Natl Acad Sci USA 77:3855–3859
10. Dignam JD, Lebovitz RM, Roeder RG (1983) Accurate transcription initiation by RNA polymerase II in a soluble extract from isolated mammalian nuclei. Nucleic Acids Res 11:1475–1489
11. Abmayr SM et al (2006) Preparation of nuclear and cytoplasmic extracts from mammalian cells. Curr Protoc Mol Biol 75:12.11.11–12.11.10

Chapter 23

Simple and Efficient Identification of Chromatin Modifying Complexes and Characterization of Complex Composition

Jeong-Heon Lee and David Skalnik

Abstract

Affinity purification and mass spectrometry analysis have been used to identify and characterize protein complexes. Wdr82-associated chromatin modifying complexes were purified by single-step FLAG affinity purification from human cells induced to express FLAG-tagged Wdr82. Purified proteins were analyzed by SDS-PAGE and specific protein bands were identified by mass spectrometry. Subsequently, purified proteins were fractionated on sucrose gradient equilibrium centrifugation to determine overall composition of each identified complex. We describe here simple and efficient approaches for the identification of chromatin modifying complexes and subsequent characterization of complex composition.

Key words: Protein complex, FLAG affinity purification, Mass spectrometry, Wdr82, Sucrose gradient equilibrium centrifugation, Complex composition

1. Introduction

The combination of affinity purification and mass spectrometry (MS) analysis has been favored to identify and characterize protein complexes in various experimental systems (1–3). In general, single-step affinity purification and tandem affinity purification (TAP) provide efficient ways to purify protein complexes. The TAP approach tends to yield higher purity of protein complexes with less background, but transiently or weakly interacting proteins can be lost due to the multiple purification steps. Single-step affinity purification has the tendency to co-purify nonspecific proteins that bind to the affinity matrix or the antibodies utilized, but this approach is still widely used to identify and characterize chromatin modifying complexes due to simple and efficient purification. This approach often provides high yields and biologically active complexes.

The WD40 domain protein Wdr82 is a component of two similar but distinct human histone H3-Lys4 methyltransferase

Minou Bina (ed.), *Gene Regulation: Methods and Protocols*, Methods in Molecular Biology, vol. 977,
DOI 10.1007/978-1-62703-284-1_23, © Springer Science+Business Media, LLC 2013

complexes (Setd1a and Setd1b) and functions as a targeting protein of Setd1 complexes by tethering to early elongating RNA pol II at transcription start sites of expressed genes (4–6). To further understand the in vivo functions of Wdr82, Wdr82-associated complexes were isolated from T-Rex HEK293 cells induced to express FLAG-tagged Wdr82. Wdr82 was found to associate with multiple protein complexes including three histone methyltransferase complexes (Setd1a, Setd1b, and MLL3/4), early elongating RNA pol II, a Pnuts/Tox4/Wdr82 (PTW) PP1 phosphatase complex, as well as additional uncharacterized proteins (7). We describe here a simple and efficient method for identification and characterization of chromatin-associated protein complexes by the combination of single-step affinity purification and MS analysis. We also describe how the overall composition of each identified complex can be determined by fractionation on linear sucrose gradients.

2. Materials

Prepare all solutions using ultrapure water and analytic grade reagents. Prepare and store all reagents at 4°C (unless indicated otherwise).

2.1. Establishment of Inducible T-Rex Stable Cells

1. Dulbeco's Modified Eagle's Medium with high glucose (DMEM) (Mediatech, Manassas, VA; Cat. # 10-017-CV).
2. Bovine calf serum (Hyclone, Cat # SH30109.03).
3. Blasticidin (Invitrogen, Cat. # R21001).
4. The T-REx HEK293 cell line (Invitrogen cat. No. R71007). These cells constitutively express the tetracycline repressor and are maintained in Dulbecco's modified Eagle's medium supplemented with 10% bovine calf serum and 5 μg/ml blasticidin.
5. pcDNA5/TO vector (Invitrogen, Cat. # V103320).
6. 2× HEPES-buffered saline (HBS): 50 mM HEPES pH 7.05, 280 mM NaCl, 1.5 mM Na_2HPO_4. Filter sterilized.
7. 2.5 M $CaCl_2$ (183.7 g $CaCl_2$ $2H_2O$ in 500 ml water). Filter sterilized.
8. Hygromycin B (Sigma, Cat. # H0654).
9. 6 mm cloning cylinder (Corning, Cat. # 3166-6).
10. 0.05% Trypsin-EDTA solution (Invitrogen, Cat. # 25300).
11. Doxycycline (Clontech, Cat. # 631311).
12. CSK buffer: 10 mM PIPES pH 7.0, 300 mM NaCl, 300 mM sucrose, 3 mM $MgCl_2$, 1 mM EGTA, and 0.5% Triton X-100.
13. Anti-FLAG M2 antibody (Sigma, Cat. # F1804).

2.2. Purification of FLAG-Tagged Proteins

1. Phosphate buffered saline (PBS) pH 7.4: 8 g NaCl, 0.2 g KCl, 1.44 g Na_2HPO_4, and 0.24 g KH_2PO_4 in 1 l water.
2. CSK buffer: 10 mM PIPES pH 7.0, 300 mM NaCl, 300 mM sucrose, 3 mM $MgCl_2$, 1 mM EGTA, and 0.5% Triton X-100.
3. Protease inhibitor cocktail (Sigma, Cat. # P8340).
4. Dounce homogenizer.
5. Anti-FLAG M2 agarose bead (Sigma, Cat. # A2220). Store at −20 °C.
6. TBS pH 7.4: 8 g NaCl, 0.2 g KCl, and 3 g Tris base in 1 l water.
7. FLAG peptide (Sigma, Cat. # F3290). Reconstitute in 1× Tris-buffered saline (TBS) and store at −20 °C.
8. ProSieve precast gel (4–20%) (Lonza, Cat. # 59520).
9. 4× SDS sample buffer (10 ml): 4 ml glycerol, 0.4 g SDS, 5 ml 1 M Tris pH 6.8, 0.01% bromophenol blue.
10. SimplyBlue SafeStain (Invitrogen, Carlsbad, CA, Cat # LC6060).
11. Parafilm M laboratory film (Fisher Scientific, Cat. # 13-374-10).
12. Industrial razor blades (VWR Scientific, Cat. # 55411-055).

2.3. Sucrose-gradient Analysis of Purified Complexes

1. Hoefer SG 15 gradient mixture (Hoefer, Cat. # SG15).
2. CSK buffer with 10% sucrose: 10 mM PIPES pH 7.0, 300 mM NaCl, 3 mM $MgCl_2$, 1 mM EGTA, 0.5% Triton X-100, and 10% sucrose.
3. CSK buffer with 50% sucrose: 10 mM PIPES pH 7.0, 300 mM NaCl, 3 mM $MgCl_2$, 1 mM EGTA, 0.5% Triton X-100, and 50% sucrose.
4. Beckman ultracentrifuge and SW41 rotor.
5. Native protein molecular weight markers (GE Healthcare, Cat. # 17-0445-01).
6. Bio-Rad Econo Pump Peristaltic Pump (Model # EP-1).
7. Disposable micropipets (VWR Scientific, Cat. # 53432-921).
8. Transparent ultracentrifuge tube (tube capacity 12 ml) (Beckman, Cat. # 5030).

3. Methods

3.1. Establishment of Inducible T-Rex HEK293 Stable Cells

Affinity purification of protein complexes can be performed using stably transfected cells that express an epitope-tagged form of the target protein to increase purification efficiency and sensitivity.

Cells that express a FLAG-tagged target protein were established using T-Rex HEK293 cells and the pcDNA5/TO expression vector. T-Rex HEK293 cells constitutively express the tetracycline repressor, which represses transgene expression downstream of tetracycline operator elements in the pcDNA5/TO vector (8). Transgene expression is induced in the presence of doxycycline, which blocks binding of the tetracycline repressor to tetracycline operator elements. This approach permits abundant expression of a target protein, even one which might produce toxicity, in a regulated manner.

1. Maintain T-REx HEK293 cells in DMEM medium supplemented with 10% bovine calf serum and 5 μg/ml blasticidin.
2. Seed 2×10^4 cells into each well of 6-well tissue culture plate and incubate at 37 °C and 5% CO_2 overnight.
3. Transfect pcDNA5/TO expression construct or empty vector using the calcium phosphate co-precipitation method (9) and incubate overnight (see Note 1).
4. Change medium and incubate for 1 or 2 days.
5. Split cells into 15-cm tissue culture dish and incubate overnight.
6. Remove medium and replace with selection medium (DMEM supplemented with 10% bovine calf serum and 5 μg/ml blasticidin, and 200 μg/ml hygromycin B) (see Note 2). Incubate cells until visible clones appear. It usually takes 2 weeks.
7. Pick clones into 24-well tissue culture plate using cloning cylinders by adding 50 μl of 0.05% trypsin-EDTA solution and pipetting.
8. Culture cells in maintenance medium (DMEM supplemented with 10% bovine calf serum and 5 μg/ml blasticidin, and 50 μg/ml hygromycin B) for a week (see Note 3).
9. Split cells into 10-cm tissue culture dish.
10. Seed $2–4 \times 10^4$ cells into each well of 6-well tissue culture plate and incubate for 1 day.
11. Add 1 μg/ml doxycycline and incubate for 2 days. Include uninduced control for each clone.
12. Harvest cells in 1.5 ml microcentrifuge tubes and remove residual medium following a short spin.
13. Add 30–100 μl of CSK buffer to pellets and vortex vigorously for 5 min. Spin extracts with Eppendorf microcentrifuge ($16{,}000 \times g \times 5$ min) at 4 °C.
14. Denature supernatant with 4× SDS sample buffer and analyze expression of target protein by Western blotting using anti-FLAG antibody or antibody against target protein (see Note 4).
15. Select high-expressing clones for further analysis.

3.2. Purification of Chromatin Modifying Complexes by Co-Immunoprecipitation and Identification of Components by Mass Spectrometry

Affinity purification of a FLAG-tagged target protein removes most nonspecific proteins. However, nonspecific interactions with beads can never be completely avoided. To distinguish authentic target protein interactions from nonspecific agarose bead interactions, we usually perform a parallel purification with cells that carry the empty expression vector. Purification steps are performed at 4 °C unless otherwise specified. Purified protein complexes can be further analyzed in sucrose gradient equilibrium centrifugation to gain insights regarding the native size and composition of the complex (10, 11). Alternatively, purified materials can be directly used for biochemical characterization of the complex.

1. Grow T-Rex inducible HEK293 cells that express FLAG-tagged target protein or carry the empty vector. We usually start an initial purification with at least twenty 10 cm dishes or five 15 cm dishes, depending on the purification efficiency and expression level of the target protein (see Notes 5 and 6).
2. After 1 day of incubation (about 70% confluency), add 1 μg/ml doxycycline and incubate for another 3–4 days (see Notes 7 and 8).
3. Harvest cells and wash pellets with 1× PBS.
4. Add 10 ml CSK buffer and vortex vigorously for 30 s. Transfer extracts to a Dounce homogenizer and homogenize ten times.
5. Centrifuge in a high-speed centrifuge (17,400 × *g*, 10 min) and transfer supernatants to new 15 ml tubes.
6. Add 200 μl of anti-FLAG-IgG agarose slurry and incubate at 4 °C for 4 h or overnight with gentle shaking.
7. Wash beads with 10 ml CSK buffer for 5 min with gentle shaking (see Note 9).
8. Centrifuge at 140 × *g* for 1 min and remove buffer.
9. Repeat steps 7 and 8 three times.
10. Briefly centrifuge beads after the final wash and remove residual buffer by pipetting.
11. Add 25 μl to 100 μl of 250 μg/ml FLAG peptide in 1× TBS and incubate for 10 min at 4 °C with gentle shaking (see Note 10).
12. Centrifuge at 140 × *g* for 1 min and carefully transfer supernatants to new 1 ml tubes.
13. Denature proteins with 4× SDS sample buffer and analyze proteins by 4–12% SDS-PAGE. Alternatively, purified complexes can be analyzed by sucrose gradient equilibrium centrifugation to determine native size and component composition of the complex.
14. Stain proteins with SimplyBlue stain reagents for several hours (see Note 11).

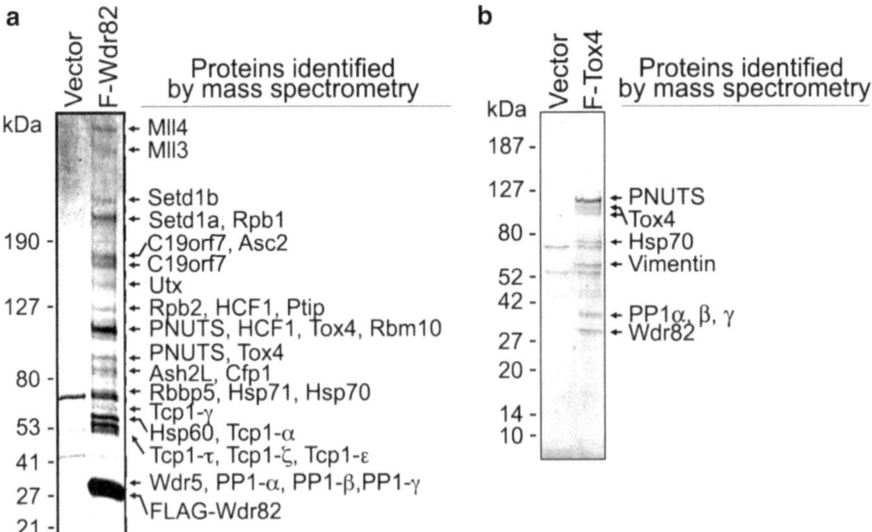

Fig. 1. One-step FLAG affinity purification of chromatin-associated proteins. T-Rex inducible HEK293 cells carrying FLAG-tagged Wdr82 (**a**) FLAG-tagged Tox4, (**b**) or empty vector were induced with 1 μg/ml doxycycline for 3 days. Nuclear extracts were prepared and subjected to one-step FLAG affinity purification. Purified proteins were analyzed by SDS-PAGE and stained with SimplyBlue stain reagent. Protein bands were excised and processed for protein identification by mass spectrometry. Proteins identified by mass spectrometry are shown by arrows. Reproduced in part from 2010 with permission from The American Society for Biochemistry and Molecular Biology (7).

15. Cut out protein bands for protein identification by mass spectrometry. We usually put the stained gel on parafilm and cut out each specific band (compared with control) on a light box using disposable razor blades. Each gel slice in a microcentrifuge tube is sent to a university-based mass spectrometry core facility.
16. One-step FLAG affinity purification of Wdr82- or Tox4-associated proteins are shown in Fig. 1. Arrows indicate proteins identified by mass spectrometry.

3.3. Analysis of Purified Complexes by Sucrose Gradient Equilibrium Centrifugation

Purified protein complexes can be further analyzed by sucrose gradient equilibrium centrifugation to gain insights of native size and composition of the complex. If the target protein is found to be a component of multiple complexes, this approach is necessary to distinguish components of each complex. Nuclear extracts are analyzed in parallel to study the endogenous complex under physiologic conditions.

1. Prepare CSK buffer containing sucrose solutions to produce the linear sucrose gradient.
2. Fill mixing and reservoir chambers of gradient mixer with 5 ml each of 50 and 10% sucrose solutions.
3. Prepare linear 10–50% sucrose gradient in an ultracentrifuge tube by opening both valves of mixing and reservoir chambers, as shown in Fig. 2a.

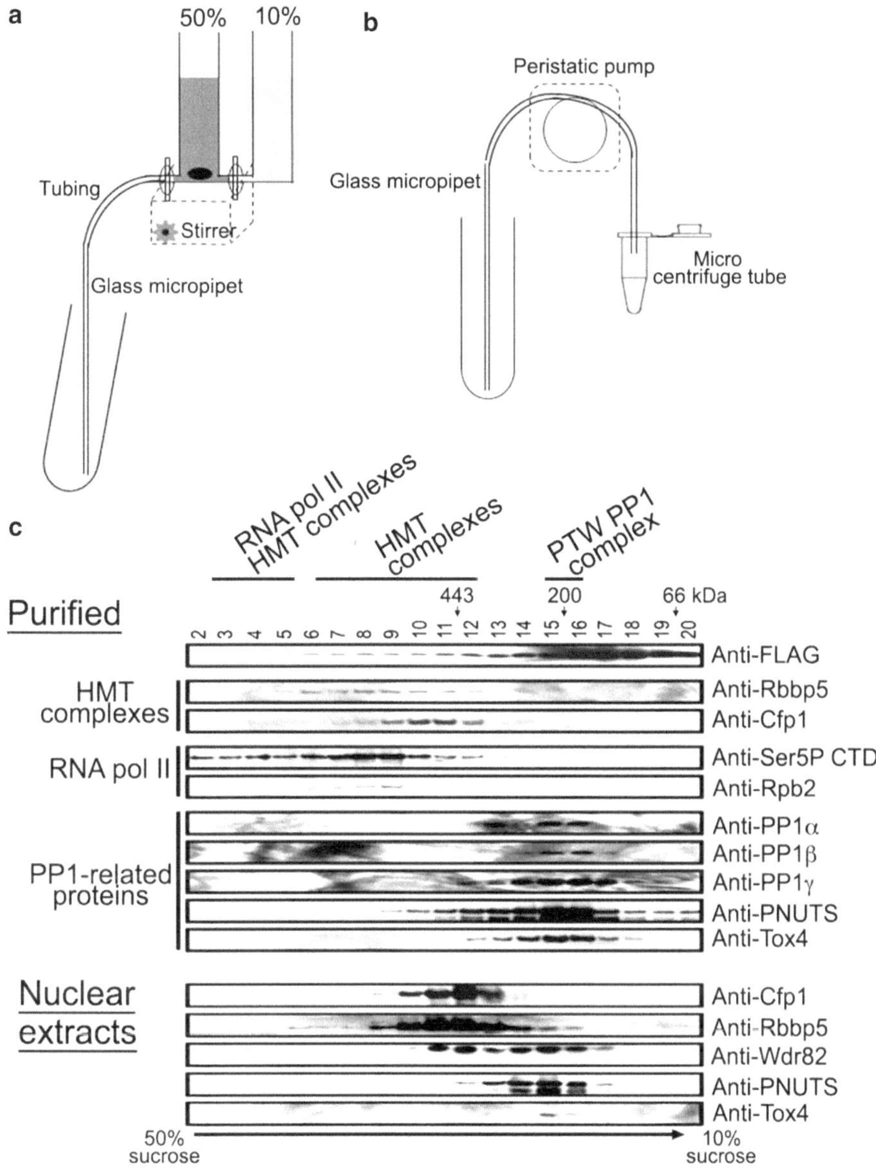

Fig. 2. Sucrose gradient equilibrium centrifugation analysis of Wdr82-associated complexes. (**a**) Preparation of a 10–50% linear sucrose gradient. After setting up sucrose solutions in gradient mixer, gently mix 50% sucrose solution by a small stirrer bar and open both valves of mixing and reservoir chambers. The linear gradient is built from bottom by releasing the solution down the side of the centrifuge tube near the accumulating solution. (**b**) Fractionation of gradient. Carefully place glass micropipette in the center of tube, slowly pump out the gradient from bottom, and collect each fraction into a microcentrifuge tube. (**c**) Purified Wdr82-associated complexes and nuclear extracts were analyzed by 10–50% sucrose gradient equilibrium centrifugation. Each fraction was denatured and an equal volume of each fraction was analyzed by Western blotting using the indicated antibodies. Native protein markers are analyzed in parallel and approximate size of complexes can be determined. Reproduced in part from 2010 with permission from The American Society for Biochemistry and Molecular Biology (7).

4. Repeat steps 2 and 3 depending on the number of samples.
5. Layer 1 ml of purified protein complex or nuclear extract over the gradient (see Note 12).
6. Place tubes into a SW 41 Beckman ultracentrifuge rotor.
7. Spin at 108,800 × *g* for 18 h at 4 °C.
8. Slowly stop centrifuge with no brake setting.
9. Take tubes out of the rotor and set up fractionation as shown in Fig. 2b.
10. Collect 0.5 ml fractions into a microcentrifuge tube from the bottom of the tube using a glass micropipette (see Note 13).
11. Denature proteins by mixing three parts of each fraction with one part of 4× SDS sample buffer.
12. Analyze fractions (fraction #1 to fraction # 20) to determine the distribution of each protein in the gradient by Western blotting. Fig. 2c shows fractionation of purified Wdr82-associated complexes and nuclear extract on 10–50% linear sucrose gradient. Each fraction was analyzed by Western blotting using the indicated antisera.

4. Notes

1. We usually utilize amino terminal FLAG or MYC epitope tags, although carboxy terminal tags may also be appropriate depending on the protein interactions being studied. 1–5 μg of DNA was transfected using the calcium phosphate co-precipitation method, but other transfection methods can be used.
2. The blasticidin-resistance gene is incorporated in the tetracycline repressor expression construct that T-Rex HEK293 cells carry, and the hygromycin-resistance gene is carried by the pcDNA5/TO vector.
3. The frequency of positive clones appears to be dependent on the size of target protein. We pick at least 10 clones for proteins that are larger than 100 kDa.
4. Initial dilution of anti-FLAG M2 antibody is 1:2,500.
5. If expression of the transgene is toxic, cells seeded to 90% confluency to obtain enough cell mass after 3 days of induction.
6. For the tight inducible expression of target protein, Tet system-approved fetal bovine serum (BD Biosciences, Cat. # 8630-1) can be used instead of regular serum. For purification purposes, we usually use regular FBS for culturing T-rex cell lines.

7. Make sure that the doxycycline is thoroughly dispersed.
8. Expression of FLAG-tagged target protein is dependent on the stability of the target protein and concentration of doxycycline in the culture medium. Various concentrations (0.1–2 μg/ml) of doxycycline can be used for expression of the transgene depending on the desired expression level of the target protein.
9. Extensive washing with bulk amounts of wash buffer in 15 ml tube decreases co-purification of nonspecific proteins as shown in Fig. 1.
10. For efficient elution, 2× SDS sample buffer can be used for SDS-PAGE or Western blotting analysis.
11. Staining of proteins in SDS-PAGE gel can be done within 2 h with SimplyBlue stain reagent.
12. To estimate the native size of each complex, additional gradients with native protein markers are analyzed in parallel (10). Each fraction is analyzed by SDS-PAGE and the gel is stained by Coomassie Blue reagent. Migration of marker proteins in the gradient is determined.
13. Mark the microcentrifuge tube for equal collection before fractionation.

Acknowledgment

This work was supported by the National Science Foundation Grant MCB-0641851 (to D. G. S.), Simmon's Clinical study funds (to J.-H. L), the Riley Childrens Foundation, and the Lilly Endowment.

References

1. Bauer A, Kuster B (2003) Affinity purification-mass spectrometry. Powerful tools for the characterization of protein complexes. Eur J Biochem 270:570–578
2. Gingras AC, Aebersold R, Raught B (2005) Advances in protein complex analysis using mass spectrometry. J Physiol 563:11–21
3. Gingras AC, Gstaiger M, Raught B, Aebersold R (2007) Analysis of protein complexes using mass spectrometry. Nat Rev Mol Cell Biol 8:645–654
4. Lee JH, Skalnik DG (2005) CpG binding protein (CXXC finger protein 1) is a component of the mammalian Set1 histone H3-Lys4 methyltransferase complex, the analogue of the yeast Set1/COMPASS complex. J Biol Chem 280: 41725–41731
5. Lee JH, Tate CM, You JS, Skalnik DG (2007) Identification and characterization of the human Set1B histone H3-Lys4 methyltransferase complex. J Biol Chem 282: 13419–13428
6. Lee JH, Skalnik DG (2008) Wdr82 is a C-terminal domain-binding protein that recruits the Setd1A Histone H3-Lys4 methyltransferase complex to transcription start sites of transcribed human genes. Mol Cell Biol 28:609–618
7. Lee JH, You J, Dobrota E, Skalnik DG (2010) Identification and characterization of a novel

human PP1 phosphatase complex. J Biol Chem 285:24466–24476

8. Yao F, Theopold C, Hoeller D, Bleiziffer O, Lu Z (2006) Highly efficient regulation of gene expression by tetracycline in a replication-defective herpes simplex viral vector. Mol Ther 13:1133–1141
9. Kingston RE, Chen CA, Rose JK (2003) Calcium phosphate transfection. In: Kingston RE (ed) Current protocols in molecular biology, Wiley, New York, p 9.1.1–9.1.11
10. Piperno G, Mead K (1997) Transport of a novel complex in the cytoplasmic matrix of *Chlamydomonas flagella*. Proc Natl Acad Sci 94:4457–4462
11. Martin RG, Ames BN (1961) A method for determining the sedimentation behavior of enzymes: application to protein mixtures. J Biol Chem 236:1372–1379

Chapter 24

Heavy Methyl-SILAC Labeling Coupled with Liquid Chromatography and High-Resolution Mass Spectrometry to Study the Dynamics of Site-Specific Histone Methylation

Xing-Jun Cao, Barry M. Zee, and Benjamin A. Garcia

Abstract

Histone lysine and arginine methylation involved in gene activation and silencing is dynamically regulated. However, partly limited to the research technologies previously available, the dynamics of global histone methylation on a site-specific basis have not been fully pursued. Heavy methyl-SILAC (Stable Isotope Labeling of Amino Acids in Cell Culture) labeling provides a remarkable signpost to distinguish the pre-existing and newly generated methyl marks on histones. Using this technology coupled with quantitative LC-MS analysis make it possible to monitor changes in the dynamics of histone site-specific methylation. In this chapter, we comprehensively describe the experimental strategy to determine the dynamics of multiple histone methylated residues including SILAC labeling, histone extraction/purification and mass spectrometry analysis.

Key words: Histone, Methylation, Dynamics, Heavy methionine, SILAC, Quantitation, Mass spectrometry

1. Introduction

Eukaryotic DNA is packaged with histone proteins into the chromatin fiber. Histone proteins not only act as the structural proteins needed to maintain chromatin architecture, but also are subjected to numerous post-translation modifications (PTMs), including acetylation, methylation, ubiquitination, sumoylation, and phosphorylation (1). Most of these modifications are located on the flexible N-terminal tails of histone proteins. These modifications serve to change histone steric properties and charges, so as to alter chromatin structure or influence the interactions among DNA, RNA, and proteins. Furthermore, the PTMs are known to act as docking sites for the recruitment of specific proteins and complexes participating in diverse gene expression

Minou Bina (ed.), *Gene Regulation: Methods and Protocols*, Methods in Molecular Biology, vol. 977,
DOI 10.1007/978-1-62703-284-1_24, © Springer Science+Business Media, LLC 2013

regulation mechanisms. Moreover, these modifications could be combinatorial and interdependent, forming a "histone code" of transcriptional activation or repression (2, 3). Methylation is unique among these PTMs due to its diverse forms and consequent variable influence over gene expression. Histone methylation can occur on lysine (mono-(me1), di- (me2), and trimethylation (me3)) and arginine (me1, symmetrical or asymmetrical me2). Multiple histone methyltransferases and demethylases are responsible for specific conversions between unmodified (me0), me1, me2, and me3 states of lysine and arginine residues on histone proteins. For example, G9a and Suv39h1 catalyze the production of H3K9me1/me2 and me3, respectively (4), whereas JHDM2A and HDM3A remove H3K9me2 (5) and me3 (6), respectively. Interestingly, various locations and degrees of methylation on specific histone residues result in different, even opposing functional outcomes. For example, methylation on H3K4 (7), H3K36 (8), and H3K79 (9) is correlated with gene activation, whereas that on H3K9 (10), H3K27 (11, 12), and H4K20 (12) is generally associated with gene silencing. In addition, H3R2me1 is associated with gene activation while H3R2me2 is linked to gene repression (13).

For a long time lysine methylation was thought to be irreversible until very recently the first histone demethylase, Lysine Specific Demethylase 1 (LSD1) was discovered (14). The dynamic regulation of methylation by methyltransferases and demethylases is an important manner for regulation of gene expression. However, most previous studies regarded histone methylation as a static condition at a particular time, and partly due to technical limitations, only a limited number of investigations pursued the dynamic change of histone methylation. Some studies focused on methylation dynamics during cell cycle progression (15–17). For example, the levels of H3K9me1, me2 and me3 were found to dynamically regulated increase from interphase to metaphase and then decrease to the initial levels at the start of the next mitotic cycle (17). Radiolabeling of methionine was used to monitor the turnover of methylation of proteins like histone H3 (18), whereas this technique only could track methylation at the protein level, not on a site-specific basis. Immunoblot analysis could specifically study dynamic change of methylation sites (17). Nevertheless, this method still cannot distinguish between "old" and "new" synthesized methyl groups.

Recently we described a new strategy to investigate the site-specific dynamics of histone methylation (19) by combining heavy methyl-SILAC (stable isotope labeling by amino acids in cell culture) labeling technology and the quantitative mass spectrometry analysis previously developed in our lab (20). Methionine is an essential amino acid that mammal cells cannot synthesize de novo, and it is also the precursor of the S-adenosyl methionine (SAM), the sole donor of the methyl group. When heavy stable isotope form of methionine, l-Methionine-methyl-^{13}C, D_3 is used in the cell culture medium, heavy-methyl groups will be introduced by methyltransferases into lysine and arginine residues on histone

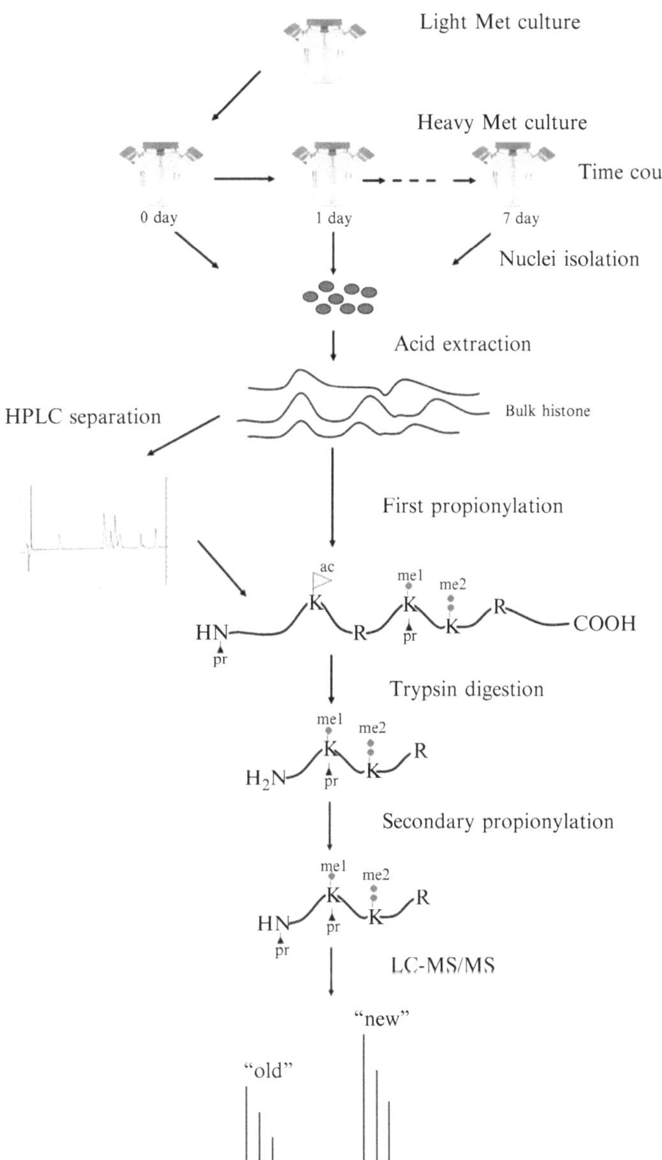

Fig. 1. Strategy for using heavy methyl-SILAC labeling to study dynamics of histone methylation. HeLa cells are introduced into the heavy methionine-labeled medium to allow for incorporation of the heavy methyl group on histones. Cells are harvested at a series of time points and subsequently histones are extracted. Bulk histone proteins or individual histone variant separated by HPLC is sequentially subjected to propionylation, trypsin digestion and LC-MS/MS analysis.

proteins. Therefore, this labeling technology coupled with quantitative mass spectrometry technology will remarkably facilitate dynamics study of site-specific methylation of histone proteins.

Our experimental scheme is shown in Fig. 1. Briefly, HeLa cells are introduced into the heavy methionine-labeled medium and harvested at a series of time points. Histone proteins are obtained by nuclei isolation and acid extraction. Bulk histone proteins or individual histone variant separated by high performance liquid chromatography (HPLC) is sequentially subjected to propionylation, trypsin digestion and LC-MS/MS analysis. Finally the abundances of light, heavy and intermediate forms of the interesting methylation sites are extracted, and normalized to construct kinetic models towards investigating their dynamics.

2. Materials

2.1. Mammalian Cell Culture and Heavy Methionine Labeling

1. Cell line: HeLa S3 cells, epithelial adenocarcinoma (ATCC).
2. Culture medium: Minimum Essential Medium Eagle, Joklik modification (SAFC Biosciences).
3. Components of homemade medium (all from Sigma-Aldrich): l-Arginine monohydrochloride; l-Cysteine dihydrochloride; l-Histidine dihydrochloride; l-Isoleucine; l-Leucine; l-Lysine monohydrochloride; l-Phenylalanine; l-Threonine; l-Tryptophan; l-Tyrosine; l-Valine; Choline chloride; Folic acid; myo-Inositol; Niacinamide; d-Pantothenic acid hemicalcium salt; Pyridoxal hydrochloride; Riboflavin; Thiamine hydrochloride; $MgCl_2$; KCl; NaCl; Na_2HPO_4; Glucose; Phenol Red sodium salt; $NaHCO_3$.
4. Heavy amino acids: l-Methionine-methyl-^{13}C, D_3 (Sigma-Aldrich); l-Lysine-$^{13}C_6$, $^{15}N_2$ dihydrochloride (Cambridge Isotope Laboratories Inc).
5. Supplements: newborn calf serum (Thermo Scientific); dialyzed fetal bovine serum (Gemini Bioproducts); GlutaMAX (Invitrogen); penicillin/streptomycin (Invitrogen).
6. MilliQ water.
7. Other: sterile Dulbecco's phosphate-buffered saline (DPBS, Invitrogen); filter units (Millipore); syringe filter (Pall); spinner flasks (Corning).

2.2. Nuclei Isolation and Histone Extraction

1. Nuclear Isolation Buffer (NIB-250): 15 mM Tris–HCl (pH 7.5), 60 mM KCl, 15 mM NaCl, 5 mM $MgCl_2$, 1 mM $CaCl_2$, 250 mM sucrose. Store at 4°C.
2. 1 M DTT: store at −20°C. Add to NIB-250 to 1 mM when use.
3. 10% NP-40.
4. Inhibitors: 5 μM Microcystin-LR (1,000×, Sigma-Aldrich), 200 mM AEBSF (400×, Sigma-Aldrich) and 5 M sodium butyrate (500×, Sigma-Aldrich).
5. 0.4 N (0.2 M) sulfuric acid (H_2SO_4), stored at 4°C.
6. Trichloroacetic Acid (TCA, Fisher Scientific), 100% stock solution stored at 4°C.
7. Acetone (Sigma-Aldrich), stored at −20°C.

8. Deionized water.
9. Bradford quantification solution (Bio-Rad).

2.3. Reverse-Phase HPLC Separation of Histone Proteins

1. C_{18} column, 4.6 mm internal diameter × 250 mm (Vydac).
2. Manual injector valve (Rheodyne).
3. System Gold HPLC system (Beckman Coulter).
4. 0.1% Trifluoroacetic acid (TFA, Sigma-Aldrich).
5. LC gradient:
   (a) Buffer A: 0.5% acetonitrile/0.2% TFA.
   (b) Buffer B: 90% acetonitrile/0.188% TFA.

2.4. Propionic Anhydride Derivatization

1. 100 mM ammonium bicarbonate (Sigma-Aldrich).
2. Ammonium hydroxide (Sigma-Aldrich).
3. Propionic anhydride (Sigma-Aldrich).
4. Isopropanol (anhydrous, Sigma-Aldrich).
5. Acetic acid (HPLC grade, Sigma-Aldrich).
6. Argon.
7. pH indicator strips (EMD Chemicals).

2.5. Trypsin Digestion

1. Sequencing grade-modified porcine trypsin (Promega).
2. 100 mM Ammonium bicarbonate (Sigma-Aldrich).
3. Acetic acid (HPLC grade, Sigma-Aldrich).

2.6. StageTip Purification

1. Empore C_{18} Disk (3 M).
2. Conditioning solution: methanol (HPLC grade, Fisher Scientific).
3. Equilibrating/loading/washing solution: 0.1% acetic acid.
4. 3: Eluting solution: 75% acetonitrile/1% acetic acid.
5. Gel loader tips.
6. Compressed air can.

2.7. LC-MS Analysis

1. Prepare an analytical column by packing a 10-cm fused silica capillary (75 μm inner diameter, 360 μm outer diameter with a 1 μm homemade laser-pulled electrospray tip) with a 50% acetonitrile/50% isopropanol slurry of reverse-phase C_{18} resin (MagicC_{18}, 5 μm particles, 200 Å pore size, Michrom BioResources) at a helium pressure (70 bar) using a bomb-loader device (Proxeon Biosystems).
2. LC gradient:
   Buffer A: 0.1 M acetic acid.
   Buffer B: 98% acetonitrile/0.1 M acetic acid.
3. AS2 autosampler (Eksigent).

4. 1200 binary HPLC system (Agilent).
5. LTQ-Orbitrap XL (Thermo Scientific).

3. Methods

3.1. Heavy Labeled Medium Preparation

1. Since there is no commercially available Joklik modified medium deficient in methionine, we need to make it according to its standard formulation.
2. Prepare 1,000× heavy methionine (15.4 mg/mL) stock solution by dissolving it in DPBS. Filter it with a 0.22 μm syringe filter (see Note 1). Store at −20°C.
3. Weigh the required amounts of all of the cell culture medium components except methionine and add them to 800 mL MilliQ water (see Note 2). Mix until completely dissolved.
4. Adjust pH to 7.2 (see Note 3) and supplement MilliQ water to 900 mL.
5. Filter the solution with a 0.22 μm membrane filter. Store at 4°C.
6. Add dialyzed FBS, heavy methionine stock solution and supplements to the medium to their working concentration when use.

3.2. HeLa S3 Cells Time-Course Labeling and Harvesting

HeLa S3 cells are maintained at 37°C in 0.2 LPM CO_2. Medium is replenished every day to maintain a cell density of $2–6 \times 10^5$ cells/mL throughout the experiment.

1. Suspension culture of HeLa S3 cells is initially maintained in normal Joklik modified medium supplemented with 10% newborn calf serum, penicillin/streptomycin, and 1% GlutaMAX.
2. Centrifuge cells at 600 × *g* for 5 min, decant the medium and wash cell pellet in sterile DPBS twice to remove the residual normal medium.
3. Resuspend cells in heavy methionine-labeled Joklik modified medium and culture at 37°C.
4. Daily aliquots are taken for continuous 7 days or as desired (see Note 4). Centrifuge cells at 600 × *g* for 5 min, decant the medium and wash cell pellet in DPBS twice.
5. Flash-freeze cell pellet in liquid nitrogen for 3 min and store at −80°C prior to further analysis.

3.3. Nuclei Isolation and Histone Extraction

1. Put frozen cell pellet on ice to thaw.
2. Add NIB-250 with 0.3% NP-40 (final concentration) to cell pellet to a final ratio of 10:1 (10 mL buffer for 1 mL cell pellet).
3. Completely resuspend cells by pipetting gently and incubate on ice for 5 min.

4. Centrifuge at 600 × *g* for 5 min to pellet nuclei at 4°C. Transfer the supernatant (cytoplasmic extract) to a different tube or simply discard.
5. Resuspend nuclei pellet gently in 10:1 NIB-250 without NP-40.
6. Centrifuge at 600 × *g* for 5 min at 4°C and decant the supernatant.
7. Repeat steps 5 and 6.
8. Slowly add 0.4 N H_2SO_4 (see Note 5) to nuclei pellet at a 5:1 final ratio (5 mL solution for a 1 mL pellet) and vortex.
9. Place the tube on ice for at least 1 h or as long as overnight by mixing intermittently or by rotation on a roller at 4°C (see Note 6).
10. Centrifuge at 3400 × *g* (see Note 7) for 10 min at 4°C and transfer the supernatant containing histone proteins to a clean tube.
11. Add 100% TCA to the supernatant to give a final concentration of 20% TCA.
12. Place the tube on ice for at least 1 h or as long as overnight (see Note 8).
13. Centrifuge at 3400 × *g* for at 10 min to pellet proteins and discard supernatant.
14. Wash protein pellet in prechilled acetone (see Note 9).
15. Centrifuge at 3400 × *g* for 10 min and discard supernatant.
16. Repeat steps 14 and 15.
17. Air-dry protein pellet at room temperature (see Note 10).
18. Resuspend dried protein pellet in deionized water (see Note 11), in 100 mM ammonium bicarbonate or in 0.1% TFA (use for HPLC separation).
19. Centrifuge at 3400 × *g* for 10 min. Slowly and carefully transfer the supernatant to a clean tube.
20. Determine the yield of the extracted histone proteins by the Bradford assay (see Note 12).
21. Aliquot histone proteins and store at −80°C prior to further analysis.

3.4. Reversed-Phase HPLC Separation of Histone Variants (see Fig. 2)

1. Load about 150 μg histone proteins dissolved in 0.1% TFA onto a C_{18} preparative column by a manual sample injector valve.
2. Run a 100 min gradient from 30 to 60% Buffer B to separate histone variants at a flow rate of 0.8 mL/min, followed by 20 min 100% Buffer B.
3. Automatically collect fractions every 2 min throughout the gradient and pool fractions spanning a single variant.

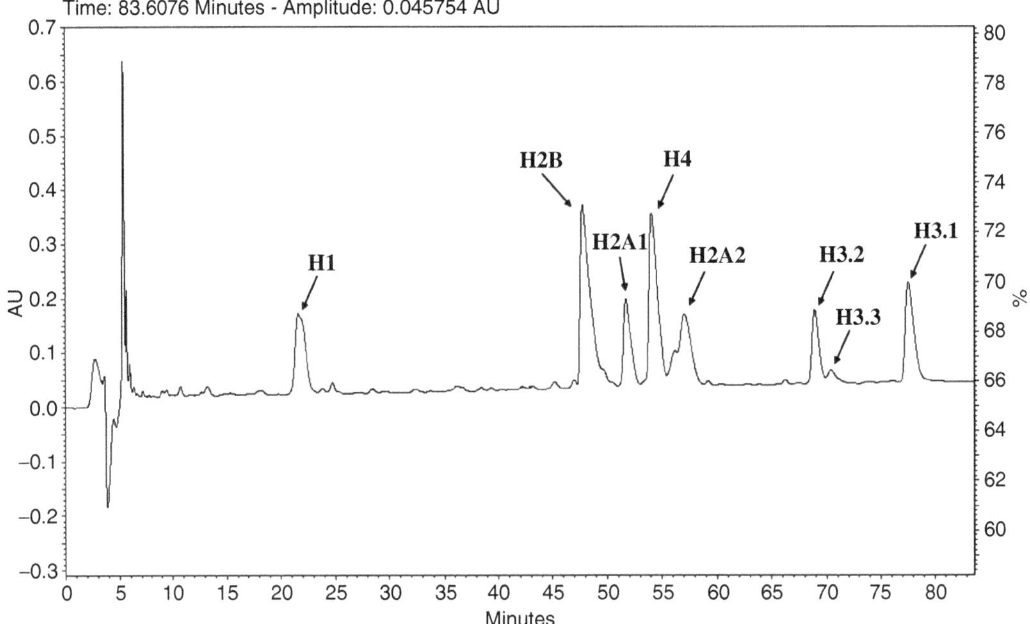

Fig. 2. Separation of histone variants with reversed-phase HPLC.

4. Lyophilize and store at –80°C prior to further analysis.

3.5. Propionic Anhydride Derivatization (see Note 13)

1. Dilute histone sample with 100 mM ammonium bicarbonate to the concentration about 0.5–1 μg/μL.
2. Take 10–20 μL histone sample. Add 0.5 μL of ammonium hydroxide to sample to make sure pH is between 7 and 9.
3. Prepare propionylation reagent by adding 25 μL of propionic anhydride to 75 μL of isopropanol (or in a 1:3 ratio of a suitable volume) (see Note 14).
4. Add about 5–10 μL propionic anhydride reagent (half relative to sample volume) to histone sample, vortex and spin down quickly.
5. Immediately add 5–7 μL ammonium hydroxide, check pH to make sure pH ~8.0. If pH <8, continue to add ammonium hydroxide until pH is approximately 8.
6. Incubate at 37°C for 15 min.
7. Dry sample down to <5 μL.
8. Repeat steps 2–7 (see Note 15).
9. Store in –80°C or use for trypsin digestion.

3.6. Trypsin Digestion

1. Dilute sample with 50–100 μL of 100 mM ammonium bicarbonate.

2. Add trypsin to sample to a 1:20 (wt/wt) ratio of trypsin to sample. Incubate at 37°C for 6–10 h.
3. Quench trypsin digestion by adding 4–8 μL glacial acetic acid.
4. Dry sample down to approximately 5 μL.
5. Repeat steps 2–8 of subheading 3.5 to perform propionic anhydride derivatization to N-termini of histone peptides.
6. Lyophilize histone peptides.

3.7. StageTip Purification

1. Pouch out a small (~1 mm diameter) piece of Empore C_{18} disk using a pipette tip with the end cut off and transfer it into a gel loader tip. Carefully use a fused silica capillary with a large outer diameter to push the disk out and jam it into the tapered part of the gel loader tip.
2. Repeat step 1 (see Note 16).
3. Reconstitute histone peptides with 20 μL 0.1% acetic acid. Check pH to make sure it acidic.
4. Condition the StageTip by loading 20 μL methanol and slowly pushing through using a compressed air can (~30 s) (see Note 17).
5. Equilibrate the StageTip by loading 20 μL 0.1% acetic acid (~30 s).
6. Load histone sample to the StageTip (~1 min).
7. Reload the sample if desired.
8. Wash the StageTip by loading 20 μL 0.1% acetic acid (~30 s).
9. Elute the StageTip by loading 20 μL 75% acetonitrile/0.1% acetic acid (~30 s).
10. Dry down to <5 μL on a SpeedVac concentrator (see Note 18) and store in −80°C prior to LC-MS/MS analysis.

3.8. LC-MS/MS Analysis

1. The sample is loaded onto the homemade C_{18} analytical column with an emission tip using an Eksigent AS2 autosampler. A 110-min LC gradient from 5 to 35% Buffer B at a flow rate of 200 nL/min (post-split) on a Agilent 1200 binary HPLC system is used to separate histone peptides.
2. LTQ-Orbitrap XL mass spectrometer is basically operated in data-dependent mode: a resolution of 30,000 for a full scan and 7 subsequent MS/MS scan in ion trap by collision-induced dissociation (CID). The selected ion monitoring (SIM) mode and target MS/MS are performed to some peptide ions with low-abundance modification states to increase their identification (such as K4 methylation). Lock mass calibration in MS mode is implemented using polysiloxane ions, 371.1012 and 445.1200.

3.9. Data Analysis

1. The RAW files are converted into mzXML files by ReAdW program (21). Histone peptide sequences and modifications are

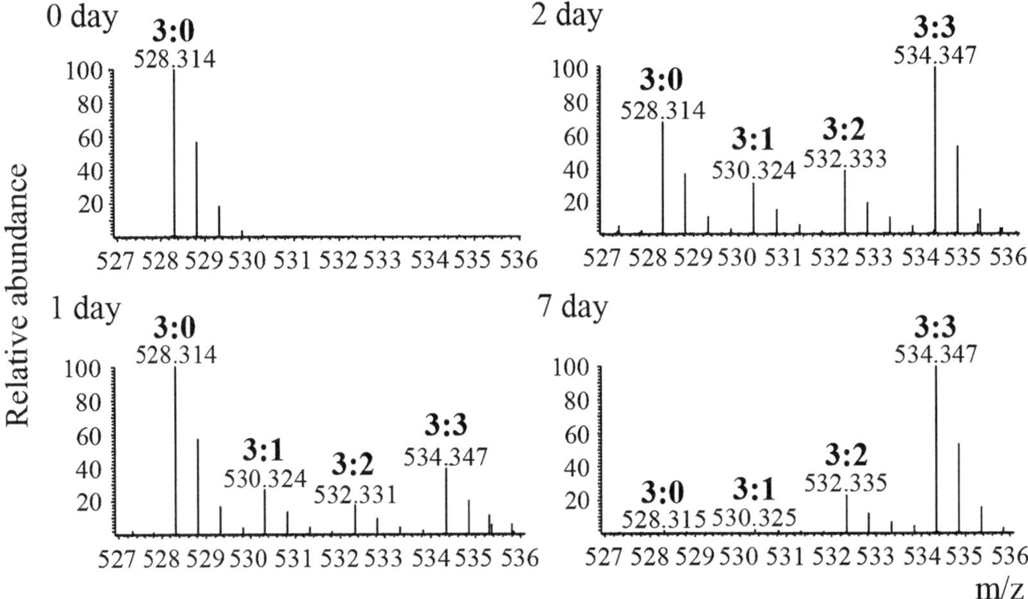

Fig. 3. Monitoring rate of H3.1K9me3. Full scan mass spectra of the $[M + 2H]^{2+}$ ions from the histone H3.1 peptide Kme3STGGKAPR extracted from the different time points (days 0, 1, 2 and 7) after pulse heavy methionine labeling. Intermediately labeled trimethylated forms (me3:1, 530.324 *m/z* and me3:2, 532.333 *m/z*) are present between the unlabeled trimethylated form (me3:0, 528.314 *m/z*) and the fully labeled trimethylated form (me3:3, 534.347 *m/z*) after day 0. Turnover of the old H3K9me3:0 can be seen in day 7, while the new H3K9me3:3 form is highest in abundance at that same time point.

confirmed by manual inspection of MS/MS spectra (see Note 19). The precursor ion tolerance is set to 10 ppm to distinguish acetylation and trimethylation, and the fragment ion tolerance is set to 0.5 Da. Propionylation of N-termini of peptides is considered as a static modification. The following variable modifications are considered: propionylation of unmodified and monomethylated lysines; dimethylation of lysine; trimethylation of lysine; acetylation of lysine; oxidation of methionine; monomethylation of H4R3.

2. To describe unambiguously a specific methylated form, we append each residue with two numbers, the first referring to the total number of methyls and the second to the number of heavy labeled methyls. For example, H3K9me3:0 refers to unlabeled trimethylated H3K9, and H3K9me3:3 refers to fully labeled trimethylated H3K9, and H3K9me3:1 and H3K9me3:2 refer to intermediately labeled trimethylated H3K9 forms containing one and two unlabeled methyls, respectively (see Fig. 3).

3. Peptide abundance is calculated by chromatographic peak integration using an in-house program or manually. The abundances of all labeled forms (light methyls, heavy methyls and possible intermediate methyls) for a specific methylation state (me1, me2

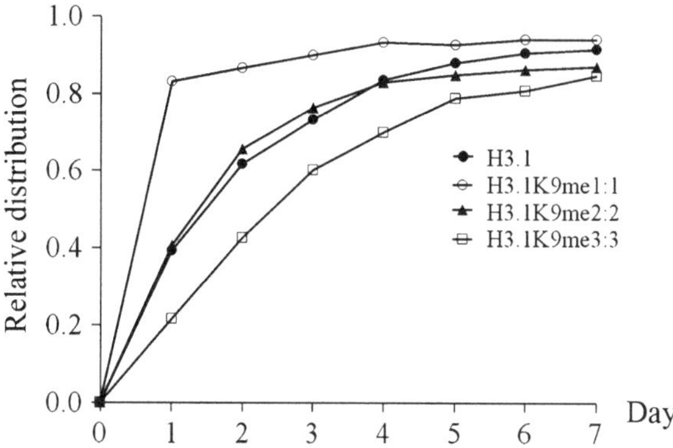

Fig. 4. Plot of the incorporation of new labels between methylation state on H3.1K9 (from Fig. 3) and general H3.1 total protein turnover.

a

$$\mathrm{Kme1{:}1(t)} = A - A \times e^{\wedge}(-t/B)$$

b

$$\mathrm{Kun} = \alpha - k_{0\to1}[\mathrm{Kun}] - k'_{0\to1}[\mathrm{Kun}] + k_{1\to0}([\mathrm{Kme1{:}0}] + [\mathrm{Kme1{:}1}]) - t_0$$
$$\mathrm{Kme1{:}0} = k_{0\to1}[\mathrm{Kun}] - k_{1\to0}[\mathrm{Kme1{:}0}] - t_1$$
$$\mathrm{Kme1{:}1} = k'_{0\to1}[\mathrm{Kun}] - k_{1\to0}[\mathrm{Kme1{:}1}] - t_1$$

Fig. 5. Kinetic modeling of histone methylation dynamics. (**a**) Logarithmic function that fits a generic fully labeled monomethylated peptide. A = the final methyl labeling efficiency of the peptide; B = time constant. (**b**) Set of first order differential equations that fits a generic unmodified histone peptide with the respective unlabeled and labeled monomethylated peptides. α = rate of synthesis; $k_{0\to1}$ = monomethylation with light SAM; $k'_{0\to1}$ = monomethylation with heavy SAM; $k_{1\to0}$ = demethylation from the monomethylation state; t_0 = degradation of unmodified state; t_1 = degradation of monomethylated state.

or me3) of a specified residue are summed and normalized to 1, hereby to obtain the relative abundance of each labeled form. From the relative abundances, one can apply various differential equations to obtain different kinetic parameters of methylation. For instance, by optimizing the relative abundances of the fully labeled methylated histone peptide over time to a single logarithmic function, one can extrapolate the time when the methylation labeling reached half of its maximum labeling efficiency, or essentially the inverse of a half-life measurement for each methylation state of a histone peptide (see Figs. 4 and 5a). While relatively direct to implement and interpret, such an approach also neglects a significant fraction of the MS data by assuming that each methylation state for a given histone peptide is formed and turned over independently of another methylation state for the same peptide. In order to take advantage of as much information about the histone peptide, one can alternatively fit all the fully labeled, partially labeled, and fully unlabeled peptides to a set of first order differential equations each representing the addition or removal of a methyl group (see Fig. 5b). By establishing a stepwise relation between all the modified peptides, one can then determine specific rate constants between specific methylation states.

4. Notes

1. The concentration of normal methionine in Joklik medium is 15 mg/L, so the corresponding concentration of heavy methionine should be 15.4 mg/L considering the molar equivalent of light and heavy methionine. To minimize loss during filtering, the stock solutions of heavy amino acids can be directly added to the medium to its working concentration, and filtered together when making the medium.
2. We can make stock solutions of almost all the required components, store them at −20°C and then add MilliQ water to make their working concentration stocks when make the medium. For histone H3 or H4, the general protein turnover rates can be estimated from the peptides containing a methionine residue, while for H1 variants without any methionine residue, other heavy essential amino acids like l-Lysine-$^{13}C_6$, $^{15}N_2$ would be necessary to be added into the medium to estimate the turnover rates.
3. The pH of the medium usually rises 0.1–0.2 U during filtration.
4. The normal doubling time of HeLa S3 cell is about 24 h.
5. 0.2 N HCl can be used as an alternative. Histone proteins are soluble in dilute H_2SO_4 or HCl buffer, while most of other nuclear proteins and nucleic acids are not and precipitate out (22, 23).
6. During the acid extraction procedure, technically you can use the maximal centrifuge force that your rotor and the associated tubes can endure.
7. The extraction time is recommended to extend to overnight when the number of cells is low (<100,000 cells).
8. Extending incubation time can increase the yield of histone proteins.
9. Acetone can remove residual TCA which would interfere with the subsequent propionic derivatization, and meanwhile can also remove TCA precipitated NP-40 which would interfere with mass spectrometry analysis.

10. Cover the tube with parafilm, poke small holes in the film and lay it at room temperature. If lyophilizing the protein pellet in a SpeedVac concentrator, usually 2–3 min is enough to completely remove the acetone.
11. Not all precipitates are dissolved in water, but histone proteins being highly basic will be dissolved.
12. At least 100–200 μg bulk histone proteins can be extracted from 1×10^7 HeLa cells according to our experience.
13. The propionic anhydride derivatization is performed twice, at the protein level and at the peptide level, respectively (24). The purpose of the chemical derivatization to histone proteins is to block unmodified and monomethylated lysine residues, and therefore, subsequent trypsin digestion results in proteolysis only C-terminal to arginine residues. In addition, the derivatization step neutralizes the charges at the N-terminal and unmodified and monomethylated lysine residues and introduces the bulky propionyl group. Propionylated histone peptides are also more hydrophobic, so they are more easily retained and resolved by standard reverse phase liquid chromatography.
14. When not in use, propionic anhydride and isopropanol should be stored under argon or nitrogen and sealed with parafilm to reduce moisture in the bottles.
15. Usually the propionylation reaction should be repeatedly performed to ensure maximum derivatization efficiency.
16. The loading capacity of a piece of Empore C_{18} disk (1 mm diameter, 0.5 mm length) is about 25 μg of total peptide (25). Therefore two pieces of the disk usually should be enough for LC-MS/MS analysis for several times.
17. Try to always keep the disk wet from this step during the procedure.
18. Be careful to not over dry the sample to reduce losses before reconstitution, as you only have to remove the acetonitrile ($20 \times 75\% = 15$ μL).
19. The labeling experiment here involves more than ten kinds of variable modifications. Therefore it is not practical to identify peptides by making a database searching (false positive rate very high). Since high pure histone proteins can be obtained by acid extraction described here and the histone peptides are subjected to high resolution mass spectrometry analysis, most peptides can be determined by their masses and their MS/MS spectra after manual inspection to determine unambiguously the localization of the modifications.
20. The light, heavy and intermediate forms of methionine and methyl groups should be considered when analyzing peptides. Methionine containing peptides may also be oxidized.

Acknowledgments

B.A.G. gratefully acknowledges funding from a National Science Foundation grant (CBET-0941143), an Agilent Thought Leader award, an NSF Early Faculty CAREER award, and an NIH Innovator award (DP2OD007447) from the Office Of The Director, NIH.

References

1. Kouzarides T (2007) Chromatin modifications and their function. Cell 128:693–705
2. Strahl BD, Allis CD (2000) The language of covalent histone modifications. Nature 403:41–45
3. Jenuwein T, Allis CD (2001) Translating the histone code. Science 293:1074–1080
4. Rice JC, Briggs SD, Ueberheide B, Barber CM, Shabanowitz J, Hunt DF, Shinkai Y, Allis CD (2003) Histone methyltransferases direct different degrees of methylation to define distinct chromatin domains. Mol Cell 12:1591–1598
5. Yamane K, Toumazou C, Tsukada Y, Erdjument-Bromage H, Tempst P, Wong J, Zhang Y (2006) JHDM2A, a JmjC-containing H3K9 demethylase, facilitates transcription activation by androgen receptor. Cell 125:483–495
6. Klose RJ, Yamane K, Bae Y, Zhang D, Erdjument-Bromage H, Tempst P, Wong J, Zhang Y (2006) The transcriptional repressor JHDM3A demethylates trimethyl histone H3 lysine 9 and lysine 36. Nature 442:312–316
7. Bernstein BE, Humphrey EL, Erlich RL, Schneider R, Bouman P, Liu JS, Kouzarides T, Schreiber SL (2002) Methylation of histone H3 Lys 4 in coding regions of active genes. Proc Natl Acad Sci USA 99:8695–8700
8. Kizer KO, Phatnani HP, Shibata Y, Hall H, Greenleaf AL, Strahl BD (2005) A novel domain in Set2 mediates RNA polymerase II interaction and couples histone H3 K36 methylation with transcript elongation. Mol Cell Biol 25:3305–3316
9. Steger DJ, Lefterova MI, Ying L, Stonestrom AJ, Schupp M, Zhuo D, Vakoc AL, Kim JE, Chen J, Lazar MA, Blobel GA, Vakoc CR (2008) DOT1L/KMT4 recruitment and H3K79 methylation are ubiquitously coupled with gene transcription in mammalian cells. Mol Cell Biol 28:2825–2839
10. Tschiersch B, Hofmann A, Krauss V, Dorn R, Korge G, Reuter G (1994) The protein encoded by the Drosophila position-effect variegation suppressor gene Su(var)3-9 combines domains of antagonistic regulators of homeotic gene complexes. EMBO J 13: 3822–3831
11. Cao R, Wang L, Wang H, Xia L, Erdjument-Bromage H, Tempst P, Jones RS, Zhang Y (2002) Role of histone H3 lysine 27 methylation in Polycomb-group silencing. Science 298:1039–1043
12. Schotta G, Lachner M, Sarma K, Ebert A, Sengupta R, Reuter G, Reinberg D, Jenuwein T (2004) A silencing pathway to induce H3-K9 and H4-K20 trimethylation at constitutive heterochromatin. Genes Dev 18:1251–1262
13. Kirmizis A, Santos-Rosa H, Penkett CJ, Singer MA, Green RD, Kouzarides T (2009) Distinct transcriptional outputs associated with mono- and dimethylated histone H3 arginine 2. Nat Struct Mol Biol 16:449–451
14. Shi Y, Lan F, Matson C, Mulligan P, Whetstine JR, Cole PA, Casero RA (2004) Histone demethylation mediated by the nuclear amine oxidase homolog LSD1. Cell 119:941–953
15. Thomas G, Lange HW, Hempel K (1975) Kinetics of histone methylation in vivo and its relation to the cell cycle in Ehrlich ascites tumor cells. Eur J Biochem 51:609–615
16. Pesavento JJ, Yang H, Kelleher NL, Mizzen CA (2008) Certain and progressive methylation of histone H4 at lysine 20 during the cell cycle. Mol Cell Biol 28:468–486
17. McManus KJ, Biron VL, Heit R, Underhill DA, Hendzel MJ (2006) Dynamic changes in histone H3 lysine 9 methylations: identification of a mitosis-specific function for dynamic methylation in chromosome congression and segregation. J Biol Chem 281:8888–8897

18. Waterborg JH (1993) Dynamic methylation of alfalfa histone H3. J Biol Chem 268: 4918–4921
19. Zee BM, Levin RS, Xu B, LeRoy G, Wingreen NS, Garcia BA (2010) In vivo residue-specific histone methylation dynamics. J Biol Chem 285:3341–3350
20. Plazas-Mayorca MD, Zee BM, Young NL, Fingerman IM, LeRoy G, Briggs SD, Garcia BA (2009) One-pot shotgun quantitative mass spectrometry characterization of histones. J Proteome Res 8:5367–5374
21. Pedrioli PG, Eng JK, Hubley R, Vogelzang M, Deutsch EW, Raught B, Pratt B, Nilsson E, Angeletti RH, Apweiler R, Cheung K, Costello CE, Hermjakob H, Huang S, Julian RK, Kapp E, McComb ME, Oliver SG, Omenn G, Paton NW, Simpson R, Smith R, Taylor CF, Zhu W, Aebersold R (2004) A common open representation of mass spectrometry data and its application to proteomics research. Nat Biotechnol 22:1459–1466
22. Murray K (1966) The acid extraction of histones from calf thymus deoxyribonucleoprotein. J Mol Biol 15:409–419
23. Shechter D, Dormann HL, Allis CD, Hake SB (2007) Extraction, purification and analysis of histones. Nat Protoc 2:1445–1457
24. Garcia BA, Mollah S, Ueberheide BM, Busby SA, Muratore TL, Shabanowitz J, Hunt DF (2007) Chemical derivatization of histones for facilitated analysis by mass spectrometry. Nat Protoc 2:933–938
25. Rappsilber J, Ishihama Y, Mann M (2003) Stop and go extraction tips for matrix-assisted laser desorption/ionization, nanoelectrospray, and LC/MS sample pretreatment in proteomics. Anal Chem 75:663–670

Chapter 25

Analysis of p300 Occupancy at the Early Stage of Stem Cell Differentiation by Chromatin Immunoprecipitation

Melanie Le May and Qiao Li

Abstract

Chromatin immunoprecipitation (ChIP) is an invaluable method to study the specific interaction of regulatory proteins with genomic DNA. Since its first development, it has been modified extensively to make it applicable to many different cell types and experimental systems. The cross-linking of regulatory proteins to genomic DNA requires monolayer cells or single cell suspensions. Here, we describe a ChIP protocol using embryoid bodies formed at the early stage differentiation of pluripotent stem cells, which we have used to determine long-range p300-dependent regulatory elements of myogenic-specific genes.

Key words: Gene regulation, Chromatin immunoprecipitation, Stem cell differentiation, Embryoid body, Transcription factor, Coactivator, DNA-binding, p300-association

1. Introduction

Chromatin Immunoprecipitation (ChIP) is a powerful technique to decipher in vivo interaction of regulatory proteins with specific genomic DNA elements (1–3). It has tremendously advanced our knowledge on protein-DNA interactions involved in the regulation of gene transcription, and on epigenetics (4). There are mainly two types of ChIP, namely, NChIP and XChIP. Whereas NChIP uses native chromatin sheared by micrococcal nuclease digestion (5), XChIP utilizes reversibly cross-linked chromatin sheared by sonication (6, 7). XChIP, first developed in 1988 (8), has since been extensively modified and simplified to much less laborious and time-consuming procedures (1, 4, 9).

Here, we focus on the use of XChIP on aggregated stem cells t embryoid bodies (EBs), which are essential for cell differentiation and early development. Reversible cross-linking of genomic DNA using formaldehyde efficiently produces both protein-nucleic acid and protein–protein cross-links (1). We describe a protocol to dis-

Minou Bina (ed.), *Gene Regulation: Methods and Protocols*, Methods in Molecular Biology, vol. 977,
DOI 10.1007/978-1-62703-284-1_25,

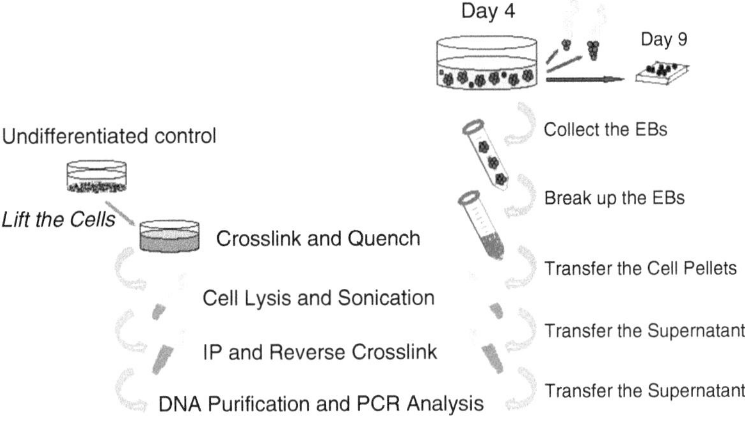

Fig. 1. Diagrammatic presentation of the ChIP procedures. Cells were cultivated into EBs in petri dishes for 4 days. Aliquot of cells were then seeded on coverslips for assessing the efficacies of myogenic differentiation on day 9, and subsequently transferred into micro tubes for input protein control and for gene expression survey by Western blot and RT-PCR analysis. The remaining EB was collected into 50 ml tubes for ChIP. Undifferentiated cells were ChIPed in parallel as control.

perse the EBs prior to the addition of formaldehyde to ensure proper cross-linking of the chromatin for the study of p300 occupancy during stem cell differentiation (Fig. 1). The cross-linked chromatin is sheared by Bioruptor® to generate DNA fragments of 200–500 base pairs in length, which are then used as substrate for p300 immunoprecipitation (IP). The antibody used in the p300 IP is coupled to Dynabeads®, and the immunoprecipitated p300 complexes are washed to remove nonspecifically bound chromatin. After removal of proteins from the immunopurified chromatin through protein digestion (10), the p300 associated DNA fragments are purified and can then be analyzed by additional methods such as real-time PCR (11), high-throughput massively parallel sequencing (ChIP-seq) (12), or hybridization to a DNA microarray (ChIP-chip) (13), versus the conventional PCR (14).

2. Materials

Prepare all solutions using DNase-free, autoclaved H_2O, and store at room temperature, unless otherwise indicated.

1. P19 cell line (ATCC).
2. α-Minimum Essential Medium, supplemented with 5% fetal bovine serum (FBS), 5% bovine calf serum (BCS) and 1× Penicillin/Streptomycin, store at 4°C.
3. Petri dishes and tissue culture dishes.

4. Phosphate-buffered saline (PBS) pH 7.4: 137 mM NaCl, 2.7 mM KCl, 10 mM Na_2HPO_4, 2 mM KH_2PO_4, store at 4°C.
5. 37% Formaldehyde.
6. 2.5 M Glycine.
7. Protease Inhibitor Cocktail Set III, EDTA-free, stored at −20°C.
8. Lysis Buffer, 50 mM Tris–HCl pH 8.0, 10 mM EDTA pH 8.0, 1% SDS.
9. Dilution Buffer, 20 mM Tris–HCl pH 8.0, 150 mM NaCl, 2 mM EDTA, 1% Triton X-100 (see Note 1).
10. TE, 10 mM Tris–HCl pH 8.0, 1 mM EDTA (see Note 2).
11. Antibody for p300 (Santa Cruz Biotechnology).
12. IgG antiserum.
13. Dynabeads® and magnetic tube rack.
14. Wash Buffer A, 0.1% SDS, 1% Triton X-100, 20 mM Tris–HCl pH 8.0, 2 mM EDTA pH 8.0, 150 mM NaCl.
15. Wash Buffer B, 0.1% SDS, 1% Triton X-100, 20 mM Tris–HCl pH 8.0, 2 mM EDTA pH 8.0, 500 mM NaCl.
16. Wash Buffer C, 1% NP-40, 1% sodium dioxycholate, 20 mM Tris–HCl pH 8.0, 1 mM EDTA, 0.25 M LiCl.
17. Elution Buffer, 1% SDS, 100 mM $NaHCO_3$ (see Note 3).
18. Proteinase K, 20 mg/ml.
19. DNA purification kit.

3. Methods

3.1. P19 Cell Culture

All cell culture procedures should be carried out in a sterilized tissue culture hood.

1. *Day 0*, plate cells in 150 mm petri dishes (see Note 4), grow and treat the cells as per differentiation protocol (11).
2. *Day 2*, plate undifferentiated control cells in 100 mm tissue culture dishes (see Note 5).

3.2. Cross-linking and Quenching

1. *Day 4*, remove three dishes per condition at a time from the CO_2 incubator and carry out the following:
   (a) Plate aliquot of EB on a coverslip coated with 0.1% gelatin placed in 6-well plate containing 3 ml of medium to determine the efficacies of myogenic differentiation on day 9.

(b) Pipette aliquot of EB into 1.5 ml micro tubes for input protein control, and for gene expression survey by Western blot and RT-PCR analysis.

(c) Collect the remaining EB into one 50 ml falcon tube (see Note 6). Place the tube upright in the incubator and repeat for other conditions.

2. Remove one tube from the incubator and aspirate the medium so that the total remaining volume is 20 ml. Break up the EBs by pipetting the cells up and down forcefully for 5 min using a 1 ml pipette (see Note 7). Place the tube upright back in the incubator and repeat this step for other conditions.
3. Vortex all tubes for 2 min vigorously.
4. Add formaldehyde to a final concentration of 1% and incubate the samples in a rotating incubator at 37°C for 15 min (see Note 8).
5. Remove the undifferentiated control cells from the CO_2 incubator. Lift the cells by using a rubber policeman. Add formaldehyde to a final concentration of 1% (see Note 9) and place dish back in the incubator for 15 min.
6. Remove both the 50 ml tubes and the undifferentiated control plates from the incubators and quench the cross-linking by adding 2.5 M glycine to a final concentration of 0.125 M. Rotate at room temperature for 5 min.
7. Transfer the undifferentiated cells from the plates to 15 ml falcon tubes.
8. Spin all the 50 ml and 15 ml tubes at 2,000 × *g* for 5 min at 4°C.
9. Aspirate the supernatant using a Pasteur pipette (see Note 10).
10. Resuspend the cell pellets in 3 ml ice-cold 1× PBS and transfer to fresh 15 ml falcon tubes.
11. Spin at 2,000 × *g* for 5 min at 4°C and aspirate the supernatant.
12. Repeat the wash with 1× PBS twice. After the final wash, remove as much supernatant as possible (see Note 11).

3.3. Cell Lysis and Sonication

1. Resuspend each cell pellet in 200 μl of Lysis Buffer supplemented with 1× protease inhibitor and transfer to 2.0 ml tubes.
2. Vortex briefly and leave the samples on ice for 10 min.
3. Vortex briefly and transfer into a Bioruptor® and sonicate on high, 30 s on and 40 s off, for 30 min (two 15 min cycles) (see Note 12).
4. Centrifuge at 14,000 × *g*, at 4°C for 15 min and transfer the supernatant to ice-cold 1.5 ml micro tube (the chromatin can be stored at −80°C for months).

3.4. Total DNA Concentration and Fragment Size

1. Transfer 5% of each sonicated sample into a fresh 1.5 ml micro tube to isolate total DNA.
2. Add Elution Buffer to a final volume of 200 μl, and then add proteinase K to a final concentration of 0.5 mg/ml. Incubate the samples in a 65°C water bath overnight.
3. Purify total DNA using a DNA purification kit and then proceed to:
   (a) Determine the DNA concentration and calculate the volume of each condition to be used for subsequent IP.
   (b) Visualize the DNA fragment size on 1% agarose gel.
   (c) Store the purified DNA at −80°C to use as the input DNA control in the final step of PCR analysis.

3.5. Immunoprecipitation and Reverse Cross-linking

1. Transfer equal amount of DNA into a 2 ml tube for all conditions and add Dilution Buffer supplemented with 1× protease inhibitor to 2 ml.
2. Add 2 μg of p300 antibody to each condition, and 5 μg of IgG to the negative ChIP control. Rotate at 4°C overnight.
3. Wash the Dynabeads® before use and resuspended in TE (see Note 13).
4. Add 40 μl of Dynabeads® to each sample and rotate at 4°C for 2 h.
5. Place the tubes in the magnetic rack, allow the beads to settle and pour off the supernatant. Add 1 ml of Wash Buffer A and rotate the tubes at 4°C for 10 min. Repeat this step for Wash Buffer B and Wash Buffer C.
6. Repeat the wash step twice with TE, but rotating the tubes for 5 min per wash. Place the tubes in the magnetic rack, pour out the supernatant, and aspirate the remaining liquid (see Note 14).
7. Add 200 μl Elution Buffer to the beads (see Note 15) and rotate at room temperature for 30 min.
8. Spin the samples at 3,000 × *g* for 1 min, transfer the supernatant to a fresh 1.5 ml micro tube, add proteinase K to a final concentration of 0.5 mg/ml and incubate the samples in the 65°C water bath overnight.

3.6. DNA Purification and PCR Analysis

1. Purify the DNA using a DNA purification kit and elute twice with 50 μl of the elution buffer supplied by the kit.
2. Analyze purified DNA with gene specific primers using Real-Time PCR (with the purified input DNA as internal control).

4. Notes

1. Dilution Buffer should be prepared in advance to allow the Triton X-100 to dissolve completely.
2. 50 mM NaCl may be added to reduce stringency.
3. Elution Buffer must be made fresh every time.
4. Mouse cells contain about 5 pg of DNA per nucleus, and human cells contain about 7.5 pg of DNA per nucleus. For optimal ChIP, aim for 25 μg DNA per IP. For P19 cells, we used three dishes per treatment, per antibody, and three million cells per 150 mm petri dish.
5. For the undifferentiated control, plate enough cells so that they will reach 80% confluence when harvested 2 days later. For P19 cells, plate one million cells per 100 mm tissue culture dish, but it will vary with specific cell type and growth rate.
6. When pooling the EBs into falcon tubes, some of the medium can be discarded by tilting the dish and removing medium from the top after the EBs have settled. However, leave as much medium as possible at this point, since some EB are small and do not readily settle.
7. Pipetting at the medium/air interface is more effective.
8. After vortexing, a thick layer of foam will form on top of the liquid. Invert the tubes several times to ensure an even distribution of formaldehyde. Optimal formaldehyde concentration and cross-linking time vary with cells types and should be determined by pilot tests.
9. Add formaldehyde slowly while swirling the dish to avoid a high concentration in one localized spot.
10. Leaving a small layer of liquid on top of the cells will minimize cell loss, and samples should be kept on ice as heat will reverse cross-linking.
11. At this point, cell pellets can be stored at −80°C for up to several weeks before proceeding with sonication.
12. Optimal sonication conditions need to be determined empirically for each cell type and the model of sonicator. Regardless, the samples need to be cooled down periodically during the sonication.
13. Do not vortex the Dynabeads® beads at any time as this will damage them. Swirl the beads in the stock bottle gently to resuspend them and pipette the desired amount into a 1.5 ml micro tube. Place the tube in the magnetic tube rack, and wait until the beads form a line against the back of the tube and the liquid turns clear. Simply pour out the liquid and resuspend the beads in TE. After two washes, the beads are ready to use and can be stored for further use.

14. Tilt the rack forward so that the remaining liquid pools away from the beads at the back of the tube, which makes it easier to aspirate.
15. Do not put Elution Buffer on ice as it will precipitate.

Acknowledgments

This work was supported by Natural Sciences and Engineering Research Council and Canadian Institutes of Health Research.

References

1. Orlando V (2000) Mapping chromosomal proteins in vivo by formaldehyde-crosslinked-chromatin immunoprecipitation. Trends Biochem Sci 25:99–104
2. O'Neill LP, Turner BM (1995) Histone H4 acetylation distinguishes coding regions of the human genome from heterochromatin in a differentiation-dependent but transcription-independent manner. EMBO J 14:3946–3957
3. O'Neill LP, Turner BM (1996) Immunoprecipitation of chromatin. Methods Enzymol 274:189–197
4. Nelson J, Denisenko O, Bomsztyk K (2009) The fast chromatin immunoprecipitation method. Methods Mol Biol 567:45–57
5. Thorne AW, Myers FA, Hebbes TR (2004) Native chromatin immunoprecipitation. Methods Mol Biol 287:21–44
6. Orlando V, Strutt H, Paro R (1997) Analysis of chromatin structure by in vivo formaldehyde cross-linking. Methods 11:205–214
7. Kuo MH, Allis CD (1999) In vivo cross-linking and immunoprecipitation for studying dynamic Protein:DNA associations in a chromatin environment. Methods 19:425–433
8. Solomon MJ, Larsen PL, Varshavsky A (1988) Mapping protein-DNA interactions in vivo with formaldehyde: evidence that histone H4 is retained on a highly transcribed gene. Cell 53:937–947
9. Flanagin S, Nelson JD, Castner DG et al (2008) Microplate-based chromatin immunoprecipitation method, Matrix ChIP: a platform to study signaling of complex genomic events. Nucleic Acids Res 36:e17
10. Solomon MJ, Varshavsky A (1985) Formaldehyde-mediated DNA-protein crosslinking: a probe for in vivo chromatin structures. Proc Natl Acad Sci USA 82: 6470–6474
11. Le May M, Mach H, Lacroix N et al (2011) Contribution of retinoid X receptor signaling to the specification of skeletal muscle lineage. J Biol Chem 286:26806–26812
12. Barski A, Cuddapah S, Cui K et al (2007) High-resolution profiling of histone methylations in the human genome. Cell 129: 823–837
13. Huebert DJ, Kamal M, O'Donovan A et al (2006) Genome-wide analysis of histone modifications by ChIP-on-chip. Methods 40:365–369
14. Higazi A, Abed M, Chen J et al (2011) Promoter context determines the role of proteasome in ligand-dependent occupancy of retinoic acid responsive elements. Epigenetics 6:202–211

Chapter 26

Mammalian Two-Hybrid Assays for Studies of Interaction of p300 with Transcription Factors

Daniela B. Mendonça, Gustavo Mendonça, and Lyndon F. Cooper

Abstract

The two-hybrid system is a powerful genetic assay that allows the interaction between two proteins to be detected in vivo. It was originally described in 1989 and since then it has been one of the main techniques used to identify interactions between proteins from different cellular organisms. Here we describe the methods to study the interaction of p300 with other transcription factors, specifically between p300 and two transcription factors related to hypoxia and inflammation, HIF-1α and NF-κB-p65, respectively.

Key words: Mammalian two-hybrid, p300, Protein–protein interaction, Gal4, VP16, Transcription

1. Introduction

One of the most important mechanisms utilized by mammalian cells to regulate a variety of cellular and molecular activities, such as transcription, translation, signal transduction and enzyme reaction is protein–protein interaction (1). The two-hybrid system is one of the various methods used to characterize protein–protein interactions. It was originally described in yeast (2), then in bacteria (3–5) and more recently in mammalian cells (1). Many eukaryotic transcription factors contain two distinct physical and functional domains: the DNA binding domain (DBD) and the transcription activation domain (TAD) (6). The DNA-binding domain specifically binds to a promoter/enhancer element and the transcriptional activation domain directs RNA polymerase II to transcribe the gene downstream of the DNA-binding domain (7). Although there are several combinations of DBDs and TADs that could be used in mammalian two-hybrid assays (6), the most commonly used DNA-binding and transcription activation domains are the Gal4 and Herpes virus

Minou Bina (ed.), *Gene Regulation: Methods and Protocols*, Methods in Molecular Biology, vol. 977,
DOI 10.1007/978-1-62703-284-1_26, © Springer Science+Business Media, LLC 2013

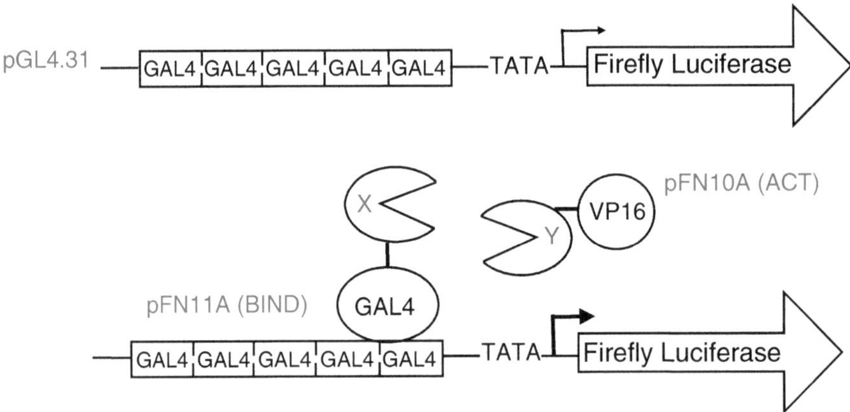

Fig. 1. Schematic representation of the CheckMate Mammalian Two-Hybrid System. The CheckMate/Flexi Vector System relies upon three plasmids that are co-transfected into mammalian cells. The pFN10A (ACT) Flexi Vector contains a herpes simplex virus VP16 transcriptional activation domain upstream of the cloning site, and the pFN11A (BIND) Flexi Vector contains the yeast GAL4 DNA-binding domain upstream of the cloning site. The pFN11A (BIND) Flexi Vector also expresses the luciferase under the control of the SV40 promoter, allowing normalization for differences in transfection efficiency. The third vector, pGL4.31, contains five GAL4 binding sites upstream of a minimal TATA box, which is upstream of a firefly luciferase gene that acts as a reporter for interactions between proteins "X" and "Y". "X" = CAD-HIF-1α or CAD-NF-κB-p65; "Y" = p300. Adapted from CheckMate Mammalian Two-Hybrid System Technical Manual (Promega).

VP16, respectively (1). For the mammalian two-hybrid analysis, at least three different types of constructs are necessary, often called "bait," "prey," and "reporter." The bait plasmid encodes a protein of interest that is fused to the DBD, while the prey plasmid encodes a target protein that is fused to the TAD. These plasmids are then co-introduced into appropriate host cells along with the reporter plasmid. Interactions between bait and prey result in association of the DBD with the TAD and the interaction between protein of interest and its potential partner can be measured by the level of reporter gene activity (8). The genes for the enzymes β-galactosidase, luciferase, and chloramphenicol acetyltransferase (CAT) are commonly used as reporter genes and allow convenient quantitation of multiple samples (6). The mammalian two-hybrid system is a straightforward method and overcome the limitations inherent to the yeast two-hybrid system (e.g., nonnuclear proteins can be difficult to detect; hybrid proteins may not be expressed stably in yeast or may not efficiently move into the nucleus; the creation of hybrid proteins may impede the natural folding patterns of the proteins, or obscure their interaction sites; and yeast cells lack complex post-translational modification mechanisms needed by many proteins for proper function) (9). Figure 1 shows a schematic representation of the Mammalian Two-Hybrid System.

2. Materials

2.1. Oligonucleotide Primers

Development of fusion constructs

The underlined sequences indicate restriction sites (for *SgfI* and *PmeI*); bold letters indicate start or stop codons.

For the C-terminal activation domain (CAD) of HIF-1α (amino acids 721-826)

HIF-1αCAD F:

GCGGCGATCGCC**ATG**GAACATGATGGTTCACTTTTTCAAGCA

HIF-1αCAD R:

AATCGTT**TAA**ACGTTAACTTGATCCAAAGCTCTGAGTAATTCTTCAC

For the C-terminal activation domain (CAD) of NF-κB-p65 (amino acids 314-550)

NF-κB-p65CAD F:

GTCGGCGATCGCC**ATG**AAGAAGAGTCCTTTCAGCGG

NF-κB-p65CAD R:

CTAGGTT**TAA**ACGGAGCTGATCTGACTCAGCA

For the Zn^{2+}-binding cysteine/histidine-rich 1 (CH1) domain of p300 (amino acids 302-418)

CH1-p300 F:

ATGGGCGATCGCC**ATG**GGTCAACAGCCAGCCCCG

CH1-p300 R:

TTGGGTT**TAA**ACTTTGAGGGGGAGACACACAGGA

2.2. PCR Reaction

1. High fidelity DNA polymerase.
2. PCR buffer.
3. dNTPs at 10 mM concentration.
4. $MgCl_2$ at 50 mM concentration.
5. PCR primers (in detail above, Subheading 2.1) at 10 μM concentration.
6. PCR template.
7. Autoclaved, distilled water.
8. Thermal cycler.
9. Thin-wall PCR tubes.

2.3. DNA Purification, Clean-Up of PCR/Digestion Products

1. QIAquick Gel Extraction Kit (QIAGEN).
2. 96–100% ethanol v/v (for the QIAquick Gel Extraction Kit).
3. 100% Isopropanol (for the QIAquick Gel Extraction Kit).

4. 1.5 or 2.0 ml microcentrifuge tubes.
5. Tabletop microcentrifuge.
6. Heating block or water bath set at 50°C.
7. Optional: Distilled water or TE buffer: 10 mM Tris–HCl, 1 mM EDTA, pH 8.0 for elution of DNA.
8. NanoDrop Spectrophotometer (Thermo Scientific).

2.4. Digestion of PCR Product and Acceptor Flexi Vectors

2.4.1. Digestion of PCR Products

1. 5× Flexi Digest Buffer (Promega).
2. Purified PCR product (up to 500 ng).
3. Flexi Enzyme Blend: *SgfI* and *PmeI* from Promega.
4. Nuclease-Free Water to a final volume of 20 μl.

2.4.2. Digestion of Acceptor Flexi Vectors

1. 5× Flexi Digest Buffer (Promega).
2. Acceptor Flexi Vectors pFN10A (ACT) and pFN11A (BIND) (200 ng) (Promega).
3. Flexi Enzyme Blend: *SgfI* and *PmeI* from Promega.
4. Nuclease-Free Water.

2.5. Agarose Gel Electrophoresis

1. Agarose.
2. 50× TAE (Tris-Acetate EDTA) Buffer Stock Solution: to 1 L of stock solution, add 242 g of Tris, 100 ml of 0.5 M EDTA pH 8.0 (this is prepared by adding 73.2 g of EDTA in 100 ml 1× PBS), 57.1 ml of glacial acetic acid, and add enough distilled water to bring final volume to 1 L. To prepare 50 ml of a 1× solution, mix one volume of 50× TAE Buffer with 49 volumes of distilled water.
3. 1 kb and 100 bp ladder.
4. Loading buffer containing tracking dies.
5. Agarose electrophoresis gel equipment and power supply.
6. Transilluminator (UV light box) and camera (optional).

2.6. Ligation of PCR Products and Acceptor Flexi Vectors

1. 2× Flexi Ligase Buffer (Promega).
2. Acceptor Flexi Vector from Subheading 2.4 (50 ng) (Promega).
3. PCR product (approximately 100 ng).
4. T4 DNA Ligase High Concentration (20 units) (Promega).
5. Nuclease-Free Water to a final volume of 20 μl.

2.7. Vector Preparation

2.7.1. Transformation and Selection

1. MAX Efficiency DH5α Competent Cells (Invitrogen).
2. Ligation reaction (1–5 μl).
3. SOC Medium per liter (Invitrogen): add 2% bacto-tryptone, 0.5% yeast extract, 10 mM NaCl, 2.5 mM KCl, 10 mM $MgCl_2$,

10 mM $MgSO_4$, 20 mM glucose, pH 7.0, and bring the final volume to 1 L with ddH_2O.

4. Chilled polypropylene tubes.
5. 37°C shaking and non-shaking incubator.
6. LB agar plates: add 15 g of agarose to 1 L of Luria Bertani (LB) media and autoclave for 20 min, cool to 45°C, add ampicillin to a final concentration of 100 μg/ml and dispense approximately 25 mL per 100-mm plate.
7. Ice bucket with ice.
8. 42°C water bath.

2.7.2. Screening for the Desired Clone

1. Polypropylene tubes.
2. Pipet tips to collect desired colonies.
3. Ampicillin.
4. LB Medium: dissolve 10 g tryptone, 5 g yeast extract, and 10 g NaCl in 800 ml distilled water. Adjust the pH to 7.0 with 1 N NaOH. Adjust the volume to 1 L with distilled water. Sterilize by autoclaving.
5. Isopropanol.
6. Ethanol.
7. Glycerol.
8. 37°C shaking incubator.
9. 1.5 or 2.0 ml microcentrifuge tubes (for minipreps).
10. 1.5 ml cryotubes.
11. 50 ml centrifuge tubes (for maxipreps).
12. Flasks (for maxipreps).
13. Tabletop microcentrifuge (for minipreps).
14. Refrigerated centrifuge with rotor (for maxipreps).

2.8. Tissue Culture and Transient Transfection

1. Human fetal osteoblastic cells—hFOB 1.19 (ATCC CRL-11372).
2. Dulbecco's Modified Eagle Medium: Nutrient Mixture F-12 (DMEM/F12).
3. Opti-MEM I Reduced Serum Media.
4. Fetal bovine serum (FBS).
5. 100× antibiotic/antimycotic solution (with 10,000 units penicillin, 10 mg streptomycin and 25 μg amphotericin B per ml) (Sigma-Aldrich).
6. 100× MEM Nonessential Amino Acids Solution 10 mM (NEAA) (Invitrogen).

7. 10× trypsin-EDTA solution (Sigma-Aldrich).
8. Phosphate buffered saline (pH 7.4): 8 g NaCl, 0.2 g KCl, 1.44 g Na_2HPO_4, 0.24 g KH_2PO_4, add double distilled H_2O to 1 L.
9. Transfection reagent: Attractene (Qiagen).
10. 150 mm tissue culture plates.
11. 48-well tissue culture plates.
12. 50 ml conical tubes.
13. Water bath set at 37°C.
14. Inverted microscope.
15. Hemocytometer with cover glass.
16. Tissue culture incubator.
17. Centrifuge.
18. Biosafety cabinet.

2.9. Mammalian Two-Hybrid Plasmids and Dual Luciferase Assay

1. CheckMate/Flexi Vector Mammalian Two-Hybrid System (Promega), which contains the following vectors:
   (a) *Functional vectors*

   Acceptor Flexi Vectors: pFN10A (ACT), which contains herpes simplex virus VP16 transcriptional activation domain upstream of the cloning site; and pFN11A (BIND), that contains the yeast GAL4 DNA-binding domain upstream of the cloning site.

   Reporter vector pGL4.31 [luc2P/GAL4UAS/Hygro] Vector, which contains a firefly luciferase gene (luc2P).

   (b) *Positive control vectors*

   pBIND-Id and pACT-MyoD that encode VP16-MyoD and GAL4-Id fusion proteins, respectively.

   (c) *Negative control vectors*

   pBIND and pACT, which lack fusion proteins, and are used as negative controls.
2. Test plasmids (fusion constructs):

   VP16-CH1-p300, GAL4-CAD-HIF-1α, GAL4-CAD-NF-κB-p65.
3. Dual Luciferase Assay Kit (Promega).
4. Glass test tubes for the Dual Luciferase Assay.
5. Ultra sensitive tube luminometer.

3. Methods

Development of fusion constructs

3.1. Oligonucleotide Primers

The primers used to construct the fusion plasmids were designed using the Flexi Vector Primer Design Tool, available at Promega website (http://www.promega.us/techserv/tools/FlexiVectorTool/default.aspx). According to the manufacturer, to facilitate cloning, the PCR primers used to amplify the protein coding region must append a *SgfI* site and a *PmeI* site to the PCR product. To append these sites, a *SgfI* site in the amino-terminal PCR primer and a *PmeI* site in the carboxy-terminal PCR primer should be incorporated, as shown on the "Subheading 2". The primer design tool described above automatically appends the restriction enzyme sites.

3.2. PCR-Mediated Amplification

Separate PCR reactions were set for each gene of interest (HIF-1α, NF-κB-p65 and p300) using the respective primer pair.

1. For PCR reaction, add:

   5 μl of 10× PCR buffer.

   1 μl of 10 mM dNTP mixture.

   2 μl of 50 mM $MgCl_2$.

   1 μl of each primer at 10 μM each.

   3 μl of 100 μg/μl of DNA of PCR template for the genes of interest.

1.0 unit (in this case, 0.2 μl) of high fidelity DNA polymerase.

Autoclaved distilled water to 50 μl.

   For CAD-HIF-1α the template DNA used was the vector HA-HIF-1α-WT obtained from Addgene (Addgene plasmid 18949); for CAD-NF-κB-p65 the template DNA used was the vector NF-κB-p65-CMV (Courtesy of Dr. Albert S. Baldwin, University of North Carolina, Chapel Hill); for CH1-p300 the template DNA used was pCMVb-p300 obtained from Addgene (Addgene plasmid 10717).

2. PCR conditions: after 2 min of the initial denaturation at 94°C, run 30 cycles of the following steps—denaturation at 94°C for 30 s, annealing at 55°C for 30 s, extension at 68°C (1 min per kb), and finish extension at 68°C for 10 min. Samples can be stored at −20°C until further use.
3. Clean-up of PCR products: using QIAquick Gel Extraction Kit (QIAGEN), according to the manufacturer's instructions and elute DNA in buffer EB. This step is important to remove the DNA polymerase and primers.

4. Determination of yield: To determine the yield, the PCR products should be quantified using UV spectrophotometry at 260 nm (NanoDrop).
5. Digestion: digest PCR products and Acceptor Flexi Vectors pFN10A (ACT) and pFN11A (BIND) (Promega), using Flexi Enzyme Blend (*SgfI* and *PmeI*) (Promega), according to manufacturer's instructions. Digestion was performed at 37°C for 30 min. For the Acceptor Flexi Vectors heat the reaction at 65°C for 20 min to inactivate the restriction enzymes. Store on ice until the PCR product and vector are ligated.
6. Clean-up of digested PCR products: using QIAquick Gel Extraction Kit (QIAGEN), according to the manufacturer's instructions and elute DNA in buffer EB. This step is important to remove the small oligonucleotides released by the restriction enzymes. According to the manufacturer's protocol, the digested Acceptor Flexi Vectors do not need to be purified.
7. 1.0% agarose gel for electrophoresis: subject the digested PCR products and Acceptor Flexi Vectors to 1.0% agarose gel prepared in 1× TAE buffer for electrophoresis, then isolate the band corresponding to the size of the desired gene.
8. Determination of yield: To determine the yield, the digested PCR products and Acceptor Flexi Vectors should be quantified using UV spectrophotometry at 260 nm (NanoDrop).
9. Ligation: ligate the digested PCR products for p300 (CH1-p300), HIF-1α (CAD-HIF-1α) and NF-κB-p65 (CAD-NF-κB-p65) to the Acceptor Flexi Vectors (VP16-encoding pFN10A or GAL4-pFN11A). CH1-p300 was ligated to VP16-encoding pFN10A; CAD-HIF-1α and CAD-NF-κB-p65 were ligated to GAL4-pFN11A. Ligation was performed for 1 h at room temperature as recommended (see Note 1).

3.3. Vector Preparation

1. Transformation and Selection: Transform MAX efficiency DH5α competent cells, as recommended by the manufacturer. 50 μl of MAX Efficiency DH5α competent cells, and 3 μl of ligation were used for the transformation. 400 μl of the transformation mixture was spread on LB agar plates containing 100 μg/ml of ampicillin. Agar plates were incubated overnight at 37°C. The remaining transformation reaction could be stored at 4°C, since additional cells may be plated out the next day, if necessary (see Note 2).
2. Screening of the desired colonies: Screen 8–12 colonies for each fusion construct (VP16-CH1-p300, GAL4-CAD-HIF1α or GAL4-CAD-NF-κB-p65). The desired colonies can be selected using pipet tips and placed into polypropylene tubes containing 3–5 ml of LB media plus 100 μg/ml of ampicillin.

These selected colonies should be put in a 37°C shaking incubator overnight (12–16 h) at 225 rpm.

3. Miniprep: Harvest the bacterial cells by centrifugation at > 8,000 rpm (6,800 × *g*) in a conventional, tabletop microcentrifuge for 3 min at room temperature (15–25°C). The DNA purification was done using the QIAprep Spin Miniprep Kit (QIAGEN), according to the manufacturer's recommendations. The DNA was eluted in 30 μl of Buffer EB.
4. Digestion: Digest the fusion constructs using Flexi Enzyme Blend (*SgfI* and *PmeI*) (Promega) to ensure that both *SgfI* and *PmeI* cleave their recognition sites flanking the protein-coding region. Digestion should be performed according to manufacturer's instructions, at 37°C for 30 min.
5. DNA electrophoresis gel: Subject the fusion constructs to 1.0% agarose gel prepared in 1× TAE buffer for electrophoresis to check the band corresponding to the desired size. The fragment size for VP16-CH1-p300 digested with the Flexi Enzyme Blend should be 5,496 and 363 bp. for GAL4-CAD-HIF1α the fragments should be 6,144 and 330 bp, and for GAL4-CAD-NF-κB-p65, the fragment sizes should be 6,144 and 729 bp.
6. Maxiprep (large scale, high purity DNA purification): From the minipreps, select one positive colony for each vector and proceed with large scale DNA purification. Inoculate the starter culture (obtained above) of 2–5 ml LB medium containing 100 μg/ml of ampicillin. Incubate for approximately 8 h at 37°C with vigorous shaking (approx. 225 rpm). Dilute the starter culture 1/500 to 1/1,000 into selective LB medium. For each fusion construct, inoculate 100 ml medium with 100–200 μl of starter culture. Grow at 37°C for 12–16 h with vigorous shaking (approximately 300 rpm). Harvest the cells by centrifugation at 4°C for 10 min at 4,000 × *g*. Isolate the desired fusion constructs (VP16-CH1-p300, GAL4-CAD-HIF-1α or GAL4-CAD-NF-κB-p65) using a Qiagen Endofree Plasmid Maxi Kit, following the manufacturer's protocol. Dissolve the DNA in a suitable volume (100 μl to 1 ml) of endotoxin-free Buffer TE, depending on the size of the DNA pellet. Make bacteria freezing stocks of the positive colonies of the fusion constructs using 50% glycerol solution (v/v) in a 1.5 ml cryotube and store at −80°C. Always use the original freezing stock for further testings.
7. Digestion: Digest the fusion constructs using Flexi Enzyme Blend (*SgfI* and *PmeI*) (Promega) to ensure that both *SgfI* and *PmeI* cleave their recognition sites flanking the protein-coding region. Digestion was performed according to manufacturer's instructions, at 37°C for 30 min.

8. Determination of yield: To determine the yield, DNA concentration should be determined by UV spectrophotometry at 260 nm (NanoDrop) and quantitative analysis on 1.0% agarose gel prepared in 1× TAE buffer. For reliable spectrophotometric DNA quantification, A260 readings should lie between 0.1 and 1.0. Prepare a 500 μl working solution of 100 μg/μl of DNA for each vector for further use.
9. Sequencing: Since PCR was used to create the original insert, the sequence of the protein-coding region in the initial clone should be verified by DNA sequencing. The T7 promoter primer should be used for sequencing of the fusion constructs.

3.4. Tissue Culture and Transfection

1. hFOB cell culture: supplement the DMEM/F12 growth media with 10% FBS and 1× penicillin/streptomycin (regular media). Cells should be grown on 150 mm tissue culture plates inside a humidified incubator at 37°C supplied with 5% CO_2.
2. Preparing hFOB cells for transient transfection: hFOBs should be grown until they reach 70–80% of confluency. For each 150 mm plates: rinse cells with 10–12 ml of 1× PBS twice, trypsinize cells with 8 ml of 1× trypsin and incubate for 5 min at 37°C incubator. After the 5 min incubation, add 12 ml of regular media to each 150 mm plate to quench the trypsin. Place detached cells into a screw-top conical tube and spin them at 200 × *g* for 4 min. Remove supernatant carefully and do not disturb cell pellet. Resuspend the cells in Opti-MEM media supplemented with 3% FBS and 1× NEAA (transfection media). The amount of media added will depend on the size of the pellet. Perform the cell counting and adjust the cell density (1.6×10^5 cells/ml).
3. Transient transfection:

   Table 1 shows recommended combinations of vectors to properly control experiments using the CheckMate™/Flexi Vector System, according to the manufacturer's recommendations. In this case, "X" represents CAD-HIF-1α or CAD-NF-κB-p65 (two different experiments will be set up), while "Y" represents CH1-p300. pBIND and pACT are Negative Control Vectors. pBIND-Id and pACT-MyoD are Positive Control Vectors.

   The method described here is for transient transfection of hFOB cells in 48-well plates, using the Fast-Forward Protocol. In the Fast-Forward protocol, plating and transfection are performed on the same day. In the Traditional Protocol, cells are plated 24 h prior to transfection. The DNA is diluted in culture medium without serum and Attractene Transfection Reagent is added to the diluted DNA to produce Attractene-DNA complexes. The cells are seeded and then complexes are added directly to the freshly seeded cells. The detailed steps for

Table 1
Experimental design for testing protein–protein interactions

Sample	ACT vector	BIND vector	pGL4.31
1	pFN10A (ACT)-Y	pFN11A (BIND)-X	+
2	pACT Vector	pFN11A (BIND)-X	+
3	pFN10A (ACT)-Y	pBIND Vector	+
4	pACT Vector	pBIND Vector	+
5	pACT-MyoD Control	pBIND-Id Control	+

Sample 1 is used to test for protein interactions between the X and Y protein-coding regions. *Sample 2* is used to test for the background level of reporter activity when one of the protein-coding regions is fused to the GAL4 DNA-binding domain. This also tests for transcriptional activation domain activity of protein X when fused to the GAL4 DNA-binding domain. *Sample 3* is used to test for the background levels of reporter activity when one of the protein-coding regions is fused to the activation domain. *Sample 4* generally provides the lowest level of background activity of luciferase activity from the reporter plasmid. *Sample 5* provides a positive control with interacting proteins fused to the activation and GAL4 DNA-binding domains

the Fast-Forward protocol are explained below, according to the manufacturer's recommendations.

(a) For each combination showed in Table 1 above, dilute 125 ng of DNA of each plasmid (amount recommended for 48-well plates) in serum-free and antibiotic-free Opti-MEM medium to a total volume of 30 μl per well. For example, if the DNA concentration is 100 μg/μl, dilute 1.25 μl DNA in 28.75 μl medium.

(b) Add 0.5 μl of Attractene Transfection Reagent per well. Mix by pipetting up and down or vortexing. Centrifuge for a few seconds to remove any liquid from the top of the tube if necessary.

(c) Incubate the samples for 10–15 min at room temperature (15–25°C) to allow transfection complex formation.

(d) During the 10–15 min incubation time, harvest the cells by trypsinization, and suspend in transfection culture medium as described above (Subheading 3.4, step 2). Count the cells, and adjust the cell density to 1.6×10^5 cells in 1 ml per well. The optimal cell density should be determined for each cell line.

(e) Add 250 μl of the cell suspension (which contains 4×10^4 cells) to each well of the 48-well plate. Next, add the transfection complexes (30 μl) to the well. The total volume per well will be 280 μl. Mix by pipetting up and down twice (or shake the plate).

(f) Incubate the cells with the transfection complexes under their normal growth conditions (37°C and 5% CO_2) for 18–24 h. Then, remove the transfection complexes and add fresh culture medium. Perform the Dual Luciferase Assay reading after 24–48 h after the removal of the transfection complexes.

3.5. Dual Luciferase Assay

1. Dual Luciferase Assay Kit Reagent preparation (according to the manufacturer's recommendations:
   (a) 1× PLB: Add 1 volume of 5× Passive Lysis Buffer (PLB) to 4 volumes distilled water. Mix well. Store at 4°C (≤1 month).
   (b) Luciferase Assay Reagent II (LAR II): Resuspend the lyophilized Luciferase Assay Substrate in Luciferase Assay Buffer II. Store at −20°C (≤1 month) or −70°C (≤1 year).
   (c) Stop & Glo Reagent: Add 20 µl of 50× Stop & Glo Substrate to 1 ml of Stop & Glo Buffer contained in either a glass vial or siliconized polypropylene tube. This will prepare sufficient Stop & Glo Reagent for 10 assays.
2. Passive Cell Lysis:
   (a) Remove growth media from cultured cells.
   (b) Rinse cultured cells in 1× PBS. Remove all rinse solution. Dispense the recommended volume (for 48-well plate, the recommended volume is 65 µλ) of 1× PLB into each well.
   (c) Passive Lysis: Gently rock/shake the tissue culture plate for 15 min at room temperature. Transfer lysate to a tube or vial.
3. Dual Luciferase Assay:

   Figure 2 shows an example of the luciferase activity involving the fusion constructs described throughout the text.

   The method described here is for Assay with Dual-Injector Luminometer.
   (a) Before you begin, set the protocol on the machine: Set injectors 1 and 2 to dispense 50 µl of LAR II and Stop & Glo Reagent, respectively. For measurements, use a 1–2 s delay and 5–10 s read time.
   (b) Predispense enough LAR II Reagent into injector 1 of the luminometer according to the number of samples.
   (c) Transfer 20 µl PLB Lysate. Mix.
   (d) Measure firefly luciferase activity.
   (e) Predispense enough Stop & Glo Reagent into injector 2 of the luminometer according to the number of samples.
   (f) Measure *Renilla* luciferase activity.

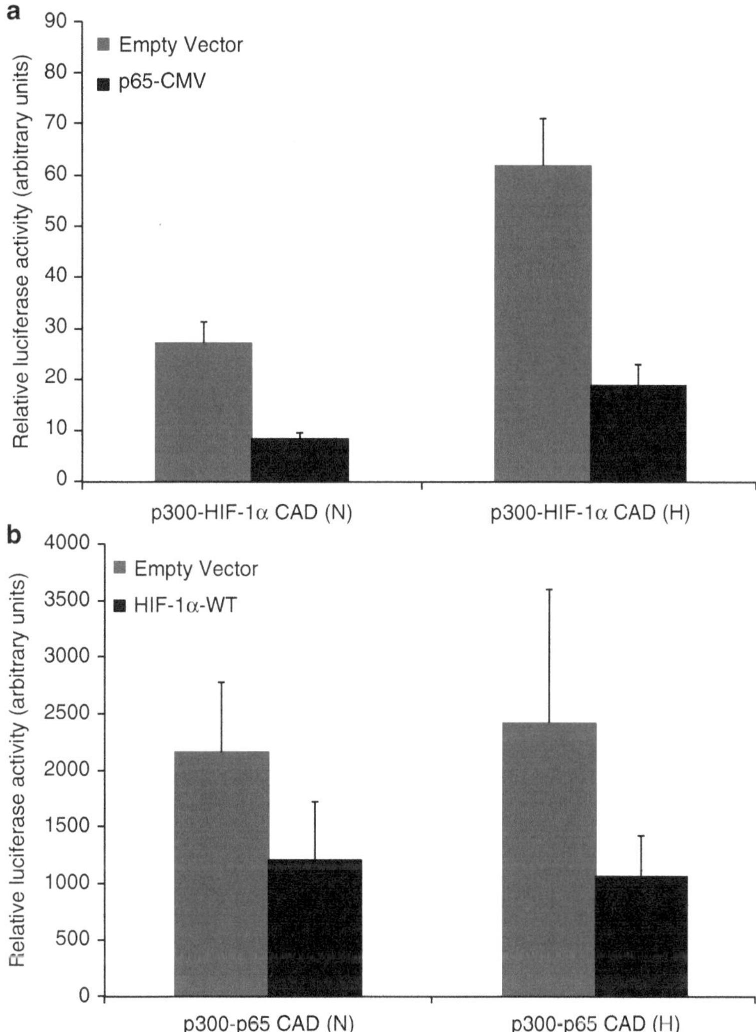

Fig. 2. p65 competes with HIF-1α for p300. hFOB cells were (**a**) co-transfected with VP16-p300, Gal4-CAD-HIF-1α fusion constructs (125 ng) and p65-CMV (375 ng) for 18 h. Cells were then exposed to Normoxia (N) or DFO-induced Hypoxia (H). Luciferase activity was measured 24 h post-treatment. Statistically significant difference compared to p300 + CAD-HIF-1α + Empty Vector (normoxia) ($p \leq 0.05$). (**b**) Co-transfected with VP16-p300, Gal4-CAD-p65 fusion constructs (125 ng) and HIF-1α-WT (375 ng) for 18 h. Cells were then exposed to N or H. Luciferase activity was measured 24 h post-treatment. Statistically significant difference compared to p300 + CAD-p65 + Empty vector (N) ($p \leq 0.05$). Data are shown as the mean ± SE of three independent experiments performed in triplicate.

4. Notes

Development of fusion constructs

1. *Ligation*

   Sometimes the ligation between the digested PCR products for p300 (CH1-p300), HIF-1α (CAD-HIF-1α) and NF-κB-p65 (CAD-NF-κB-p65) and the Acceptor Flexi Vectors (VP16-encoding pFN10A or GAL4-pFN11A) will not work the first time even though the manufacturer's recommendations are followed. In these cases, the pGEM-T Easy Vector system (Promega) should be used. These vectors contain a 3′ terminal thymidine to both ends. These single 3′-T overhangs at the insertion site greatly improve the efficiency of ligation of a PCR product into the plasmids by preventing recircularization of the vector and providing a compatible overhang for PCR products generated by certain thermostable polymerases (10). The pGEM-T Easy Vector Systems include a 2× Rapid Ligation Buffer for ligation of PCR products. According to the manufacturer, reactions using this buffer may be incubated for 1 h at room temperature. However, the incubation period may be extended to increase the number of colonies after transformation. Generally, an overnight incubation at 4°C will produce the maximum number of transformants.

   Details for ligation using the pGEM-T Easy Vector System:

   2× Rapid Ligation Buffer (Promega).

   pGEM-T Easy Vector (50 ng)

   PCR Product (insert) (amount to be used should be calculated using the equation below):

$$\frac{\text{ng of vector} \times \text{kb size of insert}}{\text{kb size of vector}} \times \text{insert : vector molar ratio}$$
$$= \text{ng of insert}$$

   Based on the size of our vectors, the calculations were done as follows:

For GAL4-CAD-HIF-1α

   ng of vector (pGEM-T Easy) = 50 ng

   kb size of insert (after digestion with Flexi Enzyme Blend—Promega) = 330 bp

   kb size of vector (pFN11A) (after digestion with Flexi Enzyme Blend—Promega) = 6,144 kb

   Insert–vector molar ratio used was 5:1
   For GAL4-CAD-NF-κB-p65

ng of vector (pGEM-T Easy) = 50 ng

kb size of insert (after digestion with Flexi Enzyme Blend—Promega) = 729 bp

Size of vector (pFN11A) (after digestion with Flexi Enzyme Blend—Promega) = 6,144 bp

Insert–vector molar ratio used was 5:1

For VP16-CH1-p300

ng of vector (pGEM-T Easy) = 50 ng

kb size of insert (after digestion with Flexi Enzyme Blend—Promega) = 363 bp

kb size of vector (pFN10A) (after digestion with Flexi Enzyme Blend—Promega) = 5,496 bp

Insert–vector molar ratio used was 5:1

Control insert DNA (used as a positive control): 2.0 μl.

T4 DNA Ligase (3 Weiss units/μl) (Promega): 1.0 μl.

Deionized water to a final volume of 10 μl.

2. *Vector preparation*
   (a) Transformation: The same protocol commented before to transform MAX efficiency DH5α competent cells can be used if using the pGEM-T Easy Vector System. However, after the transformation LB Agar plates should be prepared for blue/white screening to make it a little easier to select the colonies the next day. White colonies generally contain inserts; however, inserts may also be present in blue colonies.
   (b) Preparation of LB Agar Plates for Blue/White screening: Mix 10 μλ 100 mM IPTG with 90 μλ of ddH_2O; Mix 25 μλ X-Gal with 25 μλ ddH_2O; Add 100 μλ SOC Media and spread approximately 250 μλ per plate; Incubate the plates at 37°C and wait 30–60 min for the mixture to dry before using the plates.
   (c) Selection and Screening of the desired colonies: the white colonies should be selected if the blue/white screening was performed. Screen 8–12 colonies for each construct (pGEM-T Easy + CAD-HIF-1α, pGEM-T Easy + CH1-p300, pGEM-T Easy + CAD-NF-κB-p65). The desired colonies can be selected using pipet tips and placed into polypropylene tubes containing 3–5 ml of LB media plus 100 μg/ml of ampicillin. These selected colonies should be cultured at 37°C in a shaking incubator overnight (12–16 h) at 225 rpm.
   (d) Miniprep: Performed as reviewed in Subheading 3.3, step 3 using the QIAprep Spin Miniprep Kit (QIAGEN);

(e) Digestion: Digest the constructs using Flexi Enzyme Blend (*SgfI* and *PmeI*) (Promega), according to manufacturer's instructions. Digest from here to ligate into the new vector.

(f) DNA electrophoresis gel: Subject the fusion constructs to 1.0% agarose gel prepared in 1× TAE buffer for electrophoresis to check the band corresponding to the desired size.

(g) Maxiprep (large scale, high purity DNA purification): Done as commented on Subheading 3.3, step 6 using the Qiagen Endofree Plasmid Maxi Kit (QIAGEN).

(h) Digestion: Performed as commented before (see Note 2, step 4).

(i) Determination of yield: See Subheading 3.3, step 8.

(j) Sequencing: Inserts can be sequenced using the SP6 Promoter Primer and the T7 Promoter Primer, according to the manufacturer's recommendations.

References

1. He R, Li X (2008) Mammalian two-hybrid assay for detecting protein-protein interactions in vivo. Methods Mol Biol 439:327–337
2. Fields S, Song O (1989) A novel genetic system to detect protein-protein interactions. Nature 340:245–246
3. Hu JC, Kornacker MG, Hochschild A (2000) Escherichia coli one- and two-hybrid systems for the analysis and identification of protein-protein interactions. Methods 20:80–94
4. Ladant D, Karimova G (2000) Genetic systems for analyzing protein-protein interactions in bacteria. Res Microbiol 151:711–720
5. Dove SL (2003) Studying protein-protein interactions using a bacterial two-hybrid system. Methods Mol Biol 205:251–265
6. Feng XH, Derynck R (2001) Mammalian two-hybrid assays. analyzing protein-protein interactions in transforming growth factor-beta signaling pathway. Methods Mol Biol 177: 221–239
7. Lee JW, Lee SK (2004) Mammalian two-hybrid assay for detecting protein-protein interactions in vivo. Methods Mol Biol 261:327–336
8. Matsuzawa S, Reed JC (2007) Yeast and mammalian two-hybrid systems for studying protein-protein interactions. Methods Mol Biol 383:215–225
9. Garcia-Cuellar MP, Mederer D, Slany RK (2009) Identification of protein interaction partners by the yeast two-hybrid system. Methods Mol Biol 538:347–367
10. Mezei LM, Storts DR (1994) Purification of PCR products. In: Griffin HG, Griffin AM (eds) PCR technology: current innovations. CRC, Boca Raton, FL

Chapter 27

Fluorescence Anisotropy Microplate Assay to Investigate the Interaction of Full-Length Steroid Receptor Coactivator-1a with Steroid Receptors

Chen Zhang, Steven K. Nordeen, and David J. Shapiro

Abstract

Estrogens, acting via estrogen receptor (ER) play key roles in growth, differentiation, and gene regulation in the reproductive, central nervous, and skeletal systems. ER-mediated gene transcription contributes to the development and spread of breast, uterine, and liver cancer. Steroid receptor coactivator-1a (SRC1a) belongs to the P160 family of coactivators, which is the best known of the many coactivators implicated in ER-mediated transactivation. Binding of full-length P160 coactivators to steroid receptors has been difficult to investigate in vitro. This chapter details how to investigate the interaction of SRC1a with ER using the fluorescence anisotropy/polarization microplate assay (FAMA).

Key words: Estrogen receptor, Steroid receptor coactivator-1a, Fluorescence anisotropy, Microplate, LxxLL motif

1. Introduction

Estrogens regulate the normal growth and differentiation of reproductive tissues, bone, and nervous system and play an important role in human pathologies such as osteoporosis and breast, uterine, ovarian, liver, and lung cancers (1, 2). Estrogens, such as 17β-estradiol (E_2), execute their genomic actions mainly through estrogen receptor (ER) (3). There are two estrogen receptor isoforms, ERα and ERβ with substantial differences in the N-terminal domain and ligand binding domain (LBD) (4, 5). In the classical mechanism of ER action, the ER dimerizes when bound to E_2, and the E_2-ER complex regulates gene expression by direct binding to estrogen response elements (EREs), or closely related sequences (6–8). The consensus ERE (cERE) is a perfect palindrome, but most functional EREs are not perfect palindromes (9). For example, the

Minou Bina (ed.), *Gene Regulation: Methods and Protocols*, Methods in Molecular Biology, vol. 977, DOI 10.1007/978-1-62703-284-1_27, © Springer Science+Business Media, LLC 2013

downstream regulatory region of the proteinase inhibitor 9 (PI-9) gene called the estrogen responsive unit (ERU) consists of both an imperfect ERE palindrome and a direct repeat (10). Once bound to an ERE, the ligand binding domain of ER assumes a conformation that enables the recruitment of coactivators (11). Bound coactivators help assemble a multi-protein complex that facilitates both chromatin remodeling and formation of an active transcription complex.

The steroid receptor coactivator (SRC) or p160 family of coactivators, including SRC1/NCoA-1, SRC2/TIF2/GRIP1/NCoA-2, and SRC3/pCIP/ACTR/AIB1, represents a major class of coactivators that play key roles in ER-mediated transcription (12). The p160 coactivators interact with the hydrophobic binding cleft in the ER LBD mainly via highly conserved α-helical Leu-x-x-Leu-Leu (LxxLL) motifs, also called nuclear receptor (NR) boxes (13). All of the SRCs contain a central nuclear receptor-interacting domain (NRID) made up of three NR boxes separated by ~55 amino acids. SRC1a is an alternatively spliced form of SRC1 and contains a fourth NR box at its C-terminus (14).

We developed the fluorescence anisotropy/polarization microplate assay (FAMA) to analyze the interactions of steroid receptors with their DNA recognition sites (15) and with coactivators, such as full length SRC1a (16). In addition FAMA is a facile technology for high-throughput screening for small molecule inhibitors of receptor interactions (17–19).

In this assay, ER is pre-bound to a fluorescein-labeled ERE (flERE) in ultralow-volume microplates (16). When polarized light excites the flERE, most of the emitted light is depolarized because of rapid rotational diffusion of the flERE that results in its position being largely randomized at the time of emission. Binding of the ER protein, or of the even larger full-length SRC1a-ER complex, to the flERE slows rotation of the flERE, increasing the possibility that the complex is in the same plane at emission and excitation. These changes can be observed as an increase in fluorescence polarization, or the closely-related parameter, fluorescence anisotropy (detailed in Fig. 1). The coactivator binding assay described here is the most sophisticated application of the FAMA.

2. Material

2.1. flERE Preparation

1. Labeled sense strand: The oligonucleotide was synthesized with fluorescein (specifically: 6-FAM) at its 5′ end using phosphoramidite chemistry and PolyPak II- (Glen Research Corp, Sterling, VA, USA) and purified by the Biotechnology Center (University of Illinois). This service is provided in many bio-

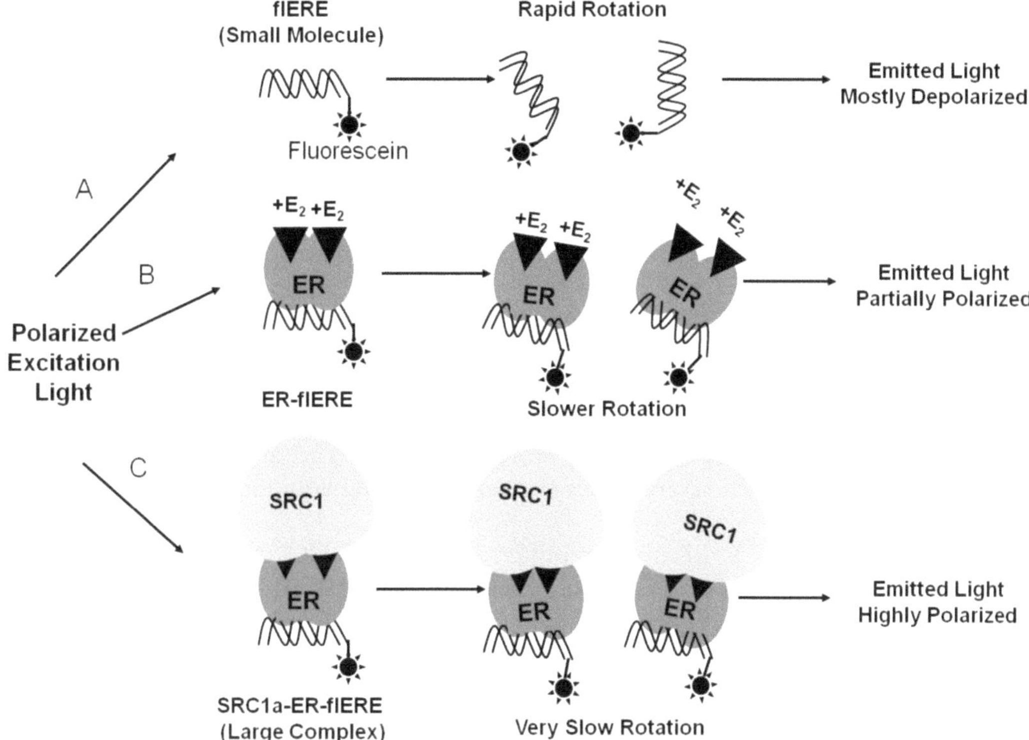

Fig. 1. Schematic of the fluorescence anisotropy/polarization microplate assay (FAMA). Excitation of the flERE with polarized light leads to excitation of only those flERE which have a similar orientation relative to the polarized excitation light. Most of the excited free flERE undergoes rapid random rotational diffusion, resulting in its position being largely randomized at the time of light emission. Therefore, the emitted light is depolarized (A). Binding of the larger ER protein to the flERE creates a high molecular weight complex. This slows rotation of the flERE, increasing the likelihood that the complex will be in the same plane at the time of emission as it was at the time of excitation. Therefore, the emitted light will be more polarized (B). When the large 160 kDa SRC1 binds to the ER-flERE, the even larger complex further slows rotation of the flERE, with the emitted light being highly polarized (C).

technology companies. The sequences of the fluorescein (fl)-labeled sense strand of cERE and PI-9 respectively are: 5′-fl- CTAGATTACAGGTCACAGTGACCTTACTCA-3′ and 5′-fl-AGATTACACTGGGGGACCCTGACCTCTAGACT TACGCTGATCACTTCGGTACGCA-3′.

2. Unlabeled antisense strand: The sequences of the oligonucleotide are complimentary to the sense strand. It can be synthesized by any company providing DNA synthesis services.
3. TE buffer: 10 mM Tris–HCl (pH 7.5), 1 mM EDTA. Annealing buffer: 100 mM NaCl in TE buffer.

2.2. Protein and Peptide

1. Full-length FLAG-epitope-tagged human ERα and FLAG-epitope-tagged human ERβ were expressed in the presence of 200 nM E_2, purified in the presence of 20 nM E_2 and quantitated as described (20). Aliquots of ER are stored at −80°C. ER can be thawed and flash frozen with liquid nitrogen once.

2. Full-length FLAG-epitope-tagged human SRC1a was purified as described (21). Aliquots of SRC1a are stored at –80°C. Use a fresh tube each time for optimal results. Used SRC1a flash frozen in liquid nitrogen exhibits slightly lower binding activity.

3. Peptides: The nuclear receptor interaction box 2 (NR-2) and box 4 (NR-4) peptides of SRC1 were synthesized by Abgent (San Diego, CA, USA). The amino acid sequences of the NR box peptides are as follows: SRC1 NR Box 2, LTERH KILHRLLQE; SRC1 NR Box 4, QAQQKSLLQQLLTE. Weigh ~1 mg of peptide and dissolve in H_2O to prepare a 10 mM stock solution. Seal the rest of the peptide in its original tube with Parafilm and store at –80°C. Aliquot the 10 mM peptide and store at –80°C. The aliquots can be thawed several times and refrozen using liquid nitrogen.

2.3. Fluorescence Anisotropy Microplate Assay

1. ER dilution buffer: 20 mM Tris–HCl (pH 7.9), 300 mM KCl, 20% Glycerol, 0.5 μg/μl BSA, 0.2 mM EDTA, 1 μM E_2.

2. SRC1 dilution buffer: 20 mM Tris–HCl (pH 7.5), 100 mM NaCl, 15% Glycerol, 0.5 μg/μl BSA.

3. FAMA buffer: 30 mM Tris–HCl (pH 7.5), 300 mM KCl, 10% Glycerol, 0.1% NP-40, 2 mM DTT, 0.2 ng/μl poly-dI:dC, 200 nM E_2. The KCl and glycerol in the FAMA buffer should be adjusted to corresponding to the salt (KCl and NaCl) and glycerol brought in by the added SRC1a and ER dilution buffer so the final concentrations of salt and glycerol in each assay are 150 nM and 5%, respectively.

4. We use black 96-well low volume, high efficiency HE microplates (Molecular Devices Corp., Sunnyvale, CA).

5. A plate-reader that can perform fluorescence anisotropy/polarization measurements in small volumes is required. The PHERAStar plate-reader (BMG Labtech) was used for the experiments described.

3. Methods

3.1. flERE Preparation

1. A_{260} values were measured to calculate the oligonucleotide concentration.

2. The fluorescein-labeled sense strand was annealed with an equimolar concentration of the unlabeled antisense strand. Add the fluorescein-labeled sense strand to an equimolar concentration of the unlabeled antisense strand in annealing buffer

to a final concentration of 1 μM each in a tube. Wrap the tube with aluminum foil.

3. Boil 400 ml of water in a large glass beaker on a hotplate and incubate the tube of oligonucleotides in the boiling water for 5 min.
4. Turn off the hotplate, leaving the oligonucleotides in the beaker on the hotplate to slowly cool to room temperature.
5. Aliquot the 1 μM flERE stock, keep it in the dark and store at −20°C. Avoid multiple freeze-thaw cycles. The flERE can be stored at 4°C for a few weeks.

3.2. ER-flcERE Binding Curves

1. Dilute the fluorescein-labeled double stranded cERE (flcERE) to 20 nM with TE buffer. The final volume of each assay in a 96-well microplate well will be 20 μl, containing 10 μl FAMA buffer, 7 μl H_2O, 1 μl flcERE and 2 μl ER. Combine H_2O, FAMA buffer, and 20 nM flcERE in proportion. The final concentration of flcERE in a 20 μl assay is 1 nM.
2. Dilute ER protein with ER dilution buffer to 10× the desired final concentrations. In the binding curves we show (Fig. 2); 200, 150, 100, 75, 50, and 25 nM ER stocks were used to achieve final ER concentrations of 20, 15, 10, 7.5, 5, and 2.5 nM. The tubes and dilution buffer should be prechilled and the diluted ER should remain on ice.
3. Add 18 μl of the complete assay mix (minus ER) prepared in step 1 to each well in a microplate.
4. Then add 2 μl of ER generated in step 2 into each well prepared in step 3. Mix by pipetting (see Note 1). In the control (ER concentration is 0), add 2 μl ER dilution buffer instead. 3 wells are assayed at each concentration.
5. Cover the microplate and incubate in the dark at room temperature for 15 min. Take off the cover and measure anisotropy with a plate-reader (see Note 2). The excitation and emission wavelengths are 485 nM and 535 nM respectively. The signal will not change significantly for at least an hour when the plate is covered and stored in the dark at room temperature.
6. Anisotropy changes resulting from ER binding are calculated by subtracting the average anisotropy values measured for flcERE alone, without added ER, from the value measured for each flcERE-ER reaction. Plot anisotropy changes vs. ER concentration and fit binding curves through the data points (Fig. 2). The K_d values of ERα and ERβ binding to flcERE are ~5 nM and ~3 nM, respectively. A concentration of ER (15 nM) that approaches saturated binding was used in the SRC1a-ER binding assays.

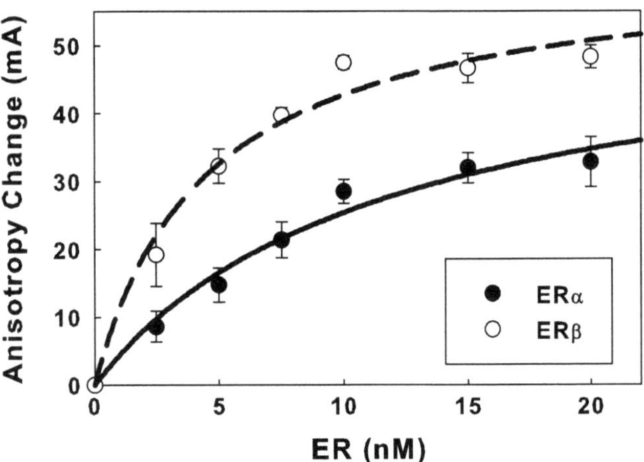

Fig. 2. Binding curves of ERα and ERβ to the flcERE.

3.3. SRC1a Binding to the ER-flcERE

1. Dilute the flcERE to 20 nM with TE buffer. Dilute the ER to 150 nM with ER dilution buffer (see Note 3). The final volume of each assay in a microplate well will be 20 μl, containing 10 μl FAMA buffer, 5 μl H_2O, 1 μl flcERE, 2 μl ER, and 2 μl SRC1a. Combine H_2O, FAMA buffer, 20 nM flcERE and 150 nM ER in proportion. The final concentrations of flcERE and ER in a 20 μl assay are 1 nM and 15 nM respectively. Add 18 μl of the combined solution to each well in a microplate. Cover the microplate and incubate in the dark at room temperature for 15–30 min to enable ER-flcERE complex to form. Measure anisotropy with the plate-reader to ensure the anisotropy values in all wells are similar. If any well appears very different than others, do not use that well in step 3.
2. While incubating the plate, dilute SRC1a protein with SRC1 dilution buffer to tenfold higher than the desired final concentrations. In the binding curves we show (Fig. 3); 800, 600, 400, 200, 100, 50 and 25 nM SRC1a stocks were used to achieve final concentrations 80, 60, 40, 20, 10, 5 and 2.5 nM. The tubes and dilution buffer should be prechilled and the diluted SRC1a should remain on ice.
3. Add 2 μl SRC1a prepared in step 2 into each well prepared in step 1. Mix by pipetting (see Note 1). In the control, there is no SRC1a and 2 μl SRC1 dilution buffer is added instead. 3 wells are assayed at each concentration.
4. Cover the microplate and incubate in the dark at room temperature for 10–20 min. Take off the cover and measure the anisotropy with a plate-reader (see Note 2).
5. Anisotropy changes resulting from SRC1a binding are calculated by subtracting the average anisotropy values measured for

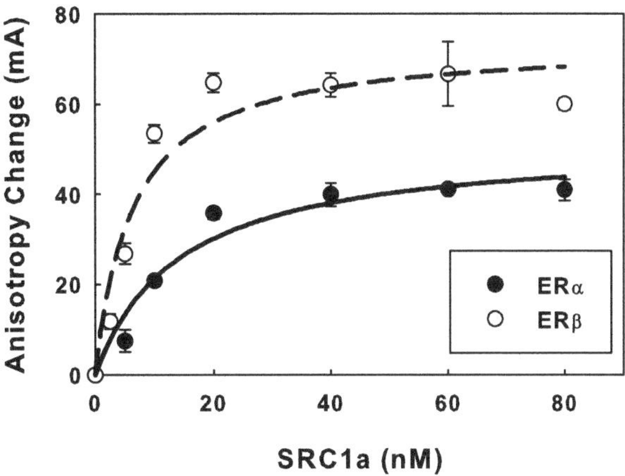

Fig. 3. Binding of full-length SRC1a to ERα-flcERE and ERβ-flcERE complexes.

ER-flcERE complex, with no added SRC1a, from the value measured for the SRC1a-ER-flcERE reactions. Plot anisotropy changes vs. SRC1a concentration and fit binding curves through the data points (Fig. 3). Robust binding curves can be obtained with FAMA. The apparent K_d values for binding of full-length SRC1a to the ERα-flcERE and ERβ-flcERE complexes were ~14 nM and ~6 nM respectively, indicating that full-length SRC1a interacts with ER-flcERE complex with very high affinity. An SRC1a concentration that results in 70–90% maximum binding was used in competition assays and kinetic assays.

3.4. Competition Assays

1. Dilute the flcERE to 20 nM with TE buffer. Dilute the ER with ER dilution buffer to 150 nM (see Note 3). The final volume of each assay in a microplate well will be 20 μl, containing 10 μl FAMA buffer, 3 μl H_2O, 1 μl flcERE, 2 μl ER, 2 μl peptide and 2 μl SRC1a. Combine H_2O, FAMA buffer, 20 nM flcERE and 150 nM ER in proportion. The final concentrations of flcERE and ER in a 20 μl assay are 1 nM and 15 nM, respectively. Add 16 μl of the combined solution to each well in a microplate. Cover the microplate and incubate in the dark at room temperature for 15–20 min to enable the ER-flcERE complex to form.
2. We used NR boxes 2 and 4 in the binding studies because previous studies suggested binding to ER is mediated primarily through these peptides (22). While incubating the plate, dilute the NR box peptide with H_2O to tenfold higher than the desired final concentrations. Add 2 μl of the peptide into each wells generated in step 1. Mix by pipetting (see Note 1). 3 wells are assayed at each concentration. Cover the microplate and incubate in the dark at room temperature for 15–20 min

to the enable the peptide-ER-flcERE complex to form. Take off the cover and measure anisotropy with a plate-reader (see Note 2).

3. While performing the incubation in step 2, dilute SRC1a with SRC1 dilution buffer to 250 nM (see Note 3). Add 2 μl SRC1a into each well prepared in step 2 so the final assay concentration is 25 nM. Mix by pipetting (see Note 1). Cover the microplate and incubate in the dark at room temperature for 15–20 min. Take off the cover and measure anisotropy with a plate-reader (see Note 2). Anisotropy can be measured several times at ~5 min intervals to ensure equilibrium is reached. Between each measurement, the plate should be covered and kept in the dark.
4. Anisotropy changes are calculated by subtracting the anisotropy values measured before SRC1a addition from the value measured after SRC1a addition. The average anisotropy changes for binding of SRC1a to the ERα-flcERE complex in the absence of added peptide is the maximum anisotropy change and defined as 100% bound. % SRC1a bound = (Anisotropy Change/Max Anisotropy Change) × 100. Plot % SRC1a bound vs. peptide concentration and fit binding curves through the data points (Fig. 4). The concentrations of peptide required to inhibit SRC1a binding to ERα-flcERE by 50% (the IC_{50}) were ~1.0 and ~3.8 μM for the SRC1a NR box 2 and NR box 4 peptides respectively. Since it requires μM NR box peptides to compete with 25 nM full-length SRC1a for binding to ER-flcERE, the LxxLL containing peptides exhibit much lower affinity than the full-length SRC1a.

Small molecules can be used instead of peptides. Competition FAMA can be used to study the ability of a small molecule to inhibit binding of SRC1a to ER.

3.5. Binding Kinetics of SRC1a

1. Dilute the flcERE to 20 nM with TE buffer. Dilute the ER with ER dilution buffer to 150 nM (see Note 3). The final volume of each assay in a microplate well will be 20 μl, containing 10 μl FAMA buffer, 5 μl H_2O, 1 μl flcERE, 2 μl ER, and 2 μl SRC1a. Combine H_2O, FAMA buffer, 20 nM flcERE and 150 nM ER in proportion. The final concentrations of flcERE and ER in a 20 μl assay are 1 nM and 15 nM respectively. Incubate the combined solution in the dark at room temperature for 15–30 min to allow formation of the ER-flcERE complex.
2. While incubating the plate, dilute SRC1a with SRC1 dilution buffer to 150 nM (see Note 3). The tubes and dilution buffer should be prechilled and the diluted SRC1a should remain on ice.

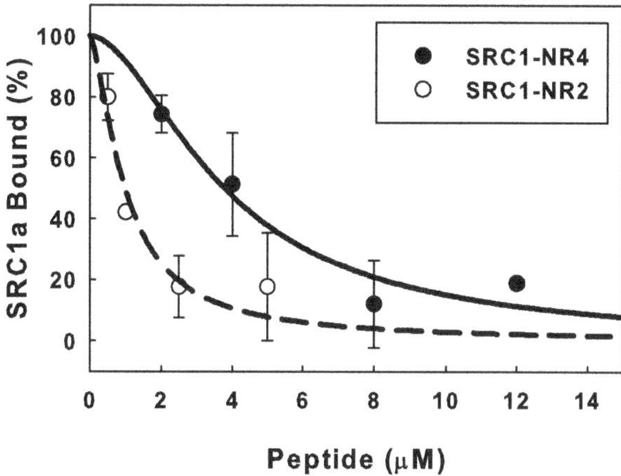

Fig. 4. NR box peptides compete for full-length SRC1a interaction with ERα-flcERE.

3. Add 18 μl in one well of the microplate, measure anisotropy of this well with the plate-reader. With the plate is still in the drawer of the plate-reader, add 2 μl SRC1a and quickly mix by pipetting up-and-down once (see Note 1). Using a kinetic mode, read the anisotropy immediately every 2 s over 5 min (see Note 4).
4. Repeat step 3 for each well and measure for at least four assays. Increase the number of assays needed depending on the quality of the experimental data.
5. Anisotropy changes are calculated by subtracting the average anisotropy values measured at time 0 (before SRC1a addition) from the value measured at time (t). The anisotropy change when binding reached a plateau at equilibrium is the maximum anisotropy change and is set equal to 100% bound. %SRC1a bound = (Anisotropy Change (time = t)/Max Anisotropy Change (time = 0)) × 100. Plot % SRC1a bound vs. time (Fig. 5). In this figure ERα is used. Binding of SRC1a to ERα-flcERE complex is extremely rapid with half time of 15–20 s. The binding reaches equilibrium with 5 min.

Similar experiments can be used to measure dissociation of SRC1a from ER-flcERE complex.

3.6. Bridging Effect

This experiments tests the ability of SRC1a to bridge, or link, two ER dimers bound at a complex DNA binding site. Linking the two DNA-bound ER dimers will increase the stability of the ER-DNA complex.

1. Perform an experiment to measure binding of ERα to the fl-PI-9 ERU as described in Subheading 3.2 for the cERE. The suggested ERα final concentrations are 400, 200, 150, 100,

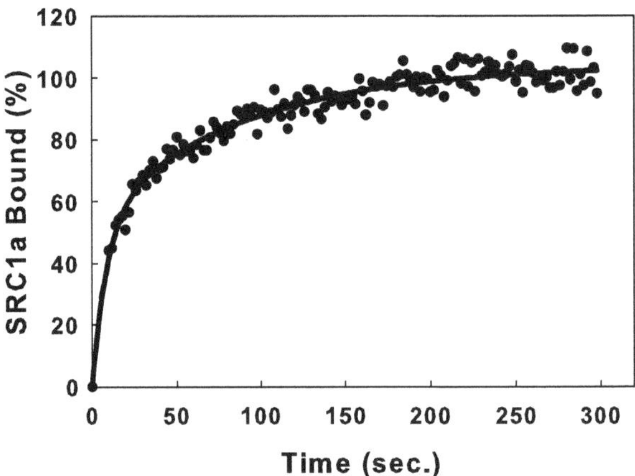

Fig. 5. SRC1a binds rapidly to ERα-flcERE complex.

50, and 25 nM since ER has much lower affinity for the PI-9 ERU than for the cERE. The K_d for binding of ERα to the fl-PI-9 ERU is ~166 nM (data not shown). The maximum anisotropy change for ERα binding to the fl-PI-9 ERU is ~90 mA, which is much higher than is seen for ERα binding to the flcERE. This is reasonable since the PI-9 ERU contains two ER binding sites, and the final complex formed includes two ERα dimers (9). Binding of 50 nM ERα to the fl-PI-9 ERU results in an anisotropy change of only ~10 mA, while binding of 2.5 nM ERα to the flcERE gives similar change as shown in Fig. 2. To test the bridging effects of SRC1a on ERα binding to the PI-9 ERU, compared to the cERE, 50 and 2.5 nM ERα are used for the fl-PI-9 ERU and flcERE respectively.

2. Dilute the flcERE and the fl-PI-9 ERU to 20 nM with TE buffer. The final volume of each assay in a microplate well will be 20 μl, containing 10 μl FAMA buffer, 5 μl H_2O, 1 μl flERE, 2 μl ER, and 2 μl SRC1a. Combine H_2O, FAMA buffer, 20 nM flERE in proportion. Use separate tubes for flcERE and fl-PI-9 ERU. Add 16 μl combined solution to each well in a microplate. 9 wells for flcERE and 9 wells for fl-PI-9 ERU will be used.
3. Dilute the ER with ER dilution buffer to 25 and 500 nM (see Note 3). Add 2 μl ER dilution buffer to wells 1–3 for both flEREs. Then add 2 μl 25 nM ER to wells 4–9 containing flcERE and 2 μl 500 nM ER to wells 4–9 containing fl-PI-9 ERU. Mix by pipetting (see Note 1). Cover the microplate and incubate in dark at room temperature for 30 min to enable formation of the ER-flERE complex.
4. While incubating the plate, dilute SRC1a protein with SRC1 dilution buffer to 400 nM. The tubes and dilution buffer should be prechilled and the diluted SRC1a should remain on ice.
5. Add 2 μl SRC1a dilution buffer to wells 1–6 for both flEREs. Then add 2 μl 400 nM SRC1a to wells 7–9 for both flEREs. Mix

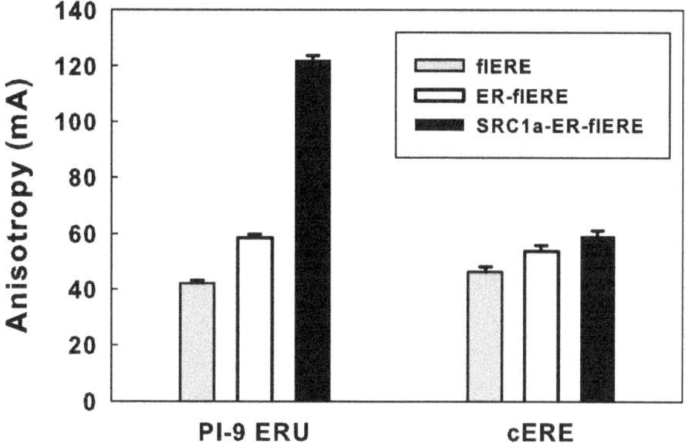

Fig. 6. Full-length SRC1a increases ERα binding to the PI-9 ERU but not to the cERE.

by pipetting (see Note 1). Cover the microplate and incubate in the dark at room temperature for 10–20 min. Take off the cover and measure anisotropy with a plate-reader (see Note 2).

6. For flcERE, the anisotropy units of SRC1a-ERα-flERE, ERα-flERE, and flERE alone are 46, 53, and 59 mA, respectively (Fig. 6). For fl-PI-9 ERU, the anisotropy units of SRC1a-ERα-flERE, ERα-flERE, and flERE alone are 42, 58, and 122 mA, respectively (Fig. 6). The very small anisotropy increase resulting from ERα binding (7 mA for flcERE and 16 mA for fl-PI-9 ERU) indicates that at a low concentration of ERα, only a very small fraction of the ER is bound to the flERE. When SRC1a is added, the minimal anisotropy increases (6 mA) with flcERE indicates the amount of ERα bound to the flcERE is still limited. The addition of SRC1a dramatically increases the anisotropy (by 64 mA) for the complex with the fl-PI-9 ERU. The results imply that SRC1a binding to the ERα-PI-9 ERU complex dramatically increased the affinity of the two ER dimers for the two binding sites of the PI-9 ERU. This likely involves bridging the two ER dimers with different LxxLL motifs so the larger complex helps to stabilize ER dimer on the low affinity binding sites.

4. Notes

1. Avoid generating bubbles when mixing the samples in the microplate. Bubbles can alter the readings.
2. Many fluorescence polarization/anisotropy plate readers lack the sensitivity required to accurately measure polarization in small 10–20 μl volumes. If the plate-reader does not directly show anisotropy as a readout, the anisotropy can be calculated

from the raw data using the formula: $FA = (I_{//} - I_{\perp})/(I_{//} + 2I_{\perp})$, where $I_{//}$ and $I_{\perp}$ are the emission light intensity (I) in the parallel ($_{//}$) and perpendicular ($_{\perp}$) directions to the excitation light, respectively. Although outcomes were independent of the instrument, gain settings vary widely between instruments; therefore the magnitude of the anisotropy change can vary significantly in experiments performed using different plate readers.

3. The activities of different batches of purified protein may vary. It is best to determine the actual concentration needed by generating a binding curve.
4. The time lag between SRC1a addition and the reading of the first data point is likely to be 5–15 s, and depends on the speed with which the microplate reader can move a microplate into place and begin reading it. A timer can be used to count. The time intervals between each reading may be limited by the plate-reader and can be more than 2 s. In that case just choose the fastest mode available.

References

1. Luconi M, Forti G, Baldi E (2002) Genomic and nongenomic effects of estrogens: molecular mechanisms of action and clinical implications for male reproduction. J Steroid Biochem Mol Biol 80:369–381
2. Katzenellenbogen BS, Katzenellenbogen JA (2000) Estrogen receptor transcription and transactivation: estrogen receptor alpha and estrogen receptor beta: regulation by selective estrogen receptor modulators and importance in breast cancer. Breast Cancer Res 2:335–344
3. Tsai MJ, O'Malley BW (1994) Molecular mechanisms of action of steroid/thyroid receptor superfamily members. Annu Rev Biochem 63:451–486
4. Evans RM (1988) The steroid and thyroid hormone receptor superfamily. Science 240: 889–895
5. Mosselman S, Polman J, Dijkema R (1996) ER beta: identification and characterization of a novel human estrogen receptor. FEBS Lett 392:49–53
6. Carroll JS, Brown M (2006) Estrogen receptor target gene: an evolving concept. Mol Endocrinol 20:1707–1714
7. Carroll JS, Liu XS, Brodsky AS, Li W, Meyer CA, Szary AJ, Eeckhoute J, Shao W, Hestermann EV, Geistlinger TR, Fox EA, Silver PA, Brown M (2005) Chromosome-wide mapping of estrogen receptor binding reveals long-range regulation requiring the forkhead protein FoxA1. Cell 122:33–43
8. O'Lone R, Frith MC, Karlsson EK, Hansen U (2004) Genomic targets of nuclear estrogen receptors. Mol Endocrinol 18:1859–1875
9. Klinge CM, Jernigan SC, Smith SL, Tyulmenkov VV, Kulakosky PC (2001) Estrogen response element sequence impacts the conformation and transcriptional activity of estrogen receptor alpha. Mol Cell Endocrinol 174:151–166
10. Krieg SA, Krieg AJ, Shapiro DJ (2001) A unique downstream estrogen responsive unit mediates estrogen induction of proteinase inhibitor-9, a cellular inhibitor of IL-1beta-converting enzyme (caspase 1). Mol Endocrinol 15:1971–1982
11. Glass CK, Rosenfeld MG (2000) The coregulator exchange in transcriptional functions of nuclear receptors. Genes Dev 14:121–141
12. Xu J, O'Malley BW (2002) Molecular mechanisms and cellular biology of the steroid receptor coactivator (SRC) family in steroid receptor function. Rev Endocr Metab Disord 3:185 192
13. Spiegelman BM, Heinrich R (2004) Biological control through regulated transcriptional coactivators. Cell 119:157–167
14. Kalkhoven E, Valentine JE, Heery DM, Parker MG (1998) Isoforms of steroid receptor coactivator 1 differ in their ability to potentiate transcription by the oestrogen receptor. EMBO J 17:232–243

15. Wang SY, Ahn BS, Harris R, Nordeen SK, Shapiro DJ (2004) Fluorescence anisotropy microplate assay for analysis of steroid receptor-DNA interactions. Biotechniques 37:807-8, 810-7
16. Wang S, Zhang C, Nordeen SK, Shapiro DJ (2007) In vitro fluorescence anisotropy analysis of the interaction of full-length SRC1a with estrogen receptors alpha and beta supports an active displacement model for coregulator utilization. J Biol Chem 282:2765–2775
17. Mao C, Patterson NM, Cherian MT, Aninye IO, Zhang C, Montoya JB, Cheng J, Putt KS, Hergenrother PJ, Wilson EM, Nardulli AM, Nordeen SK, Shapiro DJ (2008) A new small molecule inhibitor of estrogen receptor alpha binding to estrogen response elements blocks estrogen-dependent growth of cancer cells. J Biol Chem 283:12819–12830
18. Kretzer NM, Cherian MT, Mao CJ, Aninye IO, Reynolds PD, Schiff R, Hergenrother PJ, Nordeen SK, Wilson EM, Shapiro DJ (2010) A noncompetitive small molecule inhibitor of estrogen-regulated gene expression and breast cancer cell growth that enhances proteasome-dependent degradation of estrogen receptor alpha. J Biol Chem 285:41863–41873
19. Shapiro DJ, Mao CJ, Cherian MT (2011) Small molecule inhibitors as probes for estrogen and androgen receptor action. J Biol Chem 286:4043–4048
20. Zhang CC, Krieg S, Shapiro DJ (1999) HMG-1 stimulates estrogen response element binding by estrogen receptor from stably transfected HeLa cells. Mol Endocrinol 13:632–643
21. Thackray VG, Nordeen SK (2002) High-yield purification of functional, full-length steroid receptor coactivator 1 expressed in insect cells. Biotechniques 32(260):262–263
22. Bramlett KS, Wu Y, Burris TP (2001) Ligands specify coactivator nuclear receptor (NR) box affinity for estrogen receptor subtypes. Mol Endocrinol 15:909–922

Chapter 28

Use of Histone Deacetylase Inhibitors to Examine the Roles of Bromodomain and Histone Acetylation in p300-Dependent Gene Expression

Jihong Chen and Qiao Li

Abstract

The bromodomain is an evolutionarily conserved motif harbored by many transcription regulators and nearly all nuclear histone acetyltransferases including the transcriptional coactivator p300. The function of p300 is required for the expression of an array of genes, in part through histone acetylation. Here, we describe an experimental approach to examine the role of either the wild-type or a bromo-deficient p300 in the expression of p300-dependant genes. The role of histone acetylation in the expression of p300-dependent genes can also be assessed by targeting histone deacetylase activities using an inhibitor approach.

Key words: Gene regulation, Histone acetylation, Acetyltransferase, Transcriptional coactivator, Bromodomain, Histone deacetylase inhibitor

1. Introduction

The histone acetyltransferase (HAT) p300 is a master regulator of gene transcription and contains multiple functional domains for histone acetylation and the integration of different regulators at the regulatory region of chromatin (1, 2). One of these domains is the bromodomain which is evolutionarily conserved and harbored by many transcription regulators and nearly all nuclear HAT proteins (3). HeLa cervical carcinoma cells expressing the wild-type p300, are well characterized with respect to p300 function and gene regulation (4–6). On the other hand, SiHa cervical carcinoma cells have a homozygous internal deletion of exons 15–18 in the p300 gene, resulting in a p300 mutant protein deficient in the bromodomain, as well as in the HAT activity (7). Nevertheless,

Minou Bina (ed.), *Gene Regulation: Methods and Protocols*, Methods in Molecular Biology, vol. 977,
DOI 10.1007/978-1-62703-284-1_28,

the bromo-deficient p300 protein resides mainly in the nucleus (8).

In a cellular environment, the acetylation of histones is dynamically reversible. Histone deacetylases (HDAC) are responsible for enzymatic hydrolysis of the acetylated lysine acetamido groups. Inhibitors of HDACs, such as valproic acid, butyrate, and TSA, are powerful tools to study the mechanisms of histone acetylation and gene transcription. Treatment of cells with these inhibitors leads to an accumulation of acetylated histones and alteration in gene transcription (8). Here we describe an experimental approach for the use of HDAC inhibitors and taking the advantage of the different types of p300 in the HeLa and SiHa cells to study the interplay of bromodomain and histone acetylation in p300-dependent gene expression.

2. Materials

1. HeLa and SiHa cell lines (ATCC).
2. Dulbecco's modified Eagle medium (DMEM), store at 4°C.
3. Fetal bovine serum (FBS), store at –70°C.
4. Penicillin/Streptomycin 1,000× (10^5 unit penicillin/100 μg streptomycin per ml), sterilize with a 0.2 μm syringe filter and store at –20°C.
5. Trypsin 10×, store at –20°C.
6. Phosphate-buffered saline (PBS) pH 7.4, 137 mM NaCl, 2.7 mM KCl, 10 mM Na_2HPO_4, 2 mM KH_2PO_4 (PBS may be supplemented with 1 mM $CaCl_2$ and 0.5 mM $MgCl_2$ for divalent cations), autoclave and store at room temperature.
7. Tissue culture dishes.
8. Valproic acid, dissolve in distilled H_2O to make 1 M solution, sterilize with a 0.2 μm syringe filter, aliquot and store at 4°C.

3. Methods

3.1. Cell Culture

All cell culture procedures should be carried out in a tissue culture hood.

1. Aspirate medium from T75 tissue culture flask (see Note 1).
2. Wash the cells with 10 ml of PBS (do not add PBS directly onto the cells).
3. Add 1 ml of 1× Trypsin into the flask and incubate at 37°C for 1–3 min (see Note 2).

4. Dilute the trypsinized cells with 9 ml of DMEM supplemented with 10% FBS, 1× Penicillin/Streptomycin, and 1% of nonessential amino acids (see Notes 3 and 4).
5. Seed the cells directly in tissue culture dishes to be at low confluence (see Notes 5 and 6).
6. Culture the cells at 37°C in humidified air containing 5% CO_2 (cells should be seeded at least 16 h prior to treatments).
7. Change the medium and treat the cells with 2–5 mM of valproic acid for 4–8 h (see Note 7).

3.2. Harvesting Cells

1. Aspirate medium and wash the cells with PBS.
2. Add 1 ml of PBS and harvest the cells with a rubber policeman.
3. Transfer the cells into 1.5 ml micro tubes.
4. Centrifuge the cells at 2,000 × *g*, at 4°C for 5 min.
5. Aspirate the PBS (the cell pellets can be store at −70°C for multiple weeks).

3.3. Western Blot Analysis

1. The cell pellets can be used to prepare whole cell extracts or nuclear extracts for Western blot analysis (see Note 8).
2. The percentage of SDS gel required for the Western blot analysis is dependent on the molecular weight of the protein of interest.

3.4. RT-PCR Analysis

1. The cell pellets can be used to prepare total RNA for RT-PCR analysis.
2. Alternatively, 12-well or 6-well tissue culture plate can be used to grow cells. Following treatments, the cells can be lysed directly in the wells to isolate total RNA for RT-PCR analysis.

4. Notes

1. For tissue culture dishes, T75 flask takes 10 ml medium, usually used for cell maintenance, whereas different sizes of dishes are available to grow cells for Western blot or RT-PCR analysis.
2. Incubation time for trypsin depends on the cell type. It takes more time for SiHa cells to come off the flask. It is essential to inactivate trypsin after cell lifting. FBS is rich in trypsin inhibitors.
3. Different types of medium are available for various cell lines, but most cells grow well in DMEM.

Table 1
Characterization of the wild-type and bromo-deficient p300 in HeLa and SiHa cells. Quantitative Western blot analysis of bromo-deficient p300 protein, RT-PCR analysis of the relative transcript level of bromo-deficient p300, and pulse labeling analysis of the relative protein synthesis of bromo-deficient p300 in the SiHa cells are presented as the fold changes in relation to the wild-type p300 in the HeLa cells. Apparent half-life (hours) of the wild-type or bromo-deficient p300 protein was determined by pulse-chase procedures

Fold changes	p300
Western (SiHa/HeLa)	4.86 ± 2.68
RT-PCR (SiHa/HeLa)	12.46 ± 2.41
Pulse (SiHa/HeLa)	9.17 ± 4.41
Half-life (SiHa)	9.51 ± 1.13
Half-life (HeLa)	11.30 ± 1.56

4. Cells often benefit from the addition of 1% nonessential amino acids in the medium.
5. For cell maintenance, the frequency of splitting and confluence of seeding depend on the cell types. For HeLa cells, the splitting should be done every other day with seeding of 15% confluence. For SiHa cells, the splitting can be done twice a week with seeding of 25% confluence. SiHa cells grow poorly if seeded at too low confluence.
6. If the cells were to be seeded at high confluence, dilute trypsinized cells with DMEM containing FBS and centrifuge at 2,000 × g, at 4°C for 5 min. Then aspirate the supernatant and resuspend the cell pellet in medium for seeding.
7. It is important to grow and treat the two cell lines in parallel with the same batch of medium and reagents in the same environment for the gene expression pattern to be compared between these two cell lines. Care should be taken to avoid cell line cross-contamination.
8. It is imperative to carefully characterize the properties of transcriptional coactivator of interest in these two cell lines, p300 in this case (Table 1), to ensure the correct interpretation of the roles of bromodomain and histone acetylation in p300-dependent gene expression.

Acknowledgments

This work was sponsored by operating grant from the Canadian Institutes of Health Research.

References

1. Ogryzko VV, Schiltz RL, Russanova V, Howard BH, Nakatani Y (1996) The transcriptional coactivators p300 and CBP are histone acetyltransferases. Cell 87:953–959
2. Chen J, Li Q (2011) Life and death of transcriptional co-activator p300. Epigenetics 6:957–961
3. Sanchez R, Zhou MM (2009) The role of human bromodomains in chromatin biology and gene transcription. Curr Opin Drug Discov Devel 12:659–665
4. Chen J, Halappanavar SS, St-Germain JR, Tsang BK, Li Q (2004) Role of Akt/protein kinase B in the activity of transcriptional coactivator p300. Cell Mol Life Sci 61:1675–1683
5. Chen J, St-Germain JR, Li Q (2005) B56 regulatory subunit of protein phosphatase 2A mediates valproic acid-induced p300 degradation. Mol Cell Biol 25:525–532
6. Chen J, Halappanavar S, Th' ng JP, Li Q (2007) Ubiquitin-dependent distribution of the transcriptional coactivator p300 in cytoplasmic inclusion bodies. Epigenetics 2:92–99
7. Suganuma T, Kawabata M, Ohshima T, Ikeda MA (2002) Growth suppression of human carcinoma cells by reintroduction of the p300 coactivator. Proc Natl Acad Sci USA 99:13073–13078
8. Chen J, Ghazawi FM, Li Q (2010) Interplay of bromodomain and histone acetylation in the regulation of p300-dependent genes. Epigenetics 5:509–515

Chapter 29

Histone Deacetylase Inhibitor Valproic Acid as a Small Molecule Inducer to Direct the Differentiation of Pluripotent Stem Cells

Jihong Chen, Natascha Lacroix, and Qiao Li

Abstract

Pluripotent stem cells can be directed into myogenic differentiation by small molecular inducers, which preferentially activate muscle-specific transcription networks. Here we describe how to efficiently direct the differentiation of pluripotent P19 cells into skeletal muscle lineage by using histone deacetylase inhibitor, valproic acid and the ligand of retinoic acid receptor, all-*trans* retinoic acid.

Key words: Valproic acid, Stem cell differentiation, Embryoid body, Gene regulation, Histone acetylation, Transcription factor, Histone deacetylase inhibitor

1. Introduction

Pluripotent P19 embryonal carcinoma cell line is an excellent system to study the mechanisms of cellular differentiation and to identify small molecule inducers for cell fate determinations (1–3). Similar to embryonic stem cells, the P19 cells respond well to cellular cues differentiating into cell types of all three germ layers (4–6). In tissue cultures, the P19 cells can be induced into differentiation with an aggregation protocol, which involves the formation of cell aggregates, called embryoid bodies (EBs, Fig. 1). Aggregation of the P19 cells in the absence of exogenous stimuli results in mesoderm differentiation, but not myogenic specification (7). Treatment of the cells with dimethyl sulfoxide (DMSO) during EB formation generates small percentages of skeletal myocytes, whereas co-treatment with ligand of nuclear receptors, such as all-*trans* retinoic acid (RA), or bexarotene synergistically enhances myogenic differentiation (1, 3). The differentiation of P19 cells into myocytes also depends on the choice of serum (8).

Minou Bina (ed.), *Gene Regulation: Methods and Protocols*, Methods in Molecular Biology, vol. 977,
DOI 10.1007/978-1-62703-284-1_29, © Springer Science+Business Media, LLC 2013

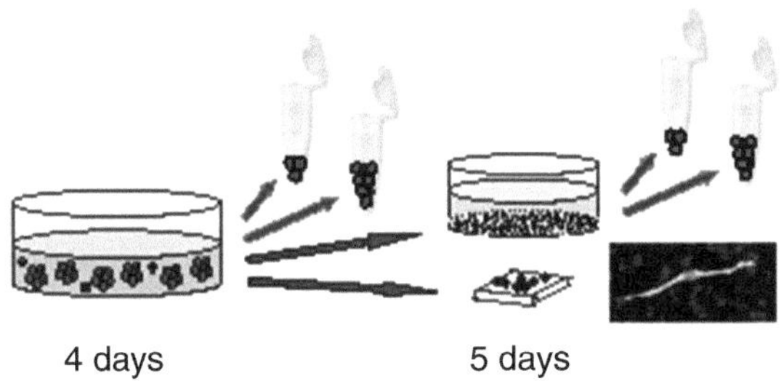

Fig. 1. Use of small molecule inducers to direct the differentiation of P19 pluripotent stem cells into skeletal myocytes. The P19 cells were first cultivated into EBs in petri dishes in the presence of valproic acid and RA, then transferred into tissue culture dishes and maintained without treatments to develop skeletal myocytes. Specific gene expression and the efficacies of myogenic differentiation can be determined by harvesting the cells at different stages for qRT-PCR and Western blot analysis, and for immunofluorescence microscopy.

Histone acetylation can be increased by inhibition of histone deacetylase (HDAC) activities, which are interesting avenues for activating gene expression and regulating stem cell fate determinations. Valproic acid has been used for many years to treat epilepsy and bipolar disorders (9, 10). It is now known as a HDAC inhibitor and has been used to induce pluripotent stem cells from primary human fibroblasts (11, 12). Here, we describe a protocol which uses valproic acid, instead of DMSO, in concert with RA, to induce the commitment of skeletal muscle lineage, underpinning the significance of histone acetylation in muscle-specific gene activation.

2. Materials

1. P19 cell line (ATCC).
2. α-Minimum Essential Medium (α-MEM), store at 4°C.
3. Fetal bovine serum (FBS) and bovine calf serum (BCS), store at −70°C.
4. Penicillin/Streptomycin 1,000× (10^5 unit penicillin/100 μg streptomycin per ml), sterilize with a 0.2 μm syringe filter and store at −20°C.
5. Trypsin 10×, store at −20°C.
6. Phosphate-buffered saline (PBS) pH 7.4: 137 mM NaCl, 2.7 mM KCl, 10 mM Na_2HPO_4, 2 mM KH_2PO_4 (PBS may be supplemented with 1 mM $CaCl_2$ and 0.5 mM $MgCl_2$ for divalent cations), autoclave and store at room temperature.
7. Petri dishes and tissue culture dishes.

8. Valproic acid, dissolve in distilled H_2O to make 1 M solution, sterilize with a 0.2 μm syringe filter, aliquot, and store at 4°C.
9. RA, dissolve in ethanol, aliquot into light-safe tubes, and store at –20°C.

3. Methods

3.1. Cell Culture

All cell culture procedures should be carried out in a tissue culture hood.

1. Aspirate medium from T75 tissue culture flask by positioning Pasteur pipette in the bottom corner of tilted flask, neck upwards (see Note 1).
2. Wash the cells with 10 ml of PBS (adding PBS onto the side walls).
3. Add 1 ml of 1× Trypsin into the flask and incubate at 37°C for 2 min.
4. Dilute the trypsinized cells with 9 ml of α-MEM supplemented with 5% FBS, 5% BCS and 1× Penicillin/Streptomycin (see Note 2).
5. Seed the cells in a new T75 flask containing 9 ml of medium (1:10 split).
6. Culture the cells at 37°C in humidified air containing 5% CO_2.

3.2. Cell Differentiation

Day 0

1. Seed 1/10 of cells from 80% confluent T75 flask (~1 million cells) into each 150 mm petri dish containing 19 ml of fresh medium (see Note 3).
2. Culture the cells at 37°C in humidified air containing 5% CO_2.

Day 1

1. Check petri dishes for EB formation, tapping the dishes lightly to lift the EBs prior to treatment.
2. Add valproic acid and RA to treat the cells (see Note 4).
3. Culture the cells at 37°C in humidified air containing 5% CO_2.

Day 2

1. Tilt the backside of the petri dish to keep the EBs at the surface of the medium. Discard 14 ml of the medium (keep the EBs in the dish).
2. Transfer the EBs with remaining medium into a new 150 mm petri dish containing 20 ml of fresh medium. Add valproic acid and RA for the 20 ml of fresh medium to maintain the treatment condition.

Day 4

1. Tilt the backside of the petri dish to keep the EBs at the surface of the medium. Discard 14 ml of medium (keep the EBs in the dish).
2. Seed 100 μl of EB directly on a coverslip coated with 0.1% gelatin placed in 6-well plate containing 3 ml of medium (see Note 5).
3. Seed 1 ml of EB in a 100 mm tissue culture dish containing 10 ml of fresh medium.
4. Culture the cells at 37°C in humidified air containing 5% CO_2.
5. Transfer the remaining EB in a 15 ml tube and centrifuge at 2,000 × *g* for 5 min. Aspirate supernatant, and resuspend the cell pellet in 1 ml of PBS. Transfer the cell suspension into 1.5 ml micro tube (depending on experimental design, the cells can be split into different tubes for qRT-PCR and Western blot analysis to examine myogenic expression-induced by valproic acid and RA at the early stage of differentiation).
6. Centrifuge the cells at 2,000 × *g*, at 4°C for 5 min. Aspirate the PBS (the cell pellets can be store at −70°C for multiple weeks).

3.3. Analysis of Myogenic Expression

Day 9

1. Fix and stain the cells on coverslips with specific antibodies for markers of skeletal myocytes to perform immunofluorescence microscopic analysis and to determine the efficacies of myogenic differentiation (see Note 6).
2. Harvest the cells in tissue culture dishes for qRT-PCR or Western blot analysis to examine myogenic expression induced by valproic acid and RA at the late stage of differentiation (see Note 7).

4. Notes

1. P19 cells should be split every other day when confluence approaches 80%.
2. Stem cell differentiation is affected by unknown factors in FBS, thus serum testing should be carried out to determine if a particular batch is suitable for differentiation studies.
3. P19 cells grow as a monolayer in tissue culture dishes, but as a suspension in regular petri dishes, which is required for EB formation and differentiation studies.
4. Pilot testing should be carried out to determine the optimal concentrations of a particular batch of valproic acid (0.1–1 mM) and RA (0.2–20 nM) for myogenic differentiation. Treatment

of the cells after 24 h of EB formation (day 1–4) increases the efficiency of myogenic differentiation by about twofold compared to an immediate treatment (day 0–4).

5. The volume of cells required for seeding on the coverslips, or in the tissue culture dishes is dependent on the efficiency of EB formation.
6. Cells can be stained simultaneously with myosin heavy chain antibody (mouse) and MyoD antibody (rabbit), followed with corresponding secondary antibodies for different filters, as well as with Hoechst for DNA (3).
7. RT-PCR analysis of Pax3 and MyoD transcripts, and Western blot analysis of Myogenin protein are well established for examining myogenic differentiation at different stages (3).

Acknowledgments

This work was sponsored by operating grant from the Natural Sciences and Engineering Research Council of Canada.

References

1. Edwards MK, McBurney MW (1983) The concentration of retinoic acid determines the differentiated cell types formed by a teratocarcinoma cell line. Dev Biol 98:187–191
2. Rudnicki MA, Reuhl KR, McBurney MW (1989) Cell lines with developmental potential restricted to mesodermal lineages isolated from differentiating cultures of pluripotential P19 embryonal carcinoma cells. Development 107: 361–372
3. Le May M, Mach H, Lacroix N, Hou C, Chen J, Li Q (2011) Contribution of retinoid X receptor signaling to the specification of skeletal muscle lineage. J Biol Chem 286:26806–26812
4. Wobus AM, Rohwedel J, Maltsev V, Hescheler J (1994) In vitro differentiation of embryonic stem cells into cardiomyocytes or skeletal muscle cells is specifically modulated by retinoic acid. Rouxs Arch Dev Biol 204:36–45
5. Kennedy KA, Porter T, Mehta V et al (2009) Retinoic acid enhances skeletal muscle progenitor formation and bypasses inhibition by bone morphogenetic protein 4 but not dominant negative beta-catenin. BMC Biol 7:67
6. Yu J, Thomson JA (2008) Pluripotent stem cell lines. Genes Dev 22:1987–1997
7. Ridgeway AG, Petropoulos H, Wilton S, Skerjanc IS (2000) Wnt signaling regulates the function of MyoD and myogenin. J Biol Chem 275:32398–32405
8. Wilton S, Skerjanc I (1999) Factors in serum regulate muscle development in P19 cells. In Vitro Cell Dev Biol Anim 35:175–177
9. Loscher W (1999) Valproate: a reappraisal of its pharmacodynamic properties and mechanisms of action. Prog Neurobiol 58:31–59
10. Johannessen CU, Johannessen SI (2003) Valproate: past, present, and future. CNS Drug Rev 9:199–216
11. Huangfu D, Osafune K, Maehr R et al (2008) Induction of pluripotent stem cells from primary human fibroblasts with only Oct4 and Sox2. Nat Biotechnol 26:1269–1275
12. Phiel CJ, Zhang F, Huang EY, Guenther MG, Lazar MA, Klein PS (2001) Histone deacetylase is a direct target of valproic acid, a potent anticonvulsant, mood stabilizer, and teratogen. J Biol Chem 276:36734–36741

Chapter 30

Sedimentation and Immunoprecipitation Assays for Analyzing Complexes that Repress Transcription

Ping Lu, Bruce S. Hostager, Paul B. Rothman, and John D. Colgan

Abstract

Co-repressor proteins function as platforms for the assembly of multi-subunit complexes that mediate transcriptional repression. Common components of such complexes are histone deacetylases, which catalyze the removal of acetyl groups from the tails of histones within nucleosomes, resulting in chromatin compaction and gene repression. In addition, co-repressor complexes generally interact with sequence-specific DNA-binding proteins that direct association with regulatory elements in the genome. Thus, identifying proteins that stably associate with co-repressors can provide insights regarding the biochemical function and target gene specificity of these molecules. Here, we describe a density gradient fractionation method for determining whether a co-repressor is incorporated into high-molecular complexes within cells and for identifying potential constituents of these complexes. We also describe a co-immunoprecipitation assay for confirming and further studying interactions between co-repressors and other proteins that may represent functional partners.

Key words: Sucrose gradient sedimentation, Co-immunoprecipitation, Immunoblots, Co-repressor proteins, DNA-binding proteins, Histone deacetylases

1. Introduction

A central role for transcriptional activation in regulating gene expression has long been appreciated, but more recently it has become clear that transcriptional repression is equally important. The application of biochemical methods has greatly advanced our understanding of the factors and mechanisms that mediate transcriptional repression (1–4). This approach helped establish that proteins called co-repressors function as "nucleators" of complexes containing enzymes which alter chromatin conformation or histone modification state, resulting in transcriptional repression (5–7). For example, common constituents of co-repressor complexes are

Minou Bina (ed.), *Gene Regulation: Methods and Protocols*, Methods in Molecular Biology, vol. 977, DOI 10.1007/978-1-62703-284-1_30, © Springer Science+Business Media, LLC 2013

histone deacetylases (HDACs), a class of enzymes that remove acetyl groups from the tails of histones within nucleosomes. In turn, the absence of acetylated histones within a gene locus is a hallmark of a transcriptionally repressed state.

We have studied the Gon4-like protein, which appears to be a novel co-repressor that is required for B cell development (8). To gain insight into the function of this molecule, we used biochemical tools to demonstrate that Gon4-like associates with another co-repressor, Sin3a, and with HDAC1 (9). One method we employed was separation of cell lysates in a continuous sucrose gradient followed by immunoblot analysis of gradient fractions. This approach separates proteins and complexes through a density gradient as driven by centrifugation. Individual molecules and stable complexes are resolved in the gradient based on sedimentation velocities, which are dictated by mass, shape, density, and hydration state (10). Following centrifugation, gradient fractions are collected and those containing the target protein are identified. Comparison to control gradients in which protein standards were separated determines whether a target protein has migrated into fractions containing high-molecular weight species, signifying association with other factors. The relevant fractions are then analyzed for the presence of other proteins that may stably interact with the target. For co-repressors, such proteins would likely be HDACs. In terms of methodology, sucrose gradient fractionation involves five major steps. First, an ultracentrifuge tube is filled with a continuous sucrose density gradient as generated by a gradient maker. Second, cell lysate is prepared and loaded on top of the gradient. Third, the sample is centrifuged to separate lysate components. Fourth, gradient fractions are collected. Lastly, fraction aliquots are analyzed by immunoblot. Using this particular approach, we obtained evidence that, within cells, Gon4-like forms complexes with other molecules and that Sin3a and HDAC1 may be constituents of these complexes (Fig. 1).

A complementary biochemical method for defining the composition of protein complexes is immunoprecipitation. In this assay, antibodies specific for a target protein are linked to a bead support and then incubated with cellular lysate containing the target. Bead-bound antibodies will bind and capture the target and other proteins stably bound to it, which are then recovered for analysis. Co-immunoprecipitation of other proteins with the target supports the hypothesis that these molecules interact in functionally relevant manner. This assay involves four major steps. First, whole cell lysates or nuclear extracts are prepared. Second, control or protein-specific antibodies are bound to beads and added to protein samples followed by incubation. Third, the bead–antibody–protein complexes are separated from unbound lysate components

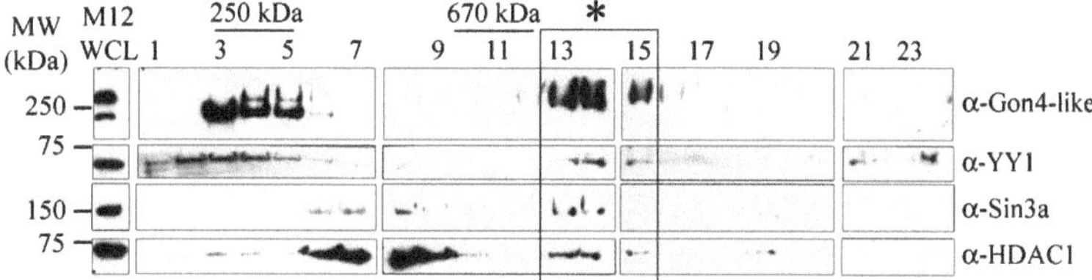

Fig. 1. Detection of the putative co-repressor protein Gon4-like and potential interacting factors in sucrose gradient fractions containing high molecular weight species. Whole cell lysates were prepared from M12 B cells and separated through a 20–50 % continuous sucrose gradient via ultracentrifugation as described in Subheading 3. After centrifugation, the fractions indicated across the *top of the panel* (*odd numbers only are shown*) were collected and analyzed by immunoblot using the antibodies listed on the *right side of the panels*. M12 whole cell lysate (WCL) was used as a positive control for immunoblots (*left side of panels*). Protein standards were separated in parallel gradients and fractions containing these are indicated at the *top of the panel*. *The box with the asterisk on top* denotes fractions that contain Gon4-like, the co-repressor Sin3a, HDAC1, and the DNA-binding protein YY1, suggesting these factors associate as components of high-molecular weight complexes. Data shown are representative of three independent experiments.

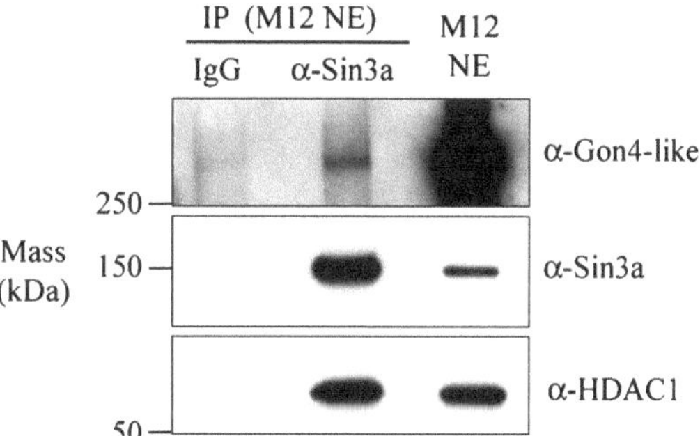

Fig. 2. Co-immunoprecipitation of Gon4-like, Sin3a and HDAC1 from M12 cell nuclear extracts (NE). Nuclear extracts were prepared from M12 B cells and used for immunoprecipitation experiments as described in Subheading 3. Affinity-purified rabbit polyclonal antibodies specific for Sin3a or nonspecific rabbit IgG (both obtained from Santa Cruz Biotechnology) were used to perform immunoprecipitations. Recovered material was immunoblotted with antibodies specific for Gon4-like (*top panel*), Sin3a (*middle panel*), or HDAC1 (*bottom panel*). M12 nuclear extract (NE) was used as a positive control for immunoblots. Data shown are representative of three independent experiments.

and washed to remove material bound nonspecifically. Finally, the recovered material is analyzed by immunoblot to confirm target recovery and to detect other proteins that may associate with it. Using immunoprecipitation analysis, we obtained data confirming that Sin3a and HDAC1 associate with the Gon4-like protein (Fig. 2).

2. Materials

Make all solutions using ultrapure, deionized water with a resistivity of 18 MΩ cm at 25 °C. Sterilize solutions by autoclaving or filtering through a 0.2 μ pore-size filter. Wash and autoclave all glassware prior to use. Unless indicated otherwise, store all reagents at room temperature (RT). Adhere to all federal and local regulations for disposal of waste materials.

2.1. Cell Culture Reagents

1. Tissue culture incubator for maintaining 37 °C temperature and 5 % CO_2 atmosphere.
2. Tissue culture grade filter screw cap flasks or plates.
3. Transformed cell line: For our studies, we use the mature B cell line M12.4.1 (11).
4. Appropriate culture media.
5. Eppendorf 5804C centrifuge, or equivalent, for spinning down cells.
6. Delbucco's Phosphate-buffered Saline (PBS) pH 7.4, lacking Mg^{2+} and Ca^{2+}.
7. Hemocytometer and coverslip.

2.2. Reagents for Determining Protein Concentrations

1. Bradford assay reagent (Bio-Rad, Hercules, CA, USA) prepared and used according to manufacturer's instructions.
2. A spectrophotometer capable of measuring visible light absorbance.

2.3. Sucrose Gradient Components

1. 10× Phosphate-buffered Saline (PBS) pH 7.4, lacking Mg^{2+} and Ca^{2+}: 100 mM $Na_2HPO_4{\cdot}2H_2O$, 17.6 mM KH_2PO_4, 1.37 M NaCl, and 27 mM KCl.
2. Sucrose gradient cell lysis buffer (Buffer S): PBS containing 1 mM EDTA, 0.5 % Triton X-100, and EDTA-free protease inhibitor cocktail (Roche Applied Science, Indianapolis, IN, USA) (see Note 1). Store at −20 °C; use ice-cold.
3. 70 % (w/v) sucrose stock solution: Dissolve 350 g of sucrose in water added up to 500 mL and filter-sterilize. The stock can be stored at 4 °C for up to 3 months.
4. 20 % (w/v) sucrose solution: For 100 mL, mix 60.73 mL of water, 10 mL of 10× PBS, 28.57 mL of 70 % sucrose stock solution, 0.2 mL of 500 mM EDTA pH 8 and 0.5 mL of Triton X-100. Chill to 4 °C prior to use.
5. 50 % (w/v) sucrose solution: For 100 mL, mix 17.87 mL of water, 10 mL of 10× PBS, 71.43 mL of 70 % sucrose stock solution, 0.2 mL of 500 mM EDTA pH 8, and 0.5 mL of Triton X-100. Chill to 4 °C prior to use.

6. Hoefer SG 15 sucrose gradient maker (Fisher Scientific, Dallas, TX, USA), tubing to connect to outlet of gradient maker for pouring the gradient, glass capillary tubes that fit snugly into the end of the tubing, Vaseline for use as a sealant.
7. Stir plate and ½-in. stir bar.
8. Molecular weight standards: Mix 100 μL of Precision-plus standards (Bio-Rad, Hercules, CA, USA) with 100 μL Buffer S and load onto a gradient. Resuspend Thyroglobulin (Sigma Chemical Company, St. Louis, MO, USA) to 20 mg/mL in a reconstitution solution that contains 50 mM Tris–HCl pH 7.5, 100 mM KCl, and H_2O. Add 50 μL of this stock to 150 μL Buffer S and load onto a gradient. Store Thyroglobulin stock at −20 °C.
9. Centrifuge tubes: 14×89 mm polyallomer tubes (Beckman, Brea, CA, USA).
10. SW41Ti swinging bucket rotor (Beckman). Chill to 4 °C prior to use.
11. Free-standing ultracentrifuge (Beckman). Chill ultracentrifuge chamber to 4 °C prior to use.
12. Ring stand and clamp for holding a 14×89 mm polyallomer centrifuge tube.

2.4. Whole Cell Lysate Reagents

1. 200 mM sodium orthovanadate ($NaVO_4$) stock solution: the tyrosine phosphatase inhibitor $NaVO_4$ (Sigma Chemical Company) must be depolymerized prior to use. Add 368 mg of $NaVO_4$ to 8 mL of water and incubate at 37 °C while occasionally mixing until dissolved. Adjust the pH to 10 by adding drops of 1 N NaOH, which will generate an orange-colored solution. Seal the tube and incubate in boiling water until the solution becomes colorless (5–10 min) and then cool to RT. Repeat the cycle of pH adjustment to 10 followed by boiling and cooling to RT until a colorless liquid at pH 10 is generated (see Note 2). Usually, two cycles are required. Add water to 10 mL, filter-sterilize, and store as aliquots at −20 °C.
2. 200 mM phenylmethylsulfonyl fluoride (PMSF) stock solution: dissolve 348.4 mg of the serine protease inhibitor PMSF (Sigma Chemical Company) in 10 mL of ethanol (or isopropanol or methanol). Incubate at 37 °C while occasionally mixing to dissolve. Store as aliquots at −20 °C and warm to RT immediately before use (see Note 3).
3. 0.5 M sodium fluoride (NaF) stock solution: Add 210 mg of the phosphatase inhibitor NaF (Sigma Chemical Company) to 10 mL of water. Incubate at 37 °C while occasionally mixing until dissolved. Filter-sterilize and store as aliquots at −20 °C.

4. Whole cell lysis (WCL) Buffer: 50 mM Tris–HCl pH 8.0, 120 mM NaCl, 0.5 % NP-40 (Roche Applied Science), 0.2 mM $NaVO_4$, 50 mM NaF, 1 mM PMSF (add immediately before use), and EDTA-free protease inhibitor cocktail (Roche Applied Science). Store WCL Buffer lacking PMSF at −20 °C (see Note 4). Use ice-cold.

2.5. Nuclear Extract Reagents

This methodology is based on that originally described by Dignam and Roeder (12).

1. Buffer A: 20 mM Hepes pH 7.5, 10 mM KCl, 1.5 mM $MgCl_2$, 1 mM EDTA pH 8, 1 mM DTT, 10 % glycerol, and EDTA-free protease inhibitor cocktail (Roche Applied Science). Store at −20 °C; use ice-cold.
2. Buffer B: 20 mM Hepes pH 7.5, 10 mM KCl, 420 mM NaCl, 1.5 mM $MgCl_2$, 1 mM EDTA pH 8, 1 mM DTT, 25 % glycerol, and EDTA-free protease inhibitor cocktail (Roche Applied Science). Store at −20 °C; use ice-cold.
3. Buffer D: 20 mM Hepes pH 7.5, 100 mM KCl, 1.5 mM $MgCl_2$, 0.2 mM EDTA pH 8, 0.5 mM DTT, and 25 % glycerol. Prepare immediately before use and cool to 4 °C (see Note 5).
4. Kontes glass dounce with Kontes Pestle "B" (Fisher Scientific). Chill on ice prior to use.
5. DNase I, grade II (Roche Applied Science): Resuspend to 10 mg/mL in 10 mM Tris–HCl pH 7.5, 0.5 mM $CaCl_2$, 0.5 mM $MgSO_4$, and 50 % (v/v) glycerol. Store at −20 °C.
6. Slide-A-Lyzer G2 Dialysis cassettes with 7 kDa mass cutoff (Thermo Scientific/Pierce, Rockford, IL, USA).
7. 10 mL syringes and 1.5 in., 18-gauge needles.
8. 1 L glass beaker.
9. Stir plate and 1.5-in. stir bar.

2.6. Immunoprecipitation Components

1. Antibodies: polyclonal or monoclonal antibodies specific for target protein and species-matched control IgG for negative control samples (see Note 6).
2. Dynal® magnetic beads conjugated to Protein A or G (Invitrogen, Carlsbad, CA, USA) (see Note 7).
3. DynaMag-2™ magnet for pulling out magnetic beads from solutions in 1.5–2.0 mL tubes.
4. 1× PBS lacking Ca^{2+} and Mg^{2+}.
5. Wash buffer: WCL buffer, prepared exactly as described in Subheading 2.3, is used as high-stringency wash buffer. Low-stringency wash buffer is of the same composition except that NP-40 is added to a final concentration of 0.1 % (rather than 0.5 %).
6. Rotator that can hold 1.5 mL tubes for mixing samples.

2.7. SDS Polyacrylamide Gel Components

This methodology is based on information in the NuPAGE® Technical Guide, which can be obtained as a PDF from the Invitrogen Web site. Consult the guide for additional details.

1. XCell SureLock® mini-cell electrophoresis system (Invitrogen): includes a chamber for running gels, a 2-gel holder (Buffer Core) with electrodes, Gel Tension Wedge to lock gels in place and a lid with wire leads for connecting to a power supply.
2. Power supply: Bio-Rad PowerPac (300 V maximum) or equivalent.
3. SDS polyacrylamide gels: 1.0 mm, 10-well, NuPAGE® Novex 3–8 % Tris-Acetate gels (Invitrogen).
4. Running Buffer: For 1 L, add 50 mL of 20× NuPAGE® Tris-Acetate SDS Running Buffer (Invitrogen) into 950 mL of water and mix thoroughly.
5. NuPAGE® antioxidant reagent (Invitrogen).
6. NuPAGE® reducing reagent (Invitrogen).
7. 10 mL syringe and a 1.5 in. 18-gauge needle.
8. 1× Sample Buffer: For 100 μL (enough for one sample), mix 25 μL of 4× NuPAGE® LDS Sample Buffer (Invitrogen), 10 μL of 10× NuPAGE® Reducing Agent, and 65 μL of water.
9. Heat block or water bath set at 70 °C.
10. Gel-loading tips (Invitrogen).

2.8. Protein Transfer to Membrane Components

This methodology is based on information in the Mini Trans-blot® Electrophoretic Transfer Cell Instruction Manual, which can be obtained as a PDF from the Bio-Rad Web site. Consult the manual for additional details.

1. Wet transfer apparatus: Mini Trans-Blot Cell System (Bio-Rad) (see Note 9).
2. Power supply: Bio-Rad PowerPac (300 V maximum) or equivalent.
3. Blunt-end tweezers (Millipore, Billerica, MA, USA) for handling PVDF membranes (see Note 8).
4. Methanol (ACS grade).
5. Transfer Buffer: for 1 L, mix 50 mL of 20× NuPAGE® Transfer Buffer (Invitrogen) and 100 mL of methanol into 849 mL of water. Chill prior to use.
6. NuPAGE® antioxidant reagent (Invitrogen).
7. Sponge blotting pads: provided with Bio-Rad's Mini Trans-Blot® Cell system.
8. 3 mm-thick filter paper cut into 7 cM × 8 cM pieces. Precut paper can be obtained from Invitrogen or Bio-Rad. Alternatively,

large sheets of paper can be obtained from Whatman/GE Healthcare (Piscataway, NJ, USA) and cut prior to use.

9. Immobilon-P PVDF transfer membrane with 0.45 μm pore size (Millipore). Cut membrane into 7 cM × 8 cM pieces prior to use.

2.9. Immunoblot Components

1. Blunt-end tweezers (Millipore) (see Note 8).
2. Methanol (ACS grade).
3. 3 mm-thick filter paper (Whatman/GE Healthcare) cut into 10 cm × 10 cm squares.
4. 10× Tris-buffered saline (TBS): 250 mM Tris–HCl, pH 7.4, 1.5 M NaCl, 20 mM KCl. Dilute to 1× with water.
5. Blocking Buffer: add 2.5 g nonfat dry milk powder into 25 mL of 1× TBS. Mix well and adjust volume to 50 mL. Incubate at 37 °C while mixing occasionally until dissolved.
6. TBS-T: Add 0.5 mL Triton X-100 to 1 L of 1× TBS (0.05 % Triton X-100). Mix using a stir bar and stir plate. Use at RT and discard after 3 days (microbes will grow within this solution).
7. 8 cm × 10 cm plastic boxes with lids.
8. Primary antibodies specific for target proteins that have been verified to work for immunoblot analysis. These can be obtained from several vendors.
9. Secondary (or detection) antibodies specific for immunoglobulin G constant regions from mouse or rabbit and conjugated to horseradish peroxidase (HRP) (Santa Cruz Biotechnology, Santa Cruz, CA, USA).
10. Enhanced chemiluminescence (ECL) HRP substrate mixes: SuperSignal West Pico Chemiluminescent Substrate (Pierce/Thermo Scientific). To increase assay sensitivity, obtain SuperSignal West Femto Chemiluminescent Substrate (Pierce/Thermo Scientific) (see Note 10).
11. Two sheets of overhead transparency film, taped together down the length of one side.
12. Glogos II Autorad Markers (Agilent Technologies, Santa Clara, CA, USA).
13. X-ray film cassette, X-ray film, and a film developer.

2.10. Coomassie® Blue Staining Reagents

This methodology is similar to that described in Molecular Cloning: A Laboratory Manual (13).

1. Glacial acetic acid (ACS grade).
2. Methanol (ACS grade).
3. Glycerol.

4. Coomassie® Brilliant Blue R-250 powder (Bio-Rad).
5. Coomassie® Staining Buffer: dissolve 0.1 g of Coomassie® Brilliant Blue R-250 powder in a solution consisting of 10 mL of glacial acetic acid, 30 mL of methanol and water up to 50 mL. Stir the solution with a stir bar on a stir plate for at least 1 h. Adjust the final volume to 100 mL with water and filter through Whatman filter paper.
6. Destaining Buffer: mix 200 mL of methanol, 100 mL of glacial acetic acid and water to a final volume of 1,000 mL.
7. Glycerol Fix Solution: mix 50 mL of methanol and 15 mL of glycerol with water up to a final volume of 500 mL.
8. Gel drying frame, membranes and clips for frame (Research Products International, Mount Prospect, IL, USA).

3. Methods

All procedures can be carried out at RT. However, keep all reagents and protein-containing samples on ice as much as possible.

3.1. Culture of Cells in Suspension for Whole Cell Lysate or Nuclear Extract Preparation

1. M12 cells are grown in suspension at 37 °C in 5 % CO_2. Stocks are maintained in T-25 cm^2 flasks and split 1:10 (cell suspension–fresh media) every 2–3 days as judged by cell concentration and media color.
2. To expand cells for harvest, transfer 5 mL of stock at high concentration to a T-75 cm^2 flask and add 25 mL of fresh media. Depending on the numbers of cells needed, T-75 cm^2 flasks are split 1:3 every 2–3 days by transferring into fresh T-75 cm^2 flasks.
3. A single T-75 cm^2 flask of M12 cells grown to high density will yield ~1×10^7 cells.

3.2. Cell Harvest

1. Decant or transfer cells into 50 mL polypropylene conical tubes.
2. Pellet the cells by centrifugation at 1,500 rpm ($405 \times g$) 10 min.
3. Remove as much media as possible by aspiration.
4. Resuspend cells in 40 mL of cold PBS and pellet by centrifugation at 1,500 rpm ($405 \times g$) 10 min.
5. Resuspend cells in 5 or 10 mL of cold PBS, pool into one 50 mL polypropylene conical tube and pellet the cells by centrifugation at 1,500 rpm ($405 \times g$) 10 min.
6. Resuspend total cells obtained in 10 mL of PBS and count using a hemocytometer.
7. Pellet the number of cells required for the method to be used (see first step in Subheadings 3.3–3.5).

3.3. Sucrose Gradient Cell Separation of Protein Lysates Prepared from Cells

1. Prepare a cell pellet containing ~1×10^7 cells.
2. Resuspend the cell pellet in 0.5 mL of Buffer S by gentle pipetting.
3. Incubate cells on ice for 20 min, inverting occasionally.
4. Spin the sample at 14, 000 rpm ($18200 \times g$) for 15 min in a microfuge at 4 °C.
5. Recover supernatant.
6. Measure the protein concentration using Bradford assay. The ideal concentration is ~5 mg/mL, allowing 1 mg of protein in a 200 μL volume to be loaded onto a gradient.
7. Place sample on ice until sucrose gradient is ready for loading.
8. Put the gradient maker onto a stir plate placed on a shelf or other solid support that is ~2 ft higher than the lab bench, allowing sufficient flow of sucrose solution. Make sure that the gradient maker is level.
9. Attach tubing to the outlet connector of the gradient maker. To the other end of the tube, attach a glass capillary tube and insert it vertically to the bottom of an ultracentrifuge tube in a rack on the bench. Apply Vaseline to the outside of the capillary tube to seal the connection with the tube (see Note 11).
10. Add water to both chambers of the gradient maker. Open the connector-stopcock and the outlet stopcock. Let water flow through, removing bubbles. Determine the flow rate, which should be ~0.5–1 mL/min; this can be adjusted by elevating or lowering the gradient maker. Afterward, allow water to drain out of the chambers.
11. Add 6 mL of 20 % sucrose solution to the mixing chamber (chamber 1, which is directly connected to the outlet stopcock) of the gradient maker. Open the outlet-connector stopcock and fill the tubing with the 20 % gradient solution to prevent bubble formation. Drain almost all of the solution from the chamber.
12. Place 6.4 mL of the 20 % and 50 % sucrose solution in chamber 1 and 2, respectively. Place a stir bar in chamber 1 and set a stir of <300 rpm. High-speed stirring may cause bubbles and disrupt gradient formation.
13. Open the connector stopcock between the two chambers.
14. Open the outlet stopcock allowing the solution to flow into the ultracentrifuge tube. After tube is nearly full (~7 mm from the top), slowly remove the glass capillary tube. The poured gradient can be stored at 4 °C up to 2 h before being loaded and centrifuged.

15. Rinse the chambers with water and repeat the process as needed. At least four tubes will be required per experiment (one for the sample, two for each of the mass markers, and another for balance).
16. Ensure that all tubes are the same weight, adding 20 % sucrose solution to the top of the gradient as necessary.
17. Gently load the lysate in Buffer S onto the top of the gradient solution using a P200 pipette. Place the tip of the pipette against the wall of the tube just above the gradient surface and slowly expel the sample.
18. On to identical gradients, load 200 μL of protein markers prediluted in Buffer S.
19. Confirm that the loaded gradients are balanced, dispensing 20 % sucrose solution at the gradient–sample interface as needed (see Note 12).
20. Place the tubes into the precooled buckets of a precooled SW41Ti swinging bucket rotor. Seal the buckets and centrifuge the samples for 16 h at 40,000 rpm (274,000 × g) in an ultracentrifuge precooled to, and maintained at, 4 °C.
21. After centrifugation, transfer all the rotor buckets containing samples to a rack and store at 4 °C. Minimize shaking of the buckets during transport as much as possible.
22. Prepare a set of 1.5 mL tubes on ice to hold the gradient fractions to be collected. Typically, 25 tubes, each filled with a 500 μL fraction, is required to collect a single gradient.
23. Set up a support stand holding a clamp that can securely hold a polyallomer centrifuge tube containing a gradient in a vertical position while minimizing movement. Remove a tube from the centrifuge bucket by pulling out with a blunt-end tweezer and secure it with the clamp.
24. Use a 1 mL pipette and tips to collect gradient fractions. Place the tip just beneath the surface of the gradient and slowly withdraw 500 μL. Transfer the aliquot to a tube on ice. Repeat this process until the entire gradient is collected.
25. Store sample tubes at −80 °C until you are ready to perform SDS-PAGE and immunoblot analysis (see Subheadings 3.7–3.9). To minimize protein degradation, avoid subjecting samples to multiple freeze–thaw cycles.
26. Aliquots (15–18 μL) of gradient fractions containing protein standards will be resolved on SDS-polyacrylamide gels and then stained with Coomassie Blue.
27. Aliquots (15–18 μL) of gradient fractions containing cell lysate components will be resolved on SDS-polyacrylamide gels and subjected to immunoblot analysis (see Fig. 1).

3.4. Making Whole Cell Lysate for Immunoprecipitation and Immunoblot Analysis

1. Prepare a cell pellet containing ~2×10^7 cells.
2. Resuspend cells in 800 μL of WCL Buffer (200 μL per 5×10^6) by pipetting. Incubate cells on ice for 20 min, inverting occasionally.
3. Spin the sample at 14,000 rpm (18200 × g) for 15 min at 4 °C.
4. Recover the supernatant.
5. Determine the protein concentration using the Bradford assay.
6. Set aside an aliquot of lysate equivalent to ~200 μg of protein as a positive control for immunoblot analysis.
7. Immediately begin immunoprecipitation assay as described below (see Note 13).

3.5. Making Nuclear Extract for Immunoprecipitation and Immunoblot Analysis

1. Prepare a cell pellet containing ~3×10^7 cells.
2. Determine the packed cell volume (PCV) of the cell pellet and resuspend cells in 5× PCV of Buffer A.
3. Spin cells at 1,500 rpm (405 × g) for 5 min and remove the supernatant.
4. Resuspend cells in 3× PCV of Buffer A and incubate on ice for 10 min.
5. Add NP40 to 0.25 % (2.5 μl of 100 % NP40 per mL) mix by gentle pipetting and incubate ice for 10 min.
6. Transfer cells to an ice-cold Kontes dounce homogenizer that was prerinsed with Buffer A.
7. Perform 25–50 strokes with Kontes Pestle B to lyse cells (see Note 14).
8. Transfer sample to a 15 mL conical tube. Spin the sample at 2,400 rpm (1037 × g) for 10 min to pellet nuclei.
9. The resulting supernatant should be cloudy, signifying efficient cell lysis. Remove all of the supernatant by pipetting or aspiration.
10. Determine packed nuclei volume (PNV) and resuspend cells in 3× PNV of Buffer B.
11. Transfer nuclei to Kontes Dounce homogenizer that has been prerinsed with Buffer B and perform 50 strokes with Kontes Pestle B to shear nuclei.
12. Transfer sample to a 1.5 mL tube.
13. Place tube on rotator in cold room and rotate for 1 h. Visible aggregates of insoluble material will be present at the end of the incubation.
14. Spin the sample at 14,000 rpm (18200 × g) for 15 min in a microfuge at 4 °C and the recover supernatant.

15. Add 2.5 μL of DNase (10 mg/mL) per mL of supernatant and place on ice.
16. Prepare a Slide-A-Lyzer G2 dialysis cassette for incubating in 1 L of chilled Buffer D for 10 min. Afterward, open the top of the cassette and use a 10 mL syringe capped with an 18-gauge needle to withdraw all the air from the dialysis chamber.
17. Load the protein sample into a 10 mL syringe capped with an 18-gauge needle. Insert the needle into the loading port of the dialysis cassette and gently expel the sample.
18. Seal the dialysis cassette and place in 1 L of chilled Buffer D. Dialyze the extract for a minimum of 3 h at 4 °C while stirring the Buffer D using a stir bar and stir plate.
19. Use a 10 mL syringe capped with an 18-gauge needle to recover the extract. Place the recovered material in a 1.5 mL tube or divide between two tubes if necessary.
20. Spin the sample at 14,000 rpm (18200×*g*) for 15 min in a microfuge at 4 °C and recover the supernatant.
21. Determine the protein concentration using the Bradford assay.
22. Set aside an aliquot of lysate equivalent to ~200 μg of protein as a positive control for immunoblot analysis.
23. Immediately begin immunoprecipitation assay as described below (see Note 13).

3.6. Immuno-precipitation Assay

1. Completely resuspend Dynabeads conjugated to Protein A or G by pipetting.
2. Remove 50 μL (~1 mg) of beads for each immunoprecipitation (see Note 15) to be done plus one extra sample and transfer to a 1.5 mL tube.
3. Wash the beads twice with 1 mL of PBS: add 1 mL of PBS to the bead slurry, place the tube in the DynaMag-2™ magnet and let stand for ~1 min. The beads will concentrate on the wall of the tube facing the magnet. Aspirate the PBS, avoiding contact with the beads. Repeat.
4. Resuspend the beads in 200 μL of PBS per immunoprecipitation and transfer 200 μL to the appropriate number of 1.5 mL tubes.
5. Add 5 μg of isotype control or target-protein-specific antibody to each tube and adjust the volume of the sample to 500 μL using PBS.
6. Incubate the samples at 4 °C for 20 min while rotating.
7. Wash beads twice with 1 mL PBS.
8. Resuspend the beads in 200 μL of the Wash Buffer to be used for the experiment (i.e., high or low stringency buffer).

9. Add to the bead suspension a volume of whole cell lysate or nuclear extract containing a total of 500–1,000 μg of protein.
10. Adjust the final volume of the sample to 1 mL by adding Wash Buffer.
11. Incubate the samples at 4 °C for 2–16 h while rotating (see Note 16).
12. Place the samples in the DynaMag-2™ magnet and let stand for ~1 min, allowing the beads to concentrate on the wall of the tube.
13. Recover the supernatant and save for immunoblot analysis if desired (see Note 17).
14. Remove the samples from DynaMag-2™ magnet and place on ice. Gently resuspend the beads in 1 mL of Wash Buffer and incubate on ice for 1–5 min.
15. Place the samples in the DynaMag-2™ magnet, allowing the beads to concentrate on the wall of the tube facing the magnet.
16. Repeat steps 14 and 15 so that the bead–protein complexes are washed a total of three or four times.
17. Perform a final wash with 1 mL of PBS and remove as much of the supernatant as possible.
18. Resuspend beads in 100 μL of 1× Sample Buffer and proceed to SDS-PAGE and immunoblot analysis (see Subheadings 3.7–3.9).

3.7. SDS-PAGE Analysis

1. Prepare 1 L of 1× Running Buffer.
2. Transfer 200 mL of 1× Running Buffer to a separate container. To this, add 500 μL of NuPAGE® antioxidant reagent and mix thoroughly just prior to loading samples onto a gel.
3. Remove 3–8 % gradient gel from packaging. Remove the plastic strip from the bottom of the gel and the comb from the top of the gel. Gently rinse the wells with 1× running buffer dispensed from a syringe capped with an 18-gauge needle.
4. Place the gel onto the "Buffer Core" so that the well side of the gel (notched plate) faces inward. On the other side, place either another gel or the "Buffer Dam".
5. Insert the gel/Buffer Core assembly into the Xcell Mini-Cell (the clear plastic container that holds the gel(s) and buffer) and lock the assembly into place using the Gel Tension Wedge.
6. Fill the inner chamber formed by the gel/Buffer Core assembly with 1× Running Buffer containing antioxidant reagent. Inspect the assembly to ensure buffer is not leaking out.
7. Fill the outer chamber with 1× Running Buffer lacking antioxidant reagent. The buffer level must exceed that of the gel wells.

8. Heat tubes containing protein samples at 70 °C for 10 min immediately prior to loading.
9. Insert tubes containing Dynabeads into the DynaMag-2™ magnet and let stand for 1 min to remove beads from the sample.
10. In one well of the gel, load pre-stained protein standards of choice.
11. Load 20–25 μL of protein samples into wells of the gel.
12. Attach the lid for the Mini-Cell and attach the wire leads from the lid to a power supply set to provide a constant voltage of 150.
13. Depending on the mass of the proteins to be detected by immunoblot, the gel run may take between 50 and 120 min.

3.8. Protein Transfer to Membrane for Immunoblotting

1. Prepare 1 L of 1× NuPAGE® Transfer Buffer chilled to 4 °C.
2. Turn off and disconnect the power supply and disassemble the Mini-Cell to obtain the gel(s).
3. Using a gel knife, carefully pry apart the gel plates. Ensure that the gel sticks to one plate; if not, soak the gel and plates in 1× Transfer Buffer to help separate.
4. Remove the top of the gel spanning the wells and then soak in 1× Transfer Buffer for ~10 min.
5. Completely submerge the PVDF membrane in 100 % methanol. Decant the methanol and add water. Incubate with gentle agitation for 5 min, exchange the water for fresh and incubate again. Exchange the water for Transfer Buffer and incubate with gentle agitation for >5 min.
6. Soak four pieces of filter paper and the blotting sponges in Transfer Buffer.
7. Place the electrode module into the buffer tank and pour 500 mL of Transfer Buffer into the tank.
8. Assemble a "sandwich" on black half of the gel holder cassette by placing the listed materials on top of each other in this order: (a) sponge pad, (b) two pieces of filter paper, (c) gel, (d) PVDF membrane, (e) two pieces of filter paper, and (f) sponge pad. When the PVDF is laid on top of the gel, make sure to remove all bubbles between them. Close and lock the gel holder.
9. Immediately slide the gel holder into the electrode module in the tank so that the clear half of the holder faces the red half of the electrode module.
10. Fill the tank near to the top with Transfer Buffer and attach the lid so that the appropriate leads and electrodes are connected. Transport the assembly to a cold room and place on a stir plate. Set the plate so that the stir bar within the chamber spins slowly to moderately.

11. Attach wire leads from the lid to a power supply set to provide a constant voltage of 50.
12. Allow transfer to proceed from 4 h to overnight.
13. Recover the PVDF membrane and wash with water and then methanol. Allow membrane to dry at RT.

3.9. Immunoblot Analysis of Protein Samples

1. Prepare TBS-T and Blocking Buffer.
2. Rehydrate the membrane by submerging in 100 % methanol. Decant the methanol and add water. Incubate with gentle agitation for 5 min, exchange the water for fresh and incubate again. Exchange the water for TBS-T and incubate with gentle agitation for >5 min.
3. Decant the TBS-T and add 10 mL of Blocking Buffer. Incubate at RT for 1 h with modest shaking.
4. Decant the Blocking Buffer and rinse the membrane twice with TBS-T. Keep the membrane in TBS-T while preparing primary antibody solution.
5. Dilute 1 mL of Blocking Buffer into 9 mL of TBS-T. Add protein-specific primary antibody to the concentration recommended by the manufacturer.
6. Add the TBS-T/antibody solution to the membrane and incubate with modest shaking. Incubate at either RT for 1 h or 4 °C overnight.
7. Remove the TBS-T/antibody solution and wash the membrane three times with TBS-T, incubating for 5–10 min with modest shaking for each wash.
8. Dilute 1 mL of Blocking Buffer into 9 mL of TBS-T. Add species-specific secondary anti-IgG antibodies conjugated to HRP to the concentration recommended by the manufacturer.
9. Add the TBS-T/antibody solution to the membrane and incubate at RT for 1 h with modest shaking.
10. Wash the membrane as described for step 7 above.
11. Add ECL substrate mix (see Note 10) to the membrane and incubate for 1 min.
12. Transfer the membrane to a piece of filter paper and allow to briefly dry.
13. Insert the membrane between two pieces of overhead transparency film and place into an X-ray cassette containing a Glogos Autorad II Marker fixed to a site, thus serving as a position marker for the X-ray film.
14. In a dark room, overlay a piece of X-ray film onto the transparency film, incubate for 10–60 s, and develop.

15. After the desired exposure has been attained, use Glogos Autorad II Marker to position the developed film over the membrane. Mark the positions of pre-stained protein standards on the film so that the mass of species detected by ECL can be estimated.

3.10. Coomassie® Blue Staining of Gels

1. After electrophoresis, put the gel into a plastic container (the top from a box of 1 mL pipette tips works well) containing Staining Buffer and incubate with gentle shaking for 1 h at RT.
2. Decant the Staining Buffer into a bottle and store for reuse.
3. Add Destaining Buffer so that the gel is submerged. Incubate with gentle shaking for at RT until bands are visible and background staining is minimal. If the Destaining Buffer becomes dark blue during incubation, replace with fresh buffer and continue incubating.
4. Wash the gel with water twice for 5 min at RT.
5. Submerge the gel in Glycerol Fix Solution and incubate 60 min at RT.
6. Mount and dry the gel using a drying frame and membrane according to the manufacturer's instructions.

4. Notes

1. Add protease inhibitor cocktail to buffers immediately before use. Avoid subjecting buffers containing the inhibitors to multiple (>2) freeze/thaw cycles.
2. In our hands, addition of NaOH is required to adjust the pH of $NaVO_4$ stock solution to 10. In some cases, it may be necessary to add 1 N HCl to adjust the pH.
3. When stored at −20 °C, PMSF will fall out of solution. Prior to adding to lysis buffer, redissolve the PMSF by incubating at 37 °C while mixing occasionally.
4. PMSF is unstable in aqueous solutions. Always add PMSF to lysis buffer immediately before use. Discard any unused lysis buffer containing PMSF.
5. Use prechilled water to prepare cold Buffer D as dialysis buffer. Prior to making nuclear extract, prepare 1 L bottles of autoclaved water and store at 4 °C overnight or longer.
6. Polyclonal antibodies are considered to be optimal for immunoprecipitation analysis because they likely recognize multiple epitopes on the target protein. However, in general monoclonal antibodies also work well. As a negative control, obtain "nonspecific" IgG isolated from the same species (usually rabbit or mouse) in which the target-specific antibodies were raised.

7. The choice of Protein A or Protein G beads depends on the species of origin and isotype of the target-specific antibodies. Consult Invitrogen's "Product Selection Guide for Immunoprecipitation" for more information. In most cases, Protein G is the correct reagent to use.
8. Never directly touch PVDF membranes with gloves or any other materials except those specified in Subheading 3; this will preserve the protein-binding capacity of the membrane.
9. We find that wet (rather than semi-dry) transfer of proteins from SDS-polyacrylamide gels provides optimal immunoblot results, particularly when proteins of relatively large mass (>100 kDa) are being analyzed. Some investigators may find that semi-dry transfer is suitable for their studies.
10. In some instances, a weak ECL signal may be obtained when the "conventional" SuperSignal West *Pico* substrate mix is used. To amplify the signal, add 100 μl (50 μl of each solution) of SuperSignal West *Femto* substrate per 2 mL of SuperSignal West *Pico* substrate.
11. The diameter of the glass capillary tube and the tubing must be very similar to minimize turbulence at the interface between them, which can disrupt gradient formation. Sealing with Vaseline will prevent airflow and bubble formation, minimizing turbulence.
12. It is critical to fill the centrifuge tube with liquid to within 5 mm or less from the top of the tube. Otherwise, the tube may collapse during ultracentrifugation.
13. Freezing and thawing can alter the structure of proteins, causing loss of functional conformation and thus affecting the outcome of immunoprecipitation experiments. To prevent this, perform the assay using freshly prepared whole cell lysates or nuclear extracts. If time is required, store protein samples on ice in an insulated bucket with a lid in a cold room.
14. The number of strokes required depends on cell size, with more strokes needed to lyse smaller cells. Cell lysis can be judged by using a microscope to view 5 μL aliquots of cells taken before and after douncing. A large amount of debris with few whole cells apparent signifies efficient lysis. If this is not observed, perform more strokes.
15. Each immunoprecipitation experiment should consist of a sample in which lysate/extract is incubated with isotype control antibodies bound to beads (i.e., a negative control sample) and a sample in which lysate/extract is incubated with protein-specific antibodies bound to beads. For all subsequent steps, these samples should be processed in parallel. If the isotype control produces a confounding signal, try a control antibody mix from a different vendor.

16. Although immunoprecipitations are commonly incubated overnight, we find that a 2–4 h incubation is effective and may help preserve protein integrity.
17. Immunoblot analysis of supernatant aliquots in parallel with material recovered by immunoprecipitation provides an assay for judging the immunoprecipitation efficiency (see Fig. 2).

References

1. Laherty CD, Yang WM, Sun JM, Davie JR, Seto E, Eisenman RN (1997) Histone deacetylases associated with the mSin3 corepressor mediate mad transcriptional repression. Cell 89:349–356
2. Zhang Y, Iratni R, Erdjument-Bromage H, Tempst P, Reinberg D (1997) Histone deacetylases and SAP18, a novel polypeptide, are components of a human Sin3 complex. Cell 89:357–364
3. Lutterbach B, Westendorf JJ, Linggi B, Patten A, Moniwa M, Davie JR, Huynh KD, Bardwell VJ, Lavinsky RM, Rosenfeld MG, Glass C, Seto E, Hiebert SW (1998) ETO, a target of t(8;21) in acute leukemia, interacts with the N-CoR and mSin3 corepressors. Mol Cell Biol 18:7176–7184
4. Fleischer TC, Yun UJ, Ayer DE (2003) Identification and characterization of three new components of the mSin3A corepressor complex. Mol Cell Biol 23:3456–3467
5. Cunliffe VT (2008) Eloquent silence: developmental functions of Class I histone deacetylases. Curr Opin Genet Dev 18:404–410
6. Grzenda A, Lomberk G, Zhang JS, Urrutia R (2009) Sin3: master scaffold and transcriptional corepressor. Biochim Biophys Acta 1789:443–450
7. Perissi V, Jepsen K, Glass CK, Rosenfeld MG (2010) Deconstructing repression: evolving models of co-repressor action. Nat Rev Genet 11:109–123
8. Lu P, Hankel IL, Knisz J, Marquardt A, Chiang MY, Grosse J, Constien R, Meyer T, Schroeder A, Zeitlmann L, Al-Alem U, Friedman AD, Elliott EI, Meyerholz DK, Waldschmidt TJ, Rothman PB, Colgan JD (2010) The justy mutation identifies Gon4-like as a gene that is essential for B lymphopoiesis. J Exp Med 207:1359–1367
9. Lu P, Hankel IL, Hostager BS, Swartzendruber JA, Friedman AD, Brenton JL, Rothman PB, Colgan JD (2011) The developmental regulator protein Gon4l associates with protein YY1, co-repressor Sin3a, and histone deacetylase 1 and mediates transcriptional repression. J Biol Chem 286:18311–18319
10. Völkl A (2010) Ultracentrifugation. In: Finazzi Agrò A (ed) Encyclopedia of life sciences (ELS). Wiley, Chichester. doi: 10.1002/9780470015902.a0002969.pub2
11. Hamano T, Kim KJ, Leiserson WM, Asofsky R (1982) Establishment of B cell hybridomas with B cell surface antigens. J Immunol 129:1403–1406
12. Dignam JD, Lebovitz RM, Roeder RG (1983) Accurate transcription initiation by RNA polymerase II in a soluble extract from isolated mammalian nuclei. Nucleic Acids Res 11: 1475–1489
13. Sambrook J, Russell DW (2001) Molecular cloning: a laboratory manual, 3rd edn. Cold Spring Harbor Press, Cold Spring Harbor

Chapter 31

Methods for Studies of Protein Interactions with Different DNA Methyltransferases

Jianchang Yang

Abstract

There are now many methods available for studying protein interactions between DNA methltransferases (DNMTs) and their binding partners. Here we describe a step-by-step procedure to identify whether proteins of interest interact with DNMTs by co-immunoprecipitation (co-IP) assay in transiently transfected cells. Though one mammalian cell is described, investigators can use the same method with other cell lines or primary cells for in vivo protein interaction studies.

Key words: Protein interactions, DNA methyltransferases, Co-immunoprecipitation, Transfection, Epitope tag, SALL4, Cytosine methylation

1. Introduction

Mammalian cells possess the capacity to epigenetically modify their genomes via DNA methylation. This process is carried out by a group of enzymes called DNA methyltransferases (DNMTs) (1, 2). To date, at least three active DNMTs have been identified and they are named DNMT1, DNMT3A, and DNMT3B (3, 4). During genetic modifications, the DNMTs usually associate with other cellular events which may contribute greatly to their enzymatic activities, subnuclear localizations, as well as sequence specificities (5–7). Indeed, identifying the individual partners for each of the DNMTs has become a hot topic in epigenetic studies (8, 9).

In laboratories, a large number of methods can be used for investigating binding proteins of DNMTs (for reviews see refs. 10–12). Two commonly used techniques are yeast two-hybrid and co-immunoprecipitation (co-IP). Both have their strengths and weaknesses. Investigators may choose the one that best answers

Minou Bina (ed.), *Gene Regulation: Methods and Protocols*, Methods in Molecular Biology, vol. 977,
DOI 10.1007/978-1-62703-284-1_31, © Springer Science+Business Media, LLC 2013

their biologically relevant questions. The yeast two-hybrid system, which requires preparation of a cDNA library, allows a versatile screening for interacting proteins inside the nucleus of yeast and detection of specific binding sites. The co-IP assays, as further described below, can be used for studying interactions between DNMTs and suspected binding partners in cultured cells, and posttranslational modifications required for these interactions are ensured. In a typical co-IP experiment, an antibody (Ab) specific to the tested protein is firstly immobilized on protein G coated magnetic beads (or agarose resins). Growing cultured cells that express the interest proteins are lysed and total proteins are prepared in non-denaturing conditions, followed by incubation with the Ab-Protein G complex. If interaction does exist, the tested protein, DNMT(s), and Ab-Protein G beads will form a new larger complex. By a series of washes, proteins that do not bind will be removed. The new tested protein-DNMT(s)-Ab-Protein G complex is then eluted from the beads and can be tested by Western blotting with individual DNMT-specific antibodies.

In certain cases, the antibodies for tested proteins may not be available or the endogenous protein levels are very low, then it will be necessary to adopt recombinant vectors that encode the tested proteins which are fused with small epitope tags (such as Flag, hemagglutinin [HA], and Myc). By transfection procedures, proteins can be exogenously expressed in cells, and the tags can be recognized by high quality antibodies that are commercially available. In other cases, investigators may seek to detect DNMT binding partners in stem cell systems such as embryonic stem cells or adult stem cells, for example, hematopoietic stem cells. It may be achievable by viral-transduction based overexpression of proteins in these cell types. In this chapter, we focus on the basic procedures for conducting co-IP in human embryonic kidney 293 (HEK293) cells, since this cell line expresses relatively high levels of all three DNMTs (13, 14). The transfection procedures for exogenous expression of interest proteins are also included, and the HA tag is described here as an example.

2. Materials

2.1. Cell Culture and Transfection Reagents

1. HEK293 cell line (American Type Culture Collection, Manassas, VA, USA).
2. Water Jacket CO_2 tissue culture incubator.
3. Aspirator or vacuum line.
4. Refrigerated centrifuges capable of centrifuging 15 ml and 2 ml centrifuge tubes.

5. Tissue culture supplies including plates, centrifuge tubes, pipettes, Eppendorf tubes, tips, etc.
6. Dulbecco's Modified Eagle Medium (DMEM) containing GLUTAMAX™ and 4,500 mg glucose (Life Technologies, Carlsbad, California, USA). Store at 4 °C.
7. Fetal bovine serum (FBS), Certified (Life Technologies). Store at −20 °C.
8. Mammalian expression vectors encoding HA-tagged protein of interest; Vectors encoding mock with HA-tag (negative control) (see Note 1).
9. Opti-MEM I Reduced Serum Media (Life Technologies).
10. Lipofectamine 2000 (Life Technologies). Store at 4 °C.
11. Dulbecco's Phosphate buffered saline (DPBS): 137 mM Sodium Chloride, 2.7 mM Potassium Chloride, 8.1 mM Disodium Phosphate, 1.1 mM Monopotassium Phosphate, pH 7.4.

2.2. Co-IP Materials

1. Cell lysis buffer: 50 mM Tris–HCl, pH 7.4, 150 mM NaCl, 0.5% Triton X100, 10% glycerol. Prepare stock solution and store at 4 °C (see Note 2).
2. Protease inhibitor cocktail: ready-to-use stock solution in DMSO, contains AEBSF, Aprotinin, Leupeptin, Bestatin, PepstatinA, and E-64 (Sigma, St. Louis, MO, USA). Store at −20 °C (see Note 3).
3. Protein G coated superparamagnetic beads: 30 mg/ml in PBS, pH 7.4, containing 0.01% Tween®-20 and 0.09% sodium azide (Life Technologies).
4. Antibody Binding and Washing Buffer; Washing Buffer (Life Technologies).
5. Monoclonal anti-rabbit HA antibody (1 mg/ml, Bethyl Laboratories Inc., Montgomery, TX, USA).
6. Control rabbit IgG (1 mg/ml, Abcam plc., Cambridge, England) (see Note 4).
7. Magnet Stand for separating magnetic particles in 1.5 ml or 2 ml tubes.
8. Labquake Tube Shaker/Rotator (Thermo Scientific, Waltham, MA).
9. Reducing sample buffer (2×): in a 15 ml conical centrifuge tube, mix 1 ml of 0.5 M Tris–HCl pH 6.8, 1 ml of 10% sodium dodecyl sulfate (SDS), 1 ml of glycerol, 0.25 ml 0.02% (w/v) bromophenol blue, and 0.1 ml 2-mercaptoethanol, store at 4 °C.
10. Heat Block.

3. Methods

3.1. Transfection of Cells with Epitope-Tagged Proteins

This step is conducted for exogenous expression of your protein of interest in cultures. It also enables co-IP to be performed using high quality epitope-specific antibodies. Design the experiment well by including proper controls to verify that immunoprecipitation has occurred (see Note 5). The brief transfection steps are depicted in Fig. 1. During cell culture procedures, use personal protective equipment such as laboratory coat and gloves at all times. Periodically clean and disinfect work surfaces with a suitable disinfectant (e.g., 70% ethanol). In addition, thermally insulated gloves, full-face visor and splash-proof apron should be worn when handling liquid nitrogen.

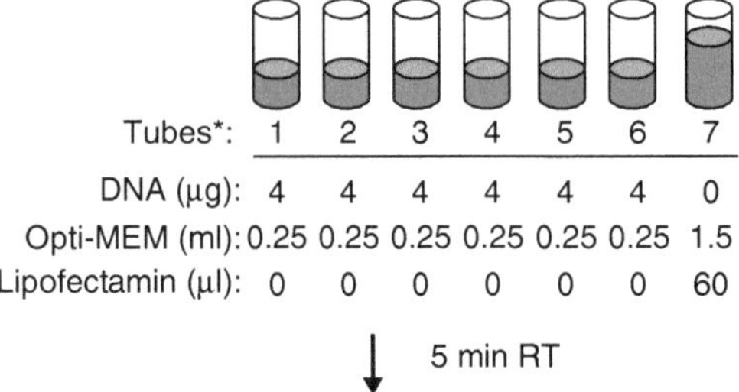

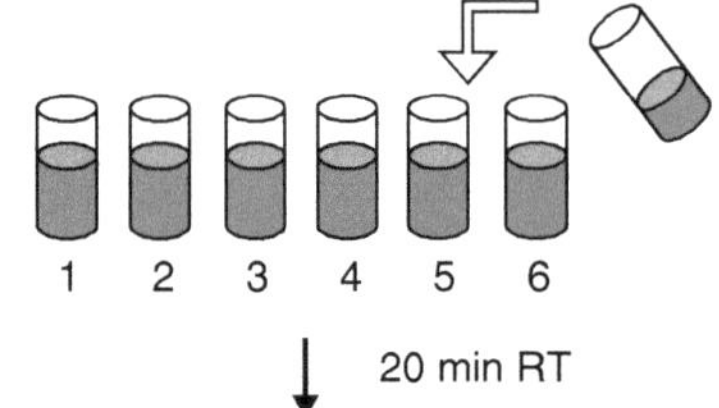

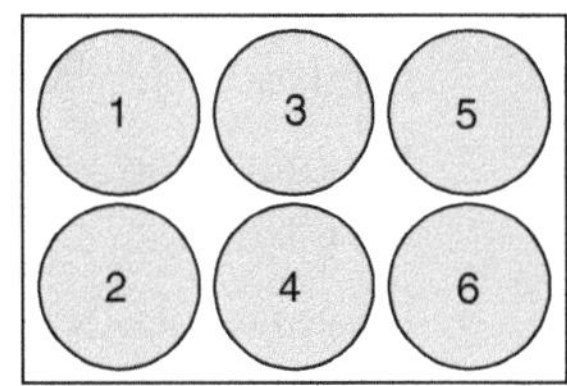

Fig. 1. Diagram of the basic transfection procedures in HEK293 cells for a co-IP experiment. This protocol is modified based on the manufacturer's user guide for Lipofectamine 2000 (Life Sciences). Samples 1, 3, and 5 are used for co-IP with an anti-HA antibody, and Samples 2, 4, and 6 are used for co-IP with control IgG.

1. Design transfection steps for co-IP. Be sure to include your controls (see Fig. 1).
2. Prepare cell growth media by adding 55 ml FBS (10% final concentration) to 500 ml DMEM media (see Note 6).
3. On the day before transfection, seed 4.0–5.0 × 10^5 HEK293 cells per well in a 6 well plate in 2 ml growth media (see Note 7). Incubate cells overnight at 37 °C in a humidified 5% CO_2 incubator.
4. On the day of transfection, check cell growth (see Note 8) and density under a microscope. The cells should be 90–95% confluent in the plate.
5. Under the tissue culture hood, mix Opti-MEM with each of the DNA expression vectors (encoding HA-tagged interest protein or controls) in sterile pre-labeled microcentrifuge tubes (0.25 ml Opti-MEM and 4 μg DNA in each tube). In a separate microcentrifuge tube, prepare the master transfection solution by mixing Lipofectamine 2000 and Opti-MEM (60 μl Lipofectamine in 1.5 ml Opti-MEM for a total of 6 samples, see Note 9).
6. Cap the tubes and gently invert eight to ten times by hand to mix. Incubate tubes for 5 min at room temperature.
7. With a P1000 pipette, distribute 0.25 ml of the mixed transfection solution into each of the six DNA-containing tubes, cap the tubes, and mix the contents thoroughly by inverting tubes eight to ten times, then incubate the mixture for another 20 min.
8. After incubation, take the cell culture plate from the incubator and place it under the hood. In each well, apply relevant combined transfection mixture (0.5 ml total) dropwise and distribute it evenly over the cell monolayers. Mix gently by rocking the plate back and forth (see Note 10).
9. Continue to incubate cells in the CO_2 incubator for 30 h prior to protein extraction and IP procedures.

3.2. Cell Lysis and Protein Extraction

1. After 30 h incubation, remove the plate containing cultured cells from the incubator and place it in the tissue culture hood. Tilt the plate slightly so that the growth medium will collect in the corners of the wells. Remove the supernatants by pipetting with a P1000 pipette (see Note 11).
2. For each well, add about 2 ml DPBS solution and pipette forcefully against the culture surface to dislodge cells. Repeat pipetting at a constant pace until all cells in the wells are detached, then collect and transfer cells in DPBS to pre-labeled 15 ml centrifuge tubes (see Note 12).

3. Wash each well once with 2 ml fresh DPBS, transfer all remaining cells in DPBS to the same centrifuge tubes, followed by centrifuge at 300 × *g* for 5 min at 4 °C.
4. Take the tubes from the centrifuge, slowly decant off all of the supernatants and resuspend the pellets by adding 1 ml fresh DPBS and pipetting. With a P1000 pipette, transfer the cell solutions to new pre-labeled 1.5 ml microcentrifuge tubes, followed by centrifuge at 300 × *g* for 5 min at 4 °C.

   The following steps can be performed on an open bench.
5. Take the tubes from the centrifuge and carefully remove the supernatants by pipetting with a P1000 pipette, being careful not to disturb the cell pellet (see Note 13).
6. In each of the tubes, add ice-cold 250 μl lysis buffer (containing 1× protease inhibitors) and briefly vortex for 20 s to resuspend the cell pellets.
7. Place the tubes on a Labquake tube rotator (rotating at 8 rpm, in a cold room) to lyse the cells for 30 min.
8. After incubation, centrifuge the lysed cells at 16,000 × *g* for 10 min at 4 °C to pellet the cellular debris.
9. Take the tubes from the centrifuge and transfer the resulted lysate (~230 μl) to new, pre-labeled and chilled microcentrifuge tubes. Put on ice for immediate use, or store at −80 °C for long term storage. Discard old tubes containing the precipitate.

3.3. Preparation of Antibody and Protein G

The purpose of this step is to ensure maximal binding of the anti-HA antibody with the Protein G beads. It is done by mixing both parts and incubation at room temperature (20–25 °C).

1. Prepare and label six sterile 1.5 ml tubes. With a P200 pipette, completely resuspend Protein G magnetic beads by gently pipetting up and down for approximately 45 s. When the liquid becomes homogeneous, distribute 50 μl solution (containing about 1.5 mg beads) into each of the tubes.
2. Place the tubes on a Magnet Stand. The beads will migrate towards the magnet and adhere to the walls of the tubes. Wait until the liquid is clear (about 30 s), then completely remove it using a gel-loading tip attached to a vacuum source, being careful not to disturb the bead-pellet on the walls of the tubes (see Note 14).
3. Take the tubes away from the Magnet Stand, add 200 μl Ab Binding & Washing Buffer along with 5 μl (5 μg) anti-HA antibody to resuspend the beads. Thoroughly mix the contents by gently pipetting up and down until the solution is homogeneous. Then, incubate the contents by leaving the tubes on a Labquake rotating shaker for 10 min at room temperature.

4. After incubation, place the tubes back to the Magnet Stand to collect the beads, and remove the supernatant as described above.
5. Take the tubes away from the Magnet Stand and wash the pelleted beads once by adding 200 μl Ab Binding and Washing Buffer. Gently pipette up and down to obtain a uniform brown suspension in the tubes.
6. Collect the beads by placing the tubes back on the Magnet Stand. Aspirate off the supernatants as described above (see Note 15).

3.4. Antigen (HA-Tagged Protein) and Antibody Interaction

In the following steps, the HA-tagged proteins (bait) along with their interacting partners (prey) in the cellular extraction will bind to the anti-HA (Ab)-Protein G complex to form a new, larger complex. This newly formed complex is linked with the magnetic beads and can be isolated from the cell lysate in the presence of external magnetic field.

1. With a P1000 pipette and fresh tips, transfer 200 μl of each cell lysate sample (prepared from step 9 of Subheading 3.2) to individual tubes containing the Ab-Protein G complex (prepared from step 6 of Subheading 3.3). Resuspend the Ab-Protein G beads by gently pipetting up and down until the solution is homogeneous (see Note 16).
2. Put the tubes on the Labquake rotating shaker, incubate with rotation for 30 min at 4 °C to allow binding of the bait proteins with the Ab-Protein G complex.
3. During the incubation time, add an equal volume of the 2× reducing sample buffer (30 μl) to each of the six remaining cell lysate samples (in the original tubes). Cap tubes loosely and heat in a 95 °C heat block (with holes filled with water) for 5 min. Then, cool tubes on ice for direct Western analysis, or transfer them to −80 °C or −20 °C freezer for long term storage.
4. After 30 min incubation, place the tubes on the Magnet Stand to collect the beads. When the supernatant liquid is clear, completely remove it using a gel-loading tip attached to a vacuum source.
5. Take the tubes away from the Magnet Stand. Wash the pelleted beads by adding 200 μl Washing Buffer and gently pipetting up and down six to eight times to obtain a uniform brown suspension in the tubes.
6. Put tubes back on the Magnet Stand to pellet the beads. Carefully remove the supernatants as described above.
7. Repeat the washing steps (from step 5 to step 6) for an additional two times (total of three washes).

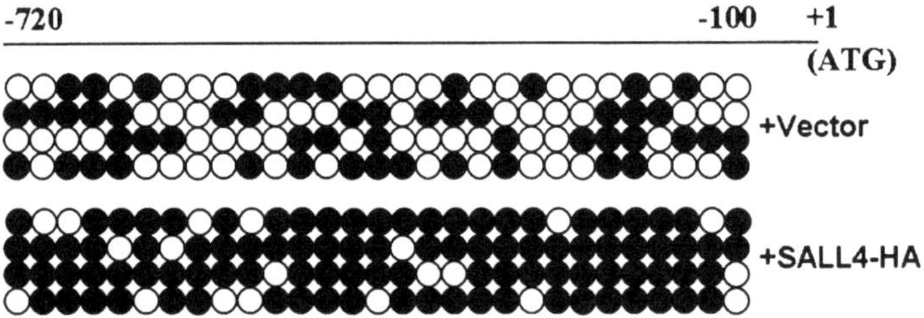

Fig. 2. IMR-90 human fibroblasts were cultured and transiently transfected with SALL4HA expressing plasmid or empty vector control. Thirty hours post transfection, cells were harvested and genomic DNAs were prepared. The purified DNA samples (500 ng) were then subjected to sodium bisulfite modification. By using specific primers, the proximal promoter region of *SALL4* gene (100–720 bp upstream of the start codon) was PCR amplified. The resulted fragments were subcloned into the pCR2.1 vector (Life Technologies) and sequenced via automated sequencing. It can be seen that the *SALL4* gene promoter is differentially methylated at specific CpGs or GC-rich sites between the two groups of cells. *Each row of circles* represents a single sequenced clone and each circle represents a single CpG or GC site. *Open circles* represent non-methylated cytosines. *Filled circles* refer to methylated cytosines. SALL4 has been found to interact with different DNMTs (14), and overexpression of SALL4 induced generally increased DNA cytosine methylation within its own promoter region.

8. With a P200 pipette, resuspend the beads in 100 μl Washing Buffer and transfer to fresh, pre-labeled tubes. This step is recommended to avoid co-elution of proteins bound to the tube walls in the following steps.

3.5. Elution of Tested Proteins

Simply add sample buffer to the pelleted beads to recover the purified proteins. The eluted proteins now can be used for direct downstream analysis (e.g., SDS-polyacrylamide gel electrophoresis followed by Western blotting). Additionally (see Note 17), DNMT enzymatic activity analysis can be conducted on co-IP products by commercially available kits, or putative DNA methylation changes of relevant genes in antigen transfected cells may be analyzed to provide further evidence of DNMT interactions. An example of the latter case is shown in Fig. 2.

1. Place the tubes (from step 8 of Subheading 3.4) on the Magnet Stand and remove all of the supernatant.
2. Take the tubes away from the Magnet Stand, add 60 μl 2×reducing sample buffer to the pelleted beads and gently pipette to resuspend the beads-Ab-protein complex.
3. Incubate the tubes in a 95 °C heat block (with holes filled with water) for 5 min.
4. Remove tubes from heat and place on Magnet Stand to pellet the beads. Load 25 μl of the supernatant containing eluted proteins onto a SDS-polyacrylamide gel, proceed with electrophoresis and Western analysis using specific DNMT(s) directed antibody (ies) (see Note 18). The remaining samples can be stored at −20 °C for further usage.

4. Notes

1. Mammalian expression vectors encoding HA-tagged proteins of interest or mock can be created using either gateway recombination technology, TOPO TA cloning, or conventional restriction enzyme-based methods, based on investigator expertise. The mock control here refers to a HA empty vector. Ideally however, constructs encoding relevant proteins lacking one or several interaction domains (to be determined) are considered the best negative controls for co-IP assays. If necessary (optional), vectors expressing known DNMT-interacting proteins may be adopted as co-IP positive controls.
2. Commercially available lysis buffers can be used, and in most cases they produce satisfactory results. Care should be taken to ensure non-denaturing conditions during protein preparations in order to maintain any interactions that occur. Denaturing lysis buffers, such as Radioimmunoprecipitation Assay (RIPA) buffer, are not recommended since they may disrupt protein interactions due to the presence of anionic detergent (e.g., SDS).
3. We routinely include protease inhibitor cocktail to minimize degradation of cell lysate proteins. However, we also obtain satisfactory cell lysate without protease inhibitors. Alternatively, phenylmethylsulfonyl fluoride (PMSF, 0.5 mM) or MG132 (30 μM) is also commonly used for cell extract preparations. Reconstitute reagents according to the manufacturer's instructions.
4. Control antibodies should be similar in nature to the specific antibody. It is recommended to use the same species as that of the primary antibody used for IP.
5. For a positive control, use the same antigen of interest for co-IP and following Western blotting assays (e.g., in addition to your samples that you run, include an input control lane that contains 8 μg of the whole cell extract used for co-IP). For negative controls, perform the entire immunoprecipitation process with an irrelevant IgG control (see Fig. 1).
6. As mentioned in the manufacturer's guide, administration of antibiotics such as penicillin and streptomycin should be avoid during cell transfection, since this may cause cell death.
7. For a successful transfection, it is important to seed the cells at the correct density so that the resulting monolayer is even and subconfluent. After adding cells, gently rock the plate back and forth and from side to side to evenly distribute cells on the surface. Set plate in incubator overnight and do not disturb.

8. We routinely check cells daily for growth and viability. Healthy cells should spread out on the surface of the wells as they proliferate in cultures. Make sure that cells are not over-confluent, look appropriate, and no signs of contamination such as cloudy media, presence of clumping cells, unexplained death, or an unusual smell.
9. Add Lipofectamine 2000 directly into the Opti-MEM and ensure that none of the reagent remains in the pipette tip.
10. Be warned that HEK293 cells are quite sensitive to perturbation. Handle gently to avoid dislodging cells in the plate.
11. Due to the potential to dislodge the cell layer, we do not recommend aspirating media from the wells using a vacuum.
12. After transfection and overnight growth, the cells may very easily detach from the plate. Simply pipette the DPBS solution directly onto the cell monolayers and then swirl the plate. The cells can then be collected by repeatedly pipetting, without the need for scraping or trypsin treatment. Change tips between samples.
13. To prevent cross-contamination, always use fresh pipette tips for each sample.
14. Aspirate off the supernatant while the tubes remain on the magnet.
15. We recommend proceeding quickly with the following steps to prevent "drying out" of the beads.
16. Mix thoroughly and change pipette tips between each transfer.
17. Reducing sample buffer is highly efficient in dissociating binding partners and is ideal for Western blotting assays. However, it is a harsh treatment and will denature all proteins in the co-IP products. For this reason, if other downstream applications (e.g., DNMT enzymatic analysis or functional assays) are considered, mild conditions such as low-pH elution with immediate neutralization should be adopted.
18. The affinity-bound Ab will co-elute with the tested antigen. This normally will not interfere with downstream Western analysis. To avoid co-elution of the Ab, it may be necessary to crosslink the Ab to Protein G beads using a cross linking reagent such as dimethyl pimelimidate (DMP).

Acknowledgment

This work was supported by the American Cancer Society-Institutional Research Grant and the Leukemia & Lymphoma Society Special Fellow award (3366-09).

References

1. Morgan HD, Santos F, Green K, Dean W, Reik W (2005) Epigenetic reprogramming in mammals. Hum Mol Genet 14 Spec No 1, R47–R58
2. Reik W, Dean W, Walter J (2001) Epigenetic reprogramming in mammalian development. Science 293:1089–1093
3. Robertson KD, Uzvolgyi E, Liang G, Talmadge C, Sumegi J, Gonzales FA, Jones PA (1999) The human DNA methyltransferases (DNMTs) 1, 3a and 3b: coordinate mRNA expression in normal tissues and overexpression in tumors. Nucleic Acids Res 27:2291–2298
4. Bestor TH (2000) The DNA methyltransferases of mammals. Hum Mol Genet 9:2395–2402
5. Bostick M, Kim JK, Esteve PO, Clark A, Pradhan S, Jacobsen SE (2007) UHRF1 plays a role in maintaining DNA methylation in mammalian cells. Science 317:1760–1764
6. Hashimoto H, Horton JR, Zhang X, Bostick M, Jacobsen SE, Cheng X (2008) The SRA domain of UHRF1 flips 5-methylcytosine out of the DNA helix. Nature 455:826–829
7. Sharif J, Muto M, Takebayashi S, Suetake I, Iwamatsu A, Endo TA, Shinga J, Mizutani-Koseki Y, Toyoda T, Okamura K et al (2007) The SRA protein Np95 mediates epigenetic inheritance by recruiting Dnmt1 to methylated DNA. Nature 450:908–912
8. Hervouet E, Vallette FM, Cartron PF (2010) Dnmt1/transcription factor interactions: an alternative mechanism of DNA methylation inheritance. Genes Cancer 1:434–443
9. Qin W, Leonhardt H, Pichler G (2011) Regulation of DNA methyltransferase 1 by interactions and modifications. Nucleus 2:392–402
10. Berggard T, Linse S, James P (2007) Methods for the detection and analysis of protein-protein interactions. Proteomics 7:2833–2842
11. Phizicky EM, Fields S (1995) Protein-protein interactions: methods for detection and analysis. Microbiol Rev 59:94–123
12. Cho S, Park SG, Lee DH, Park BC (2004) Protein-protein interaction networks: from interactions to networks. J Biochem Mol Biol 37:45–52
13. Kim GD, Ni J, Kelesoglu N, Roberts RJ, Pradhan S (2002) Co-operation and communication between the human maintenance and de novo DNA (cytosine-5) methyltransferases. EMBO J 21:4183–4195
14. Yang J, Corsello TR, Ma Y (2012) Stem cell gene SALL4 suppresses transcription through recruitment of DNA methyltransferases. J Biol Chem 287:1996–2005

INDEX

A

Abortive transcription 217, 219
dinucleotide extension assay 221
initiation 217, 220, 223
luciferase based assay 221, 222
promoter-dependent and independent 219–220, 222, 225
stimulatory effect 223
TFIIB and RNAP interactions 218–219
Acetyltransferase. *See* Histone acetyltransferases (HATs)
Affinity. *See also* Immunoaffinity purification
beads 98, 104, 281
capture 22
FLAG 97, 274, 289, 294
gel 196, 197, 201, 227, 277
matrix 282
probe 188
pulldown 185–186, 188–189
purification 137, 138, 141, 147, 183, 289, 290, 293
resins 234, 274, 281–284
selection 96, 97, 103
Antibody
affinity-bound 274
affinity-resin 274
bead-bound 97, 366
control 393
immobilized agarose beads 103, 106
mixture for kinase activity detection 260–266

B

Bromodomain 353, 354, 356

C

cAMP responsive element binding protein (CREB/CREB1) 114, 115, 229, 232, 236–239
fluorescent imaging 235
target promoters 114
CBP/CREBBP. *See* Coactivator(s)
Cellular memory 8
ChIP. *See also* HaloCHIP; Immunoprecipitation
acquisition of high quality DNA 53
-chip 95
isolation and shearing 57–58, 62, 113, 316
nucleosome-free region (NFR) 36, 38, 42, 43, 45, 47, 49
-on-chip 126
-qPCR 53, 60
-seq 35, 53, 54, 60, 62, 95, 126, 316, 319
X-ChIP 315
Chondrogenesis 194, 197
Chromatin
accessibility 3, 5, 13
assembly 193–200
associated proteins 194, 290, 294
cross-linking 39, 41, 53, 126, 315–320
DHS site cleavage assay 30
DNase I hypersensitivity 13, 14, 21–23
immunoprecipitation (*see* ChIP)
in single living cells 249
in vitro transcription 195–200
modification 4
modifying complexes 289, 293
open 4, 13, 21
remodeling 183, 194, 340
factor 196
sonication 57, 61, 62, 112, 113, 118, 315–318, 320
states 4, 8, 22
structure 3, 9, 13–15, 21, 194, 299
template 201
Cis
acting 80, 81, 84
element(s) 1, 3, 79–84, 95, 96, 193, 250
regulatory DNA 95, 96, 125
regulatory module(s) 35
regulatory sequences 125
Coactivator(s) 7, 8, 193, 194, 197–199, 315, 339, 340, 353
CBP/CREBBP 6, 7, 9
NCOA/SRC 5, 8, 339, 340
p300/EP300 .... 3, 6, 7, 9, 193, 194, 197, 198, 200, 201, 315, 316, 323–325, 329, 353–356

Minou Bina (ed.), *Gene Regulation: Methods and Protocols*, Methods in Molecular Biology, vol. 977, DOI 10.1007/978-1-62703-284-1, © Springer Science+Business Media, LLC 2013

Co-immunoprecipitation (co-IP)
DNA methyltransferases 385–388, 393
transcription factor interaction 243
CoREST 9
Gon4 like..........366, 367
HDACs 9, 354, 360, 366, 367
NCoR..........5, 9
SIN3.......... 3, 5, 9, 366, 367
Co-repressor(s) 9, 194, 365, 366
CpG 2, 5, 8, 146
islands (CGIs) 2, 8, 9, 138
methylation.......... 392
-rich promoters..........2, 4, 5, 9
Cytosine methylation 153, 385, 392

D

Dimerization between transcription factors in living cells 229
Dinucleotide extension assay.......... 221
DNA
bait.......... 97
-based affinity purification.......... 138, 141, 147–148
bending (*see* Protein-induced DNA bending)
binding
activity 200
domain(s).......... 159, 194, 239, 323, 324, 333
protein(s) 7, 53, 79, 95–97, 112, 113, 122, 125, 126, 129, 134, 365
site(s) 53, 125, 347
transcription factor(s) 193
compaction 2
element(s) 1, 13, 21, 22, 44, 97, 125, 137, 159, 315
fluorescence staining (SYBR green) 31
genomic 2, 4, 5, 13
high-quality for sequencing.......... 53
methylation..........9, 194
preparation.......... 141, 146–147
purification 26, 45, 95
DNA methyltransferases (DNMTs)..........387–392
DNase I
-chip 14, 22, 25
cleavage kit 129
digestion 21–32, 97, 99, 135
hypersensitive site (DHS)..........4, 5, 7, 13, 14, 21–24, 27, 30, 31 (*see also* Nucleosome-free region)
qPCR..........24, 30–33
-seq 15, 17, 22–27, 29, 32

E

Electrophoretic Mobility Shift Assay (EMSA)......105, 125, 137, 159, 160, 165, 169, 176, 177, 179–181
Enhancer(s) 2, 4, 13, 35–37, 42–49, 53, 66, 79, 137, 153, 194, 195, 233
Enzyme(s). *See also* DNase I
DNA modifiers (*see* DNMTs)
hisone modifiers 3, 4, 6, 8, 9
kinase(s) peptide microarrays..........259–270
Epigenetic(s). *See also* Histone modification
changes 14
DNA modification 154, 385
regulation..........193, 194
Epitope(s)
cross-linked.......... 111
peptides 274
protein 112
tag.......... 273–278, 281, 284, 291, 341, 385, 386, 388
c-myc 274
HA 274
proteins.......... 278
V5 274
Estrogen Response Element (ERE) 183, 185, 187, 188, 191, 331, 340, 342–349

F

FAIRE (Formaldehyde Assisted Isolation of Regulatory Elements)..........41, 49, 50
FLAG
affinity
beads 104
purification 289, 291, 292, 294
anti-FLAG M2 .... 97, 103, 106, 290
affinity gel.......... 197
agarose bead.......... 291
M2 peptide 97, 99
peptide..........283, 291
tag..........101, 102, 104, 196, 197, 247, 275, 276, 290, 291
mediator subunits 276, 277
p300.......... 197
smad3 .... 197, 201
sox9.......... 107
Tox4.......... 294
wdr82..........289, 290, 294
Fluorescence
anisotropy microplate assay (FAMA)..........339–350
cross-correlation spectroscopy (FCCS)..........220–240
resonance energy transfer (FRET) 203–214
staining by SYBR green.......... 32
Fluorescent
detection scanner 121
modifications.......... 206
plate reader 26
Fluorometric quantitation 60
Fluoro-PEG Puro spacer 97, 98, 102, 107
Functional assay(s). *See also* Luciferase assay
nucleosome-free region (NFR)..........35–37, 46, 50

G

General transcription factors (GTFs) 2
TFIIA 2, 60, 218
TFIIB 2, 60, 217–225
TFIID 2
TFIIE 2, 218
TFIIF 2
TFIIH 2, 218Gold nanoparticle-based DNA probe 183–192
Gold nanoparticles (AuNPs)
DNA probe 185, 191
functionalized with ERE (estrogen response element) 184, 191
solution 185, 187
synthesis 187

H

HaloCHIP and HaloTag for protein-DNA complex isolation 111–123
Heterochromatin 4, 14, 23, 194
Heterodimer 231, 240, 244
Heterooligomer 95, 96
HIF-1 323–325, 328–331, 335
Histone(s)
acetylation 4, 8, 193, 194, 198, 299, 354, 356, 360
acetyltransferases (HATs) 4, 6, 9, 194, 353 (*see also* p300; Coactivator(s))
code 300
core 2–4, 13, 193, 197
purification 196
deacetylases (HDACs) 3, 9, 354, 360, 365–367
inhibitor 353, 359–362
deacetylation 9
-DNA interactions 3, 194
methylation 4, 5, 8
dynamics 299–307
methyltransferase(s) 4, 289, 290, 300
modification(s) 4–6, 8, 9, 53, 193, 300, 307, 365

I

Immunoaffinity purification
chromatography 274
epitope tagged proteins 273, 274
protein complexes 273–275, 281–286
mammalian mediator complexes 276
Immunoblot 183, 184, 189, 300, 365–367
Immunofluorescence
myogenic differentiation 362
transcription factor interaction 245–247
Immunoprecipitation 366
In vitro virus (IVV) technology, mRNA display. *See* mRNA display, IVV technology

K

Kinase activity profiling 259–270

L

Luciferase assay
abortive transcription initiation 217, 221, 224
activity of promoter DNA 65–77
dual luciferase reporter system 79–92

M

Mammalian Two-Hybrid System 324, 328, 333–338
Mass spectrometry
identification of chromatin modifying complexes 289, 293, 294
LC and SILAC based to study dynamics of histone methylation 299–301, 310, 311
quantitative nanoproteomics 183
SILAC based to study protein-DNA interactions 137–139, 142, 148, 151–154
Mediator(s) 1–3, 6, 9, 183, 273, 276, 278
Mediator complex isolation 273–287
Microarray(s)
ChIP-chip 95, 316
DNase ChIP 14
peptides for kinase substrates 259–261
Micrococcal nuclease (MNase) digestion assay 199
MLL 1, 2, 8, 290
mRNA display for in vitro selection of DNA binding proteins 95–105
Multisubunit protein complex(es) 2, 6, 8, 183, 193, 194, 273, 278, 281, 290, 294, 340, 365
Mutagenesis 87, 218, 219

N

NF-IL6 gene 159, 165
NF-κB 68, 325, 329–332, 336, 337
Nuclear extract(s)
for affinity purification 138, 276, 278, 280, 281, 284, 285, 294–296
import 244
localization 244
preparation 140, 144, 162, 172, 183, 185, 188, 374, 376
recruitment assay 243, 244
SILAC labeling 138–140, 144, 154
Nuclear receptor(s) 1–9, 250, 339, 340, 342, 350, 359, 366, 370
Nuclear recruitment assay 243
cell culture and transfections 245
cell seeding 245
immunofluorescence 245–247
polyethylenimine, transfection 246, 247

Nuclei
DNase I digestion ... 21–32
isolation ... 13–18, 40, 280, 302–305, 376
permeabilization ... 39
restriction enzyme digestion ... 36
Nucleosome
array(s). ... 196, 198
assembly protein ... 196
formation ... 2, 13
-free-region (NFR) ... 35–51
(*see also* DNase I hypersensitive)

P

p300 (EP300) ... 1, 3, 5–7, 9, 193, 194, 197, 198, 200, 201, 315–317, 323–325, 328–350, 353, 354, 356
PCR ... 21, 114–116, 120, 134, 329, 392
ChIP ... 53, 59–61
nucleosome-free region ... 36, 44–46, 50
purification kit ... 26, 38, 42, 43, 83, 98, 99
quantitative (qPCR).. ... 24–32, 54, 105, 111, 114, 360, 362, 363
RT ... 96, 97, 103–105, 107, 111, 318, 319, 355, 356
Peptide microarrays for kinase substrates ... 259–261
Pol II. *See* RNA polymerase
Polycomb repressive complexes (PRC) 1 and 2 ... 8, 9
Post-translation modification(s) ... 4, 80, 299, 324, 386
Promoter(s). *See also* Luciferase assay
active ... 4, 5
activity ... 66, 75, 81
analysis ... 65, 114
bivalent ... 4
core ... 2, 45, 46, 203
CpG rich ... 9
deletion analysis ... 66, 79–84
distal and proximal ... 1, 3, 13, 35, 392
DNA ... 65–69
DNase I hypersensitivity ... 23, 24
elements ... 37, 323
repressed ... 4
Promoter-independent abortive initiation assays ... 219–220, 222, 225
Protein
binding region detection ... 169
-DNA
complex(es) ... 111–114, 159, 160
cross-linking ... 36, 37, 111–115, 117, 118, 218
interactions ... 53, 111, 137, 138, 159, 169, 203, 315, 365
-induced DNA bending ... 203, 205
TATA box/TATA DNA ... 2, 45, 84, 203, 206–207
-protein interactions ... 9, 96, 112, 159, 201, 218, 219, 237, 325, 333
Protein networks
CREBBP and EP300 roles ... 3, 7, 9
examples of hubs ... 3–9
HATs ... 5, 6, 9
HDACs ... 3, 9
H3K4 methyltransferases ... 8, 9
mediator ... 6
nuclear receptors ... 5, 8, 9
PRC1 and PRC2 ... 8–9
Proteomics ... 135, 138, 183, 184

Q

Quantitative
nanoproteomics ... 183, 184
PCR (qPCR) ... 24, 25, 29–30

R

Regulatory
codes.... ... 193
complexes ... 6, 9
DNA ... 95, 96, 159
element(s) ... 21, 35–37, 41, 49, 50, 53, 54, 65, 67, 79, 250, 315, 365
events ... 1
factors ... 73, 193, 249
hub(s) ... 1, 3–9
machineries ... 6
mechanisms ... 249
modules ... 35, 36, 49, 50, 159, 165
networks ... 8, 95, 97 (*see also* Protein networks)
proteins ... 4, 13, 79, 315
region(s) ... 79, 137, 340, 353
sequences ... 13, 66, 125, 137
Repress/repression/repressive ... 3, 4, 6, 8, 9, 67, 80, 159, 218, 300, 365, 366
RNA polymerase II (Pol II, RNAP) ... 2, 80, 217, 278, 323

S

SALL4 ... 392
Sedimentation analysis of complexes that repress transcription ... 365
Setd1a and Setd1b ... 3, 8, 290
SILAC (Stable Isotope Labeling of Amino acids in Cell culture)
to identify specific protein-DNA interactions ... 137
to study dynamics of site-specific histone methylation ... 299
Smad3/4 ... 194, 197, 200, 201
Sox4 ... 244
Sox9 ... 194, 197, 200
Spectroscopy. *See* Fluorescence

Stable cell lines
immunoaffinity of protein complexes 273, 278
single-cell analysis 249, 252–256
Stem cell(s)
differentiation 315, 316, 362
E14 embryonic 36, 43, 44
embryonic bodies (EBs) 315
ESC 154
neuronal NSC 65
pluripotent 359
systems 386
Steroid receptor coactivator-1a (SRC1a) 339, 340, 342–350

T

TATA
box 2, 45, 84, 203
TBP complex 203, 205–208, 218
Tbx3 244
TGF-β 194, 197
Transcription 1, 4, 6, 8, 35, 198, 217, 218, 339, 353
Transcriptional
activation 8, 45, 50, 201, 249, 323, 324, 365
domain 328, 333
dynamics 249
activators 35
activity 66, 79, 194, 197, 249
apparatus 197
coactivators(s) 194, 353, 356
complex(es) 183, 193, 194, 198, 201
component(s) 194, 199, 201
control 79
enhancers 34
machinery 1, 7, 36
regulation 80, 194
regulatory
elements 35, 37
modules 36, 49, 50
networks 95, 97
repression 365
start site 46
states 8
synergy 1, 9
Transcriptionally
competent 8
inactive 5
poised 2, 3
repressed 366
Transcription initiation/start 2, 5, 6, 24, 46, 49, 66, 80, 193, 217, 218, 220
Transcription factor(s) 1–3, 6–9, 14, 35, 36, 42, 49, 54, 67, 68, 75, 79, 80, 95–97, 111, 115, 125, 126, 159, 193, 194, 198, 200, 217, 229, 231, 243, 244, 248, 323
binding site (TFBSs). 2, 3, 14, 35, 53, 67, 68, 80, 84, 125, 126, 129, 138, 153
binding to chromatin 14
-specific antibody 38
Transfection 46–48, 66, 68–71, 81, 83, 87, 88, 115, 122, 231, 233, 243–248
Transgene array for imaging of transcriptional activation dynamics 249
Trithorax group of proteins 8

V

Valproic acid (HDAC inhibitor) 354, 355

W

Wdr82 289

Y

Yeast one-hybrid assay 125–135
Yeast two-hybrid system 386

MIX
Papier aus verantwortungsvollen Quellen
Paper from responsible sources
FSC® C105338

If you have any concerns about our products, you can contact us on
ProductSafety@springernature.com

In case Publisher is established outside the EU, the EU authorized representative is:
Springer Nature Customer Service Center GmbH
Europaplatz 3, 69115 Heidelberg, Germany

Printed by Libri Plureos GmbH
in Hamburg, Germany